Fabrication and Machining of Advanced Materials and Composites

This reference text discusses processing, structure, and properties of metal matrix composites, polymer matrix composites, and ceramic matrix composites for applications in high-end engineering equipment and biomedical and nano-biotechnology areas.

The text begins by discussing fundamentals, classification, designing, and fabrication of composite materials, followed by ultrasonic vibration-assisted machining of advanced materials, fabrication of transparent advanced composites, fabrication of composites via microwave sintering, and hybrid machining of metal-matrix composites. It covers important topics including fabrication of shape-memory polymers, additive manufacturing for the fabrication of composites, 3-D printing processes for biomedical applications, and ultrasonic vibration-assisted machining of advanced materials.

This text will be useful for undergraduate and graduate students and academic researchers in areas including materials science, mechanical engineering, manufacturing science, aerospace engineering, electronics, and communication engineering.

This book

- Covers processing, structure, and properties of metal matrix composites, polymer matrix composites, and ceramic matrix composites
- Discusses nano-materials and their potential applications in the areas of biomedical and nano-biotechnology
- Provides modern processing techniques to synthesize advance materials
- Explores applicability of the materials using mechanical, chemical, thermal, and electrical tests

Discussing advanced materials, their manufacturing techniques and applications in diverse areas including automotive, aerospace engineering, and biomedical, this text will be useful for undergraduate and graduate students and academic researchers in areas including materials science, mechanical engineering, manufacturing science, aerospace engineering, electronics, and communication engineering. It will further discuss electro discharge machining of steels using chromium alloy-based electrodes and advanced machining techniques for hard materials.

Fabrication and Machining of Advanced Materials and Composites

Opportunities and Challenges

Edited by
Subhash Singh and Dinesh Kumar

CRC Press
Taylor & Francis Group
Boca Raton London New York

CRC Press is an imprint of the
Taylor & Francis Group, an **informa** business

First edition published 2023
by CRC Press
6000 Broken Sound Parkway NW, Suite 300, Boca Raton, FL 33487-2742

and by CRC Press
4 Park Square, Milton Park, Abingdon, Oxon, OX14 4RN

CRC Press is an imprint of Taylor & Francis Group, LLC

Library of Congress Cataloging-in-Publication Data
A catalog record for this title has been requested

ISBN: 978-1-032-22455-8 (hbk)
ISBN: 978-1-032-35548-1 (pbk)
ISBN: 978-1-003-32737-0 (ebk)

DOI: 10.1201/9781003327370

Typeset in Sabon
by MPS Limited, Dehradun

Contents

Preface

Advancements in the field of materials has led to the replacement of conventional materials with new-age materials that exhibit superior properties and expedite novel creations. This significantly impacts the growth of novel strategies to meet the elevating technological demands. Composites, hybrid composites, biomaterials, functionally graded materials (FGMs), shape memory alloys and polymers, and 2-D nanostructures are some of the commonly known advanced materials. One important aspect of advanced materials is their machining and fabrication. Employing the conventional machining methods to these advanced materials would be ineffective and a complicated process owing to the diverse material properties and structures. Therefore, it is essential to understand the various advanced manufacturing techniques and appropriately use them for the advanced materials. Intense research on this and many research publications in many reputed journals related to this field are evident. But there isn't a book that entails the opportunities and challenges of fabricating and machining the advanced materials. Therefore, this book has been written with an objective to focus on various aspects of machining aforementioned advanced materials discussing method suitability, as well as the various pros and cons related to the fabrication methods. Reading this book will also give an idea about the various advanced materials and how their evolution transpired.

Adhering to the above objectives, the chapters in this proposed book are meticulously planned to cover the majority of the portion related to the topic. The initial chapter in the book covers basic topics of historical evolution and introduction to composites in Chapter 1. The subsequent chapters (Chapters 2 and 3) focus on the broad classification of composite materials and the various techniques employed in the fabrication of the composites, respectively. Chapter 2 is completely dedicated to metal, ceramic, and polymer matrix composite classifications and the classifications based on various reinforcement types. Chapter 3 focuses on the various manufacturing methods of the polymer, metal and ceramic matrix composites, and their advantages and disadvantages correlating them to the nature of composite. This basics on the fabrication techniques is followed up with ultrasonic vibration assisted machining of advanced materials in Chapter 4. This is a novel and an advanced machining technique employed for hard-to-machine materials, especially composites that possess superior mechanical characteristics. The following two chapters (Chapters 5 and 6) focus on the additive manufacturing technique. In Chapter 5, the focus is on 3-D printing of all advanced materials, but the application is limited to the biomedical field. Chapter 6 focuses specifically on the additive manufacturing of composites. Chapters 7 and 8 are related to fabrication and manufacturing potential of polymer materials. Chapter 7 specializes in fabrication of shape memory polymers, whereas Chapter 8 describes the novel, electrochemical discharge machining for the fabrication of hybrid polymer matrix composites. Sustainability is an important aspect while dealing with manufacturing of composites. Hence, in Chapter 9 this is discussed in detail for orthopedic applications. The following Chapters 10, 11, and 12 each have been allotted to the innovative fabrication techniques of advanced materials such as electrical discharge machining, vibration-assisted casting, and microwave energy-based processing, respectively. Chapter 11 has a special focus on functionally graded materials. Likewise, the remaining chapters in the book compile the topics like hybrid machining of metal matrix-based composites (Chapter 13), electrical discharge machining of shape memory alloys (Chapter 14), and machining of 2-D-based nanostructures (Chapter 15). The final chapter (Chapter 15) in the book also gives a special mention to the machining challenges pertaining to the 2-D nanostructures.

The intention behind writing this book is to make it suitable for anyone willing to study about advanced materials and manufacturing techniques. The fundamental addressees of this book would be the undergraduate and post-graduate students in diverse fields of engineering like mechanical, metallurgy, manufacturing and materials science, and biomedical and research scholars and scientists working in various research centers. Professors and lectures working in various universities will also find this book extremely useful. Experts working in various industries related to the field of manufacturing and materials along with editors, reviewers, and reading enthusiasts of numerous journals would find this book informative and interesting. Hence, it is trusted that this book would motivate and enthuse its readers to research this particular field of engineering.

Editors

Dr. Subhash Singh is currently working as associate professor in the Department of Mechanical and Automation, Indira Gandhi Delhi Technical Univeristy for Women, New Delhi. Previously, he worked as assistant professor in the Department of Production and Industrial Engineering, NIT Jamshedpur. He earned his B. Tech from KNIT, Sultanpur, 2003 and M. Tech from NIT Kurukshetra, 2007. He has completed his Ph.D. from IIT Roorkee in 2017. Dr. Singh specializes in areas such as modification of nano-materials, thin coating, fabrication of MMCs, synthesis of 2-D materials, FSP, and machining of biodegradable materials. He has published more than 30 research papers in various prestigious journals. He has also authored one book and 20 book chapters. He has completed three research projects.

Dr. Dinesh Kumar is currently working as an assistant professor in the Department of Production and Industrial Engineering, NIT Jamshedpur. He earned his B. Tech from Uttar Pradesh Technical University Lucknow (Now APJ Abdul Kalam Technical University Lucknow) in 2009 and M. Tech from the Mechanical Engineering Department, MNNIT Allahabad in 2011. He completed his Ph.D. from IIT Roorkee in 2016. He has published more than nine research articles.

Contributors

Girija Nandan Arka, National Institute of Technology, Jamshedpur, Jharkhand, India

Ashish Das, National Institute of Technology, Jamshedpur, Jharkhand, India

Akshay Dvivedi, Indian Institute of Technology, Roorkee, Uttarakhand, India

Rama Kanti, Indian Institute of Technology, Roorkee, Uttarakhand, India

Amaresh Kumar, National Institute of Technology, Jamshedpur, Jharkhand, India

Dinesh Kumar, National Institute of Technology, Jamshedpur, Jharkhand, India

Divyanand Kumar, National Institute of Technology, Jamshedpur, Jharkhand, India

K. Santhosh Kumar, National Institute of Technology, Jamshedpur, Jharkhand, India

Kundan Kumar, National Institute of Technology, Jamshedpur, Jharkhand, India

Pradeep Kumar, Indian Institute of Technology, Roorkee, Uttarakhand, India

Vikas Kumar, Federal TVET institute, Addis Ababa, Ethiopia

Mamta Kumari, National Institute of Technology, Jamshedpur, Jharkhand, India

P. Prasanna Kumari, Vignan's Institute of Engineering, for Women, Visakhapatnam, Andhra Pradesh, India

Md. Manzar Iqbal, National Institute of Technology, Jamshedpur, Jharkhand, India

Ashok Kumar Jha, National Institute of Technology, Jamshedpur, Jharkhand, India

Radha Raman Mishra, Birla Institute of Technology and, Science Pilani, Pilani, India

Rahul S. Mulik, Indian Institute of Technology, Roorkee, Uttarakhand, India

Kaushik Pal, Indian Institute of Technology, Roorkee, Uttarakhand, India

Chander Prakash, Lovely Professional University, Phagwara, India

Shashi Bhushan Prasad, National Institute of Technology, Jamshedpur, Jharkhand, India

S.V. Satya Prasad, National Institute of Technology, Jamshedpur, Jharkhand, India

Rajneesh Raghav, Indian Institute of Technology, Roorkee, Uttarakhand, India

Farhan Ahmad Shamim, Aligarh Muslim University, Aligarh, India

Sahil Sharma, Indian Institute of Technology, Roorkee, Uttarakhand, India

Ranjit Singh, Dr. B.R. Ambedkar National, Institute of Technology, Jalandhar, India

Ravi Pratap Singh, Dr. B.R. Ambedkar National, Institute of Technology, Jalandhar, India

Subhash Singh, Indira Gandhi Delhi Technical, University, New Delhi, India

Sunpreet Singh, National University of Singapore, 21 Lower Kent Ridge Road, Singapore

Tarlochan Singh, Indian Institute of Technology, Bombay, Mumbai, India

Anand Mukut Tigga, National Institute of Technology, Jamshedpur, Jharkhand, India

Rajeev Trehan, Dr. B.R. Ambedkar National, Institute of Technology, Jalandhar, India

Mohammad Uddin, South University of Australia, Australia

Ravi Verma, Institute of Electronics, Microelectronics and Nanotechnology, Villeneuve-d'Ascq-France

Chapter 1

Introduction, History, and Origin of Composite Materials

Subhash Singh

Department of Mechanical and Automation Engineering, Indira Gandhi Delhi Technical University for Women, New Delhi, India

Mohammad Uddin

Department of Mechanical Engineering, South University of Australia

Chander Prakash

School of Mechanical Engineering, Lovely Professional University, Phagwara, India

CONTENTS

1.1 INTRODUCTION TO COMPOSITES

Composites are material systems with multiple phases produced by combining diverse materials in an attempt to achieve those superior characteristics as well as performance which cannot be achieved by the individual constituting components [1]. Unlike in alloys, the obtained phases within a composite aren't due to phase transformations, natural reactions, or any alternative phenomena. An alloy is always a homogeneous mixture in which the components added don't retain their original characteristics and are generated by natural processes. This isn't the case within a composite because the added constituents restore their original

DOI: 10.1201/9781003327370-1

characteristics and also the composite needn't always be homogenous [2]. That's the basic difference between an alloy and a composite. When observed at a macroscopic scale, among the various phases, a composite would generally comprise of a continuous, weaker phase, termed "the matrix" and a much stronger, stiffer phase termed "reinforcement." But in certain situations, due to the effects of processing and chemical reactions, between the phases there would be another distinct phase termed "interphase" present among the matrix as well as the reinforcements [3]. Finally, a composite is formed with stacked layers of reinforcement fibers as well as a matrix with required characteristics in a particular direction or in multi-directions. Fig. 1.1 represents all the three possible phases present in a composite material.

The composite's characteristics are influenced by the constituents' geometry and characteristics as well as phase distributions. The reinforcement's volume/weight fraction is one of the prime factors. The material system's uniformity or the homogeneity is influenced by the reinforcement's dispersion. The higher the non-uniformity, the higher is the heterogeneity within the material. This leads to greater characteristic scattering and therefore resulting in higher probability of failure at the weaker sections of the system of material. The system's anisotropy is influenced by orientation as well as geometry of reinforcement within the composite [4]. Also, within a composite material, there are multiple roles of each of the existing phases which are related to the nature of application for which the composite is designed. For low- to medium-performance applications of composites, particulate or short fibers of reinforcement is employed which could have certain stiffening but composite's strength would be limited. It is the material's mechanical characteristics that are dependent on the nature of matrix utilized as it becomes

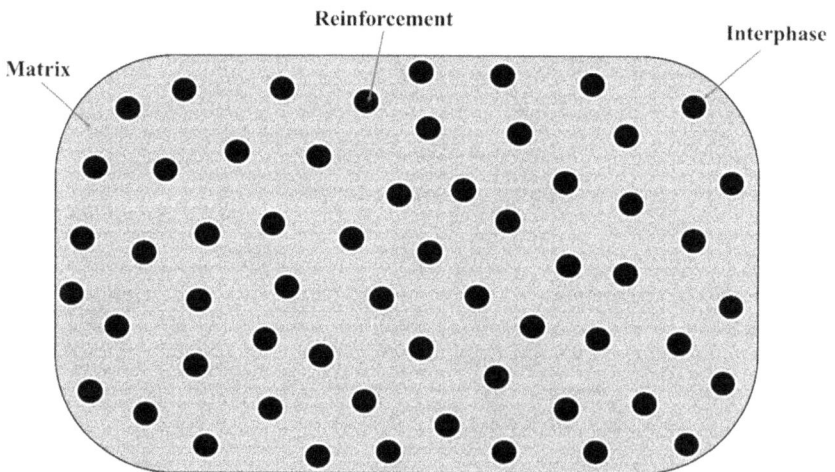

Figure 1.1 Constituent phases within a composite [3].

the major constituent that bears the load. For high-performance applications, the structure of the composites would comprise of continuous fiber–type reinforcement that becomes the composite's backbone influencing the characteristics such as material strength as well as material stiffness along the length of the reinforcement. It is the matrix's duty within the composite to bond, protect, and support the sensitive fibers, thereby looking after the localized transfer of stress in between consecutive reinforcement fibers [5]. Though smaller in size, it is the interphase that controls the mechanisms of failures, its propagations and toughness of the material against fractures, as well as the entire stress-strain behavior of the composite till the occurrence of failure. Apart from these, some on the other factors on which the characteristics of a composite material depend are characteristics of the individual constituents of the matrix as well as the reinforcement, the number of phases existing within a composite, compatibility among the matrix, as well as the reinforcement and fabrication technique of the composite material. The matrix and reinforcement types should be such that any interactions between them should not result in harmful constituents of interphases that would hamper the overall characteristics of the composites. Therefore, tailorability is one of the prime aspects of composites suitable to the application and the desired characteristics within the material. This is possible through proper selection of constituents, constituent proportions, constituent morphologies, crystallographic textures and degrees of crystallinity, material dispersions, along with the interface composition among the selected constituents [1]. This endorses the composites to meet the demands of diverse fields of engineering. Hence, composite materials hold a major share in the materials sciences and other aspects of commercial manufacturing sectors of engineering [6]. The composites formed by combination of matrices as wells as reinforcements will comprise of diverse advantages as well as disadvantages that are responsible for diverse engineering applications. It is essential to consider the pros and cons before selecting the materials to be combined in formation of a composite. The advantages and disadvantages of composites have been represented in Table 1.1 [7].

1.2 APPLICATIONS OF COMPOSITES

Due to the nature of their fabrication technique and their laminate structure, the composites exhibit exceptional characteristics like good resistance against corrosion resistance, lower weights, enhanced, strength and productivity, reasonable costs of the material, flexibility in design along with great durability. Hence, there are applications of composites in diverse fields of engineering such as aircrafts and aerospace, automotive sector, marine, infrastructures, oil and energy, biomedical, as well as sports and recreation [8]. Apart from these, there are other applications as well. The various applications of composites are discussed in detail.

Table 1.1 Advantages and disadvantages of composite materials [7]

S No.	Advantages	Disadvantages
1	High stiffness/strength to weight ratio resulting in reduced weight.	Higher production as well as material costs.
2	Easy tailorability of characteristics such as high strength or stiffness along the direction of load applied.	There might be a weaker transverse section if unidirectional fibers are used.
3	Redundant paths of load in between consecutive fibers.	Weaker matrices with lower toughness.
4	Higher durability without any corrosion.	Reusability and material disposal might be difficult.
5	Damping within a composite is inherent.	A composite is difficult to analyze.
6	Enhanced or reduced thermal/electrical conductivity.	Matrix might be subjected to degradation in testing conditions

1.2.1 Aircrafts and Aerospace

The structures like space antennae, optical instruments, and mirrors used in aerospace vehicles are fabricated from extensively stiff and lightweight composites of graphite because graphite composites can be fabricated with near-zero hydric and thermal expansions [9]. Therefore, it is possible to achieve exceptional dimensional stability in extreme conditions of the surroundings. The secondary and primary structures like fairings, bulkheads, floor beams, fuselages, wings, and empennages in civilian as well as military aircrafts predominantly make use of composites exhibiting characteristics such as exceptional strength and stiffness with lower densities [10]. One such example of composite usage can be illustrated by mentioning the fabrication of Boeing 787 Dreamliner and the Airbus A380, which happens to be the largest aircrafts in the whole world. The use of carbon/epoxy as well as graphite/titanium composites in the Boeing 787 is seen. Around 50% of the aircraft's weight comprises the composite (Fig. 1.2) [11]. Also, compared to the previous versions (Boeing 747, Boeing 757, Boeing 767, Boeing 777), the usage of composites has significantly grown in the Boeing 787, which primarily includes the wings and fuselage. Fig. 1.2 represents the evolution of composites usage in the commercial aircrafts of Boeing.

Even the Airbus A380 makes use of considerable quantities of hybrid glass/epoxy/aluminum laminate (GLARE) composites. This includes all the advantages of composites and metals and alleviates the disadvantages. Military airplanes like the B-2 bombers as well as smaller unmanned drones make use of the stealth properties of the carbon/epoxy composites. Organizations like NASA are also utilizing composites for space-related applications like in rockets and many spacecrafts. One such example can be the use of carbon composites reinforced with Kevlar in a solar-powered flying wing called the Helios, which was manufactured to analyze the

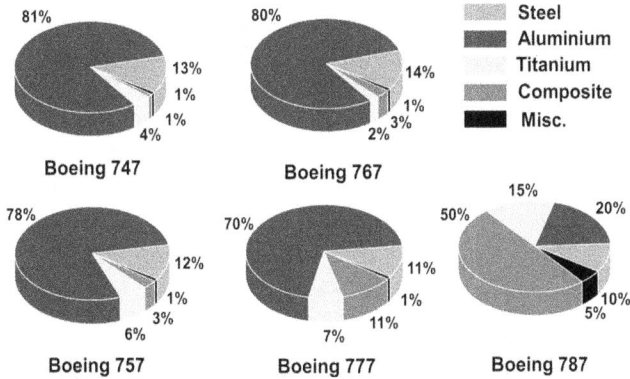

Figure 1.2 Increasing use of composite materials in Boeing commercial aircrafts [11].

environment [12]. Apart from the ones mentioned above, applications of composites in the aerospace and the aircraft sectors include compressor and fan blades, air-foil surfaces, flywheels, wing box structures, engine bay doors, radars, solar reflectors, satellite structures, jet and rocket engines, and also in transmission structures of helicopters like the turbine shafts and blades as well as the rotor shafts.

1.2.2 Automotive and Transportation

The composites and the automotive industry go hand in hand as the automotive sector is one of the largest consumers of composites. The need for reducing weight is one of the prime factors in automobiles for better fuel efficiency. Moreover, strategic vehicular designs combined with composite significantly enhance the fuel efficiency. Composite applications include transportation as well as the automotive industry for load bearing components like bodies, frames of trucks, as well as railcars. Automobile components such as engines, cylinders, pistons, crankshafts, connecting rods, and leaf springs are manufactured. In general, a leaf spring made of glass/epoxy composite has a weight that is 1/5th of that of a steel leaf spring. carbon fibre reinforced polymer (CFRP), a fiber-reinforced polymer, is utilized in automobiles [13].

1.2.3 Marine

Similar to the aircraft industry, even the marine sector strives for lightweight structures. For the same, composites reinforced by fibers of carbon are utilized for fabricating lighter and damage resistant ship hulls of diverse form with thick glass section [14]. Sandwich construction is carried out that comprises thin face sheets of composites bonded to a thick core that is lighter in weight. Examples of these are corvettes and minesweepers. The composite structures in a ship are cheaper to maintain, are cost effective for manufacturing and possess good insulation with exceptional resistance to corrosion. In the marine sector, for commercial as well as the military-related applications, the deck, bulkheads, propellers, mast, and boat hulls, as well as several other parts of a ship are fabricated through composites to make use of their corrosion-resistant characteristics. Other areas of composites' application in a water body vessel are super yachts' furniture and interior moldings [15].

1.2.4 Construction and Infrastructure

The use of composites for construction and infrastructural applications is a newer concept. Yet on a global platform, this application has a huge market for composites. The use of composites in this field is to fabricate ideal materials for construction with exceptional strength. Structural members are reinforced with composites to fabricate earthquake-resistant bridges or buildings with unique and complex designs that seem impractical and are not possible through conventional materials. Buildings that are energy efficient, aesthetically appealing, versatile, and have unique functionality with enhanced thermal performance can be constructed through composites [16]. Flexible designs with complex carvings in building designs like ribs, curves, corrugates, and free contours of changing thickness can be easily constructed by using composites. Moreover, conventional appearances such as chrome and gold, copper, stone, and marble can be obtained in an extremely cost-effective manner. Hence, there has been a significant gain in contemplating the application of composites across diverse commercial as well as residential structure by the community of architecture. Even huge pipes for transportation of water and oil are fabricated through composites of glass/polyester or Kevlar. Thermosetting polymer composites are being used in architectural components like roofs and wall panels, fixtures and moldings, doors and window frames, vanity sinks and shower stalls, swimming pools, etc. Not only for constructions, but also for repairing existing structures like buildings, bridges, rails, and roads, composites are utilized [17].

1.2.5 Corrosive Environments

Composites due to their exceptional corrosion-resistant abilities are utilized in applications involving handling chemicals and exposure to corrosive, severe, outdoor conditions. Thus, composites are found generally found in oil/gas refineries, plants of chemical processing, plants of pulp and paper, and facilities where water treatment is done [18]. Tanks, pumps, cabinets, hoods, fans, and ducts in these areas fabricated through composites achieve resistance for longer time durations against harmful, corroding fluids. The major setback of corrosion in pipes of sewer, desalination projects, and oil and gas supplies were evaded through fiber-reinforced polymer composites.

1.2.6 Energy

The use of lightweight and extremely strong carbon fibers in place of glass fibers as a reinforcing material in composites has enhanced the field of materials. The material advancements in the U.S. Department of Energy, have had a major impact in the redefinition of the energy industry. Composites enhance the efficiency of conventional suppliers of energy through wind and solar power. The use of composites of carbon fiber in wind turbine blades provide better strength, flexibility at lower weights, and costs resulting in enhanced output efficiencies [19]. There has been substantial growth in the wind industry since the use of composites and this growth is to continue further in the future with the advent of advanced composites as well as fabrication techniques. Composites are also utilized in risers for drilling in offshore oil drilling installations.

1.2.7 Electricals and Electronics

Design as well as flexibility in processing are the benefits of composite materials; hence, they can substitute the alloys of metal in electrical appliances. But the aspects of function and design are linked to the variable customer choices. Therefore, a higher trend is seen in the electrical appliances sector in comparison to other industrial organizations [20]. The use of composite is seen in panels of equipment, power tools, handles, frames as well as appliance trims and diverse other applications. In the electrical appliances like dishwashers, refrigerators, ovens, dryers, and ranges, composites are used. Certain equipment related components are fabricated by using composite materials. The components are control panels consoles, knobs, kick plates, shelf brackets, motor housings, and side and vent trims, etc. The significant advancement in the electronics field is because of the composites' usage. It is the strong dielectric characteristics with arc-track resistance properties of the composite materials such as the thermosetting polymer matrix composites that are needed for electronics-related applications. Some of the components manufactured through the composite materials are arc shields and chutes, control systems and lighting-based components, bus supports, microwave antennas, metering devices, and controls for motors. There are also standoffs, their insulators, printed boards of wiring, hardware related to pole line, equipment of the substation, switch gears, as well as blocks and boards of the terminal [21].

1.2.8 Sports/Recreational Applications

The composites reinforced by fibers exhibit exceptional properties suitable for producing sports and other recreational activities equipment. The easily moldable fiber-reinforced composites have exceptional strength, modulus of elasticity, and corrosion resistance with very low density. So composites related to sports include bicycle frames, fishing poles, jumping boards, hockey sticks, helmets, horizontal and parallel bars, kayaks, and racquets of tennis. So, the sporting equipment fabricated using composites of carbon and glass fibers would be extremely light with exceptional strength and durability [22]. This will benefit the sport enthusiasts and athletes to deliver their best performance without worrying about any fatigue while utilizing the sports gear.

1.3 HISTORICAL EVOLUTION OF THE COMPOSITES

Before discussing the historical evolution of composites, it is essential to understand the basic fact that composites have always been a part of the human history, dating back to thousands of years ago, approximately around 3400 B.C. till the present, modern times. Though there has been diverse kinds of composite materials developed, the idea behind developing them is the same.

Fig. 1.3 represents the broad history of how composites have always been an integral part of human life. All the key moments have been highlighted across a timeline spanning from the inception of composites till the modern era, the details of which will be discussed in the current section of this book chapter. As mentioned earlier, diverse composites have been utilized for diverse applications but the common idea behind fabricating composites is to achieve superior characteristics of the material in order to meet the demands of the application under testing conditions. The existence of composites within nature has occured seen since the origin of time. The presence of flora and fauna in nature, which are systems of composite fibers, are proof of this fact. Leaves of a tree; bird's wings; and the complex, musculoskeletal human body system, comprising of bones, muscles, tendons can be considered as a few examples [23].

Initially, around 5000 years back, in 3500 B.C., information on the use of composites was known to us. In the ancient Mesopotamian civilization (modern-day Syria, Kuwait, and Iraq nations), the first man-made fibrous composite was produced, and it was meant for daily usage. Different pieces of wood were stuck together in diverse directions to create the earliest plywood type. A similar kind of method was adopted by the ancient Egyptians around the same time where some fibrous strips, from the papyrus plant, were stuck together perpendicular to one another in two different layers to form the earliest kind of paper called the "papyrus." The Egyptians found this development extremely significant, so the fabrication method of the papyrus was kept extremely confidential; hence, a monopoly was created. This development of paper (papyrus) reformed the entire concept of preserving any treasured information. Hence, for many centuries, Egypt was the largest exporter of papyrus paper across the world.

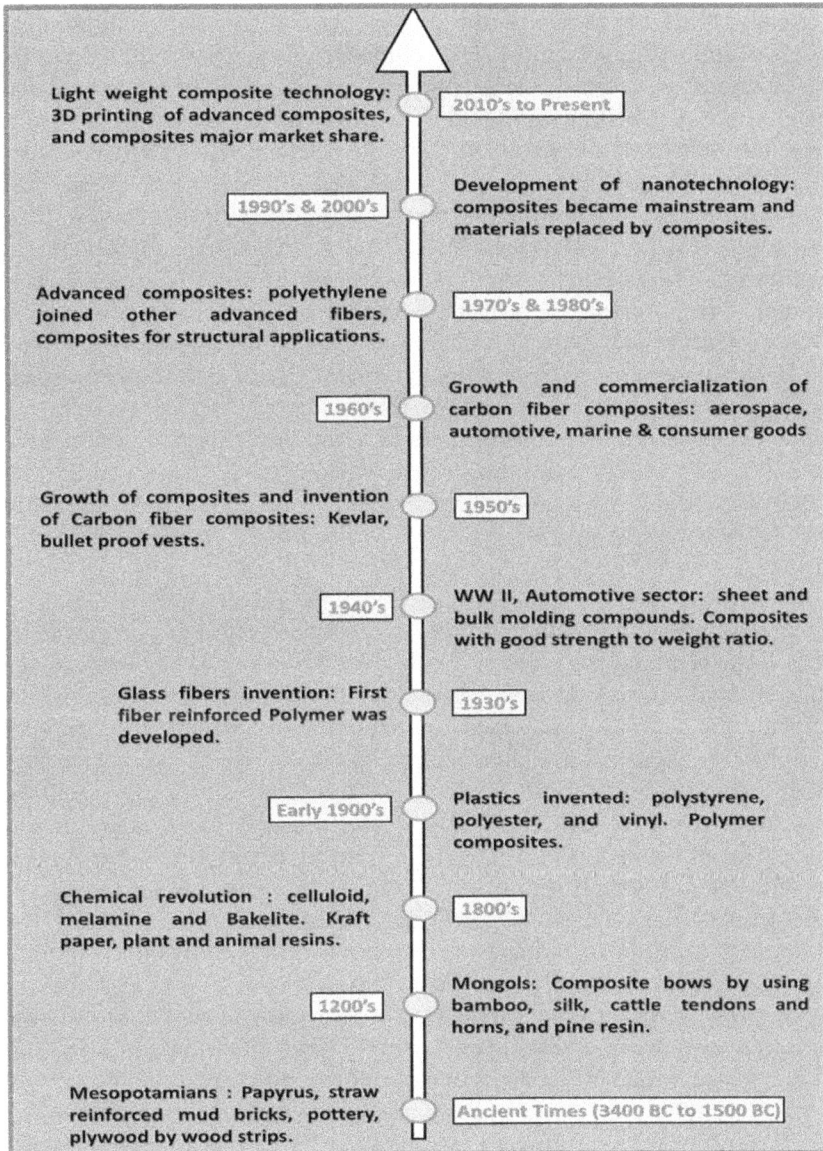

Figure 1.3 Historical evolution of composites [23].

The early Egyptians also fabricated baskets, sails, ropes, and boats using the papyrus plant's wooden chops [24]. Moreover, the papyrus/linen layers along with cartonnage were soaked within plaster and death masks were made. As the years passed by, around 1500 B.C., further development was seen from the Egyptians and the Mesopotamian civilizations in the form of composite bricks. Mud was reinforced with straws to fabricate strong, composite bricks, which were utilized in the construction of houses and religious and public structures. The composite mixture was also used in the reinforcement of boats as well as pottery. Not only this, but there was also mention of the use of metals. Gold was one of the first metals to be used in the ancient civilizations in the form of ornaments and then it was the use of copper for domestic usage as well as in weapons and other things [25]. Around 1200 B.C., upon the completion of copper age civilization, there was a surge in bronze usage, which led to the beginning of Bronze Age civilization. The end of Bronze Age resulted in the surge of Iron Age civilization. Therefore, metals have also been extremely familiar to the humans from ancient times. There was also a mention of development of diverse kinds of concrete, mortars, and lime around 25 B.C. with the preparation techniques similar to the modern-day cement. But the ancient mixture was supposed to be superior in properties in comparison to the modern-day's Portland cement. Further information on composite advancements was mentioned for diverse high-level applications, wherein the men associated with fabrication, the artisans, builders, and engineers of the ancient civilizations strived for better quality composites.

Many years later, around 1200 A.D., in the eastern part of Asia, the Mongol warriors were the first inventors of composite archery bows fabricated by including the materials such as bamboo/wood, horns/cattle tendons, and silk combined with naturally obtained pine resin. On top of a bamboo, the tendons were placed on the outer (tension) side, whereas horn sheets were placed on the inner (compressed) side of the bows that were finally wrapped tightly with silk/birch bark or both and finally sealed the entire bow using the pine resin [26]. These composite bows were used by the armies of Genghis Khan, and these were supposed to be an extremely powerful and accurate bow during that time. These were the strongest and most feared weapons used till the invention of guns/firearms happened later in the 14th century. A testing done by one of the museums on a 900-year-old surviving composite bow proved that these bows are as powerful as the modern-day bows and any target placed 490 yards away could be easily and accurately hit at a distance almost five times the length of a football field. Such was the development of composites in that era.

In the 1800s, there were many experiments carried out by the canoe builders, using diverse composite resins in order to fabricate laminates of paper. In the process, multiple, sturdy layers of wood pulp, termed Kraft paper, were stuck together using shellac (a glue-like material). The idea behind this was good. However, this laminate of paper couldn't meet the expectations because it didn't last long when submerged inside water. In the later years of the 1800s, around 1870s to 1890s, there was a chemical revolution that completely transformed the development of composites [27]. People were able to fabricate the first of its kind, synthetic/man-made, polymer resins that were capable of being converted from a liquid state into a solid state by a technique termed "polymerization," in which the molecules could be cross-linked. Until then, the resin used was a sticky substance called pine resin that could be extracted from animals or plants and used for sealing bows or boats. The earliest kind of man-made resins were melamine, Bakelite, and celluloid. These resin evolution and polymerization techniques progressed into the early years of the 1900s, which paved way to the invention of plastics through hardening (curing) of unsaturated, man-made resins.

In the beginning of the 1900s, advancements in chemicals resulted in the invention of plastics. Polyester, polystyrene, vinyl, and phenolic were some of the developed materials of synthetic resins that outshined the existing resins but there was a requirement of a suitable reinforcement to strengthen them and make them rigid to make them suitable for structural applications. In the year 1907, Leo Hendrik Baekeland, a chemist born in Belgium but hailing from New York, U.S., developed a unique, synthetic resin, polyoxybenzylmethylenglycolanhydride, familiar by the name of Bakelite. It was a thermosetting resin formed when phenol reacts with formaldehyde and an elimination reaction occurs. It was extremely brittle when developed, but it would become soft with a better strength upon combination with cellulose. It was identified as the first of its kind plastic prepared from a synthetic resin [28]. In the beginning, Bakelite was commercially used in the preparation of gearshift knobs in Rolls Royce cars in the year 1917. Due to its non-conducting and heat-resistant nature, it was widely applicable in various consumer and industry-related goods like electrical insulators, pipe stems, telephone and radio casings, children's toys, and kitchenware. There was a recognition for the significance of Bakelite as the first man-made plastic came much later in the year 1993 by American Chemical Society. This was a significant national landmark in the chemical history. In the middle part of the 19th century, between 1920s and 1930s, newer resins with enhanced properties were developed.

The 1930s are believed to be the era of resins and the field of composites in totality. Some significant resins had been developed in this duration that played an important part in the modern times. In the beginning of the 1930s, DuPont and American Cyanamid, American chemical industries, fabricated some advancements in polymer resins. In due course of time, each of the companies, for the very first time fabricated polyester resins on their own. In the year 1935, remarkable changes occurred in the field of plastics [23]. Ray Greene, an employee in the Owens Corning, an industry dealing with glass and composites, invented a novel material by mixing plastic polymer and glass. This newly invented material was extremely light in weight and had exceptional strength. It was termed "fiber-reinforced polymer (FRP)" and is familiar by the name of fiberglass. A technique was devised by the Owens-Illinois to draw glass into thin strands (fibers) and weave them into textile fabrics. Owens Corning had started a new FRP industry. The invention of FRP completely reformed the scenario of composites in the field of fabrication forever. A patent was granted to

Carleton Ellis for the unsaturated polyester resins in the year 1936, which are one of the most dominant resins in the field of fabrication as they exhibit good curing (hardening) characteristics. Apart from this, epoxy, a resin system with exceptional performance, was available by 1938. There were also fibers of asbestos reinforcing the phenolic resins being used.

In 1940s, there was World War II (WW II) and an Industrial Revolution that became a boon to the field of composites. WW II created a necessity for lightweight, higher strength, and weather-resistant composites. Due to WW II, the research within the FRP industries materialized into production due to sheer necessity. Moreover, it was found out that composites of fiber glass are transparent (invisible) to the frequencies of radio signals. Moreover, it sheltered the radar equipment at the time of war. Therefore, this resulted in the usage of fiber-reinforced composites of glass for the domes in radar and various electronic applications [23,29]. Before WW II, all the airplane wings made use of thin wooden layers along with plastic resins. This was initially practiced by Howard Hughes over Spruce Goose as a form of composite airplane wings. At the time of WW II, there was further development of the upcoming composite industries as there was intense search by the military for materials with lower weight, better strength, and a higher durability for air and water modes of transport to withstand the corrosive effects of salt present inside water and air. Hence, airplanes had fiberglass for weight reduction and corrosion resistance in all conditions of weather. In 1942, Ray Greene produced a dinghy from a glass fiber composite that later changed the boating world. There was a consumption of more than 7 million pounds of fiberglass during WW II by the year 1945 and that was only for military-based applications. Once the war ended, the decline in the consumption of fiberglass for military applications was observed. By then the public sector was complete aware of the fiberglass benefits. Therefore, the later part of the 1940s saw the growth of fiberglass composites for commercial applications and in the field of manufacturing. Immediately when FRP and fiberglass turned out to be a consumable product, the opportunity was grabbed by the boating and automobile sectors, which then turned out to be the next highest fiberglass composite consumers. In the year 1946, the boating industries commercially produced and launched boat hulls reinforced by the fiberglass-based composites. At the time of 1947, the testing and prototype building of an entire automobile body produced from fiberglass composite was complete, which resulted in the fabrication of the Chevrolet Corvette later in the year 1953 [23,30]. The car was fabricated by impregnation of preformed fiberglass took place with resin that was then molded in metal dies. With the growth of the automobile sector, various novel molding techniques came into existence. Sheet and bulk molding compounds are two such significant molding methods employed at that time, not only in the automobile sector but also across diverse manufacturing industries that came up during that time. The aspect of resistance against corrosion was adopted by the piping industries that commercially fabricated a pipe made entirely of fiberglass in the year 1948. This was later taken up by the oil industries in the subsequent years, which turned out to be a huge success.

The successful run of composites continued in the 1950s, and it expanded very quickly. Many applications of the composites became a reality, such as sports cars, trucks, boats, ducts, pipes, and storage tanks. Even surfboards made of fiberglass came into existence, boosting the surfing sport. There is also development of novel methods of manufacturing polymer composites like filament winding, vacuum bag molding, and pultrusion that are currently in use. Moldings and ladders, components which are linear, are produced through pultrusion process. Composites suitable for aerospace applications are fabricated through filament winding, such as large-scale motors of rockets. This paved the way for space exploration in the coming years of the 1960s.

The marine field was the largest composite material in the 1960s. The patent on carbon fibers was approved in the year 1961, but it took several years past that since it was been applied commercially. The carbon fiber composites enhanced the stiffness-weight ratios of the thermoset components. It was largely utilized in the fields related to consumer/sporting goods, automotive, and aerospace, which it boosted with its superior characteristics. The year 1966 was when Kevlar, an advanced, carbon fiber derived composite, was invented by Stephanie Kwolek, a chemical expert at DuPont [23,31]. Kevlar is a strong fiber that is para-aramid and is familiar for its application in stab as well as ballistic-resistant armor. This proved to be extremely beneficial for protection of police as well as the armed forces. The development of novel and enhanced resins greatly enhanced the requirements of composites, especially in the areas pertaining to corrosion and higher temperatures. Apart from this, there was also a predominant usage of metals like aluminium (Al) and steel used around this time, which has continued till the modern times. Composites comprising of the metals like Al and boron came into existence in the year 1970. Also around this time was the usage of rods of iron combined with concrete that were employed for structural applications. This resulted in the growth of steel-reinforced concretes.

In the subsequent two decades, there was tremendous advancement in the field of composites, for commercial applications, due to their exceptional strengths. The combination of novel, polyethylene with ultra-high molecular weight along with advanced fibers found their applications in components of aerospace, armors, structural applications, devices in medical fields, equipment in sports, and various other diverse fields. The automotive sector overtook the marine sector in the 1970s to become the highest consumer of composites, which is still intact today. In the early 1970s and late 1980s, advanced composites were predominantly utilized for structural applications on continents like Asia and Europe [23]. It is during this period that the first composite highway bridge in the world was constructed along with a bridge deck completely made of composite materials.

The middle of the 1990s witnessed the integration of composite materials along with the fields of construction as well as production [32,33]. The composites were the predominant materials used in these fields. Polymer composites (thermosets) now became a cheaper and more reliable option, replacing conventional metals as well as fabricated thermoplastics in the industries of electrical and appliances, transportation, and construction. A striking growth was visible in the usage of composites across consumers as well as industries. A few examples are electrical cross-arms and infrastructures, hardware of pole lines and insulators, doorknobs and handles, and weather-resistant strains. Apart from this, high strength and seismically upgrade buildings, pedestrian bridges, parking lots, and transportation structures made completely of FRP-reinforced composites were constructed across diverse areas in the world.

Nanotechnology was being employed commercially in the beginning of the 2000s [34]. Nanotechnology may be defined as fabrication and structural applications of systems of nanosize in a controlled manner at molecular or atomic scale, in order to develop materials with enhanced or novel characteristics. Nanomaterials possess enhanced characteristics due to the fact that for the same material mass they would consist of a larger area of surface in comparison to bulk materials, which will allow better material reactivity. Also, at nanoscales, matter behavior could be dominated by quantum effects after influencing the material properties. Therefore, it would be possible to manufacture composites of smaller sizes with enhanced characteristics in a cost-effective manner. The role of composites is significant in nanotubes of carbon as the nanotubes in bulk would be ideal fibers to enhance the electrical and thermal as well as mechanical characteristics of the bulk material. The integration of nanomaterials into enhanced resins and fibers was the new trend of composites formed. The advancement of the technology such as physical vapor deposition broadened the application range of the composites to diverse fields like aerospace, automotive sector, etc. Apart from strength, durability is another factor that is evident in the modern-day composites.

In the 2010s, there was further development in the field of manufacturing that resulted in the surge of the advanced 3-Dimensional (3-D) printing technique. This technique simplified the production process and now the production was possible in small-scale businesses as well as homes in which by making use of a personal computer (PC) and computer aided design (CAD) software any object of intricate geometry and size could be easily printed [35]. As a result of this, demand of advanced composites became very high and components via 3-D printing with exceptional characteristics could be prepared using diverse fibers of reinforcement in an extremely short duration. Polymer composites could be fabricated through carbon fibers or glass fibers of discontinuous strands via 3-D printing across fields like tooling, automotive, biomedical, electronics, etc. [36]. 3-D printing also resulted in the printing of metal matrix composites. It was in the year 2014 that the first carbon fiber 3-D printer in the world was invented by Mark Forged.

In the present times, research on advanced composites draws manufacturers, universities, as well as government grants. Composites in present times have exceptional characteristics such as lighter densities and enhanced strength, and are therefore needed by medical experts and sporting suppliers for enhanced safety. There has been excessive necessity of lightweight materials in the manufacturing sectors due to escalating greenhouse gas emissions and the need to protect the atmosphere. Also, escalating prices of crude oils has become worrisome to industries. This substituted the use of conventional materials with advanced composites having exceptional performance capabilities under trying situations for diverse applications in engineering fields [37]. So, in particular, it is the automotive and aircraft sectors that rely on modern-day, lightweight composites for the enhancement of automobile efficiency along with lower emissions of carbon dioxide (CO_2). The aircraft sectors make use of advanced composites in airplane turbine blades to reduce the weight and have a higher efficiency [38]. Government organizations like the defense sectors use radar transparency and the lightweight nature of advanced composites in fabricating stealth aircrafts, drones, blast/ballistic protective equipment, and firearms [39]. There can be different types of metal, polymer, and ceramic types of advanced composites used for diverse applications, since durability and sustainability are the major criteria for using composites. The idea behind striving to achieve composite durability is to cut down on manufacturing as well as maintenance costs [40]. This growth on extensive use of composites has been possible only due to the advancements in the manufacturing field, so it can be said that composites and the field of manufacturing go hand in hand. Observing the trend and this vast growth of composites usage from ancient to present times as well as the future, it is extremely evident that further composite research will take place and continue to grow for material sustainability and recyclability. This will surely provide a more secure and bright future in the field of composites.

1.4 CONCLUSIONS

Since the ancient civilization till the modern times, there has been tremendous development in the field of composite materials and the prime reasons for this progress are inventions of various polymers, fibers as well as ceramics, in different centuries in the vast history of human civilization. Also, the need for materials with better characteristics drove this development process of the composite materials field. Though at a point in time, during the World War II, the aircraft industry drove the technological development in need of lightweight materials, in the subsequent years the cost factor in comparison to the monolithic materials was added as a reason.

In today's world, other factors like assurance of quality, reproducibility, and behavioral predictions of the material in its tenure, along with cost and the need for lightweight materials drives the development in all areas. But one important aspect here is that development in manufacturing methods also resulted in the development of advanced composites. They go hand in hand. This is because the process of manufacturing is essential for quality assurance and product characteristics once it reaches the finished state. Significant research has been devoted towards producing functional as well as reliable composites in a cost-effective manner and commercially feasible methods. This is the reason why there has been invention of 3-D printing technology and fabrication of nanocomposites.

Though composite technology is under a developing stage, it has attained a maturity stage and the future looks extremely promising. The expanding market in the form of automotive, aerospace sectors, and lightweight material necessity has reduced the costs of basic material constituents, which further resulted in lowered manufacturing costs. Development in the technology has given experience in handling advanced material composites with better techniques. It is the demand for multifunctional materials that is providing novel challenges as well as development potential of a novel material system with enhanced characteristics. Therefore, composite materials continue to be a field of research.

REFERENCES

[1] Chung DDL. *Composite Materials Science and Applications*. Second ed. Springer-Verlag London, 2010.
[2] Sekar K, Vasanthakumar P. Mechanical properties of Al-Cu alloy metal matrix composite reinforced with B_4C, Graphite and Wear Rate Modeling by Taguchi Method. *Materials Today: Proceedings*. 2019 Jan 1; 18: 3150–3159.
[3] Daniel IM, Ishai O. *Engineering Mechanics of Composite Materials*. Second ed. New York: Oxford University Press, 2006.
[4] Talreja R, Varna J. *Modeling Damage, Fatigue and Failure of Composite Materials*. Woodhead Publishing, 2016.
[5] Hemath M, Mavinkere Rangappa S, Kushvaha V, Dhakal HN, Siengchin S. A comprehensive review on mechanical, electromagnetic radiation shielding, and thermal conductivity of fibers/inorganic fillers reinforced hybrid polymer composites. *Polymer Composites*. 2020 Oct; 41(10): 3940–3965.
[6] Fleischer J, Teti R, Lanza G, Mativenga P, Moehring HC, Caggiano A. Composite materials parts manufacturing. *CIRP Annals*. 2018 Jan 1; 67(2): 603–626.
[7] Peters ST, editor. *Handbook of Composites*. Springer Science & Business Media, 2013 Nov 27.
[8] Gay D. *Composite Materials: Design and Applications*. CRC Press, 2014 Jul 29.

[9] Jamir MR, Majid MS, Khasri A. Natural lightweight hybrid composites for aircraft structural applications. In *Sustainable Composites for Aerospace Applications*, Mohammad J, Mohamed T, editors (pp. 155–170). Woodhead Publishing, 2018 Jan 1.

[10] Jawaid M, Thariq M, editors. *Sustainable Composites for Aerospace Applications*. Woodhead Publishing, 2018 Apr 27.

[11] Warren AS. Developments and challenges for aluminum – A Boeing perspective. *Materials Forum*. 2004 Aug; 28: 24–31.

[12] Zhu X, Guo Z, Hou Z. Solar-powered airplanes: A historical perspective and future challenges. *Progress in Aerospace Sciences*. 2014 Nov 1; 71: 36–53.

[13] Ravishankar B, Nayak SK, Kader MA. Hybrid composites for automotive applications – A review. *Journal of Reinforced Plastics and Composites*. 2019 Sep; 38(18): 835–845.

[14] Selvaraju S, Ilaiyavel S. Applications of composites in marine industry. *Journal of Engineering Research and Studies*. 2011 Apr; 2(II): 89–91.

[15] Pemberton R, Summerscales J, Graham-Jones J, editors. *Marine Composites: Design and Performance*. Woodhead Publishing, 2018 Aug 20.

[16] Karbhari VM. Introduction: The use of composites in civil structural applications. In *Durability of Composites for Civil Structural Applications* (pp. 1–10). Woodhead Publishing, 2007 Jan 1.

[17] Erofeev V. Frame construction composites for buildings and structures in aggressive environments. *Procedia Engineering*. 2016 Jan 1; 165: 1444–1447.

[18] Hu N, editor. *Composites and Their Applications*. BoD – Books on Demand, 2012 Aug 22.

[19] Swolfs Y. Perspective for fibre-hybrid composites in wind energy applications. *Materials*. 2017 Nov; 10(11): 1281.

[20] Agarwal BD, Broutman LJ, Chandrashekhara K. *Analysis and Performance of Fiber Composites*. John Wiley & Sons, 2017 Oct 30.

[21] Chung DD, Composite materials for electrical applications. In *Composite Materials*, Deborah DLC, editor, (pp. 73 89). Springer, London, 2003.

[22] Tang DZ. The application of carbon fiber materials in sports equipment. In *Applied Mechanics and Materials* (Vol. 443, pp. 613–616). Trans Tech Publications Ltd, 2014.

[23] Ngo TD, editor. Introduction to composite materials. *Composite and Nanocomposite Materials from Knowledge to Industrial Applications*, (pp. 3–30). 2020 Feb 25.

[24] Herakovich CT. Mechanics of composites: A historical review. *Mechanics Research Communications*. 2012 Apr 1; 41: 1–20.

[25] Hartman D, Greenwood M, Miller DM. *High Strength Glass Fibers*. Agy, 1996.

[26] Johnson T. 2018. *History of Composites: The Evolution of Lightweight Composite Materials*.

[27] Nagavally RR. Composite materials-history, types, fabrication techniques, advantages, and applications. *International Journal of Mechanical and Production Engineering*. 2017; 5(9): 82–87.

[28] Seymour RB, Deanin RD, editors. *History of Polymeric Composites*. VSP, 1987.

[29] https://www.mar-bal.com/language/en/applications/history-of-composites

[30] Taub AI, Krajewski PE, Luo AA, Owens JN. The evolution of technology for materials processing over the last 50 years: The automotive example. *JOM*. 2007 Feb; 59(2): 48–57.

[31] Yadav R, Naebe M, Wang X, Kandasubramanian B. Body armour materials: From steel to contemporary biomimetic systems. *RSC Advances*. 2016; 6(116): 115145–115174.

[32] Bakis CE, Bank LC, Brown V, Cosenza E, Davalos JF, Lesko JJ, Machida A, Rizkalla SH, Triantafillou TC. Fiber-reinforced polymer composites for construction—State-of-the-art review. *Journal of Composites for Construction*. 2002 May; 6(2): 73–87.

[33] Henerichs M, Voss R, Kuster F, Wegener K. Machining of carbon fiber reinforced plastics: Influence of tool geometry and fiber orientation on the machining forces. *CIRP Journal of Manufacturing Science and Technology*. 2015 May 1; 9: 136–145.

[34] Santamaria A. Historical overview of nanotechnology and nanotoxicology. *Nanotoxicity*. 2012; 926: 1–2.

[35] Su A, Al'Aref SJ. History of 3D printing. In *3D Printing Applications in Cardiovascular Medicine* 2018 Jan 1 (pp. 1–10). Academic Press.

[36] Wang X, Jiang M, Zhou Z, Gou J, Hui D. 3D printing of polymer matrix composites: A review and prospective. *Composites Part B: Engineering*. 2017 Feb 1; 110: 442–458.

[37] Prasad SS, Prasad SB, Verma K, Mishra RK, Kumar V, Singh S. The role and significance of Magnesium in modern day research – A review. *Journal of Magnesium and Alloys*. 2021 Jun 24; 10(1): 1–61.

[38] Pradeep AV, Prasad SS, Suryam LV, Kumari PP. A comprehensive review on contemporary materials used for blades of wind turbine. *Materials Today: Proceedings*. 2019 Jan 1; 19: 556–559.

[39] Jayalakshmi CG, Inamdar A, Anand A, Kandasubramanian B. Polymer matrix composites as broadband radar absorbing structures for stealth aircrafts. *Journal of Applied Polymer Science*. 2019 Apr 10; 136(14): 47241.

[40] Tsai SW, Hahn HT. *Introduction to Composite Materials*. Routledge, 2018 May 2.

Chapter 2

Intense Classification of Composite Materials

Subhash Singh

Department of Mechanical and Automation Engineering, Indira Gandhi Delhi Technical University for Women, New Delhi, India

Sunpreet Singh

Department of Mechanical Engineering, National University of Singapore, Singapore

CONTENTS

2.1 INTRODUCTION

Composites are extremely popular and extensively used in materials science and because they accurately blend the right amounts of properties such as toughness, stiffness, lighter nature of the materials, and good resistance against corrosion [1]. Composites as defined are a combination of multiple material constituents of diverse chemical or physical characteristics that together form a material possessing distinctive and superior properties than that of the individual materials within the combination [2]. The constituents of a composite remain without losing

DOI: 10.1201/9781003327370-2

their personal characteristics, nor do they blend or get dissolved as happens in an alloy or a solution mixture. In composites, the constituents combine such that each constituent material can be observed separately when viewed via a microscope to contribute their characteristics to enhance the properties of the final material composition [3]. Composites comprise of a weaker, softer base material called the matrix that surrounds as well as binds another stronger, filler material termed the "reinforcement" and together they form a composite. Mechanical characteristics within a composite such as thermal conductivity, formability, and ductility are due to the existence of the matrix material, whereas characteristics such as enhanced strength as well as stiffness or lower thermal expansion are due to the existence of the reinforcement material. Also, the loads applied on the composite material are endured by the reinforcements. The reinforcement within a composite may be particulates, fragments, whiskers/fibres of materials that are naturally occurring or synthesized artificially [4,5]. Composites are not only artificially synthesized but a few are also naturally obtained. Wood consists of cellulose fibre chains (reinforcement) and organic lignin (polymer matrix) as well as bones consisting of hydroxyapatite reinforcement crystals surrounded by collagen matrix are few examples [6]. Based on the type of matrix and reinforcement types, there has been a broad classification of composite materials. Since it is the nature of the individual constituent materials within a composite that impact a composite's characteristics, it is essential to understand and study the categorization of these composites. Hence, in this present book chapter, this intense classification of composites is discussed in detail.

2.2 CLASSIFICATION OF COMPOSITES

The composites are basically classified in the following ways:

> The first kind of categorization is based on the nature of matrix material used in the composite. According to this the composites are metal matrix composites (MMCs), ceramic matrix composites (CMCs), polymer matrix composites (PMCs) and carbon-carbon composites (CCCs) or also called as carbon matrix composites. Sometimes, the PMCs and CCCs together are termed as organic matrix composites (OMCs) [7].

The next kind of categorization of composites is based on the type of reinforcement. As per this, the composites are segregated as fibre-reinforced composites (FRCs), particulate-reinforced composites (PRCs), and lamina-reinforced composites (LRCs). In PRCs, the particulates can be in the form of powder or flakes. Concrete and particles of wood in boards can be taken as certain examples of these PRCs. The fibres in

FRCs can be discontinuous/continuous. When embedment of fibre parti-cles is done in a matrix, it is an FRC. In the case of short/discontinuous fibres, the composite characteristics change w.r.t the fibre length. In the case of continuous fibres, the composite's properties such as the Young's modulus are constant. These long fibres with smaller diameters do possess exceptional tensile characteristics but when pushed along the axial di-rection, these bend/buckle. Therefore, sufficient support needs to be given for the individual fibres. The LRCs comprise of different material layers surrounded and placed together through matrix material. Sandwich structures can be considered an example of LRCs [8].

According to the size and scale of the material fabricated, at 10^{-9} or nanoscales, the composites comprising of reinforcements of nanoscales are termed "nanocomposites." This is an advanced version of the composites. Even the nanocomposites may be classified based on their matrix and reinforcement types just like the conventional composites. It is just that the material scale is at the nano-level changes the material properties in comparison to bulk materials [9].

All of the above composites are artificially synthesized by combining various materials to get the desired properties. But there is a class of composites, another classification, termed the "natural composites," which are readily available in nature [10]. These are not synthesized by any man-made methods. They just have the matrix and reinforcements that are part of nature. For example wood, bones, etc. come under this category. The broad classification of various composite materials is represented by a flowchart in Fig. 2.1.

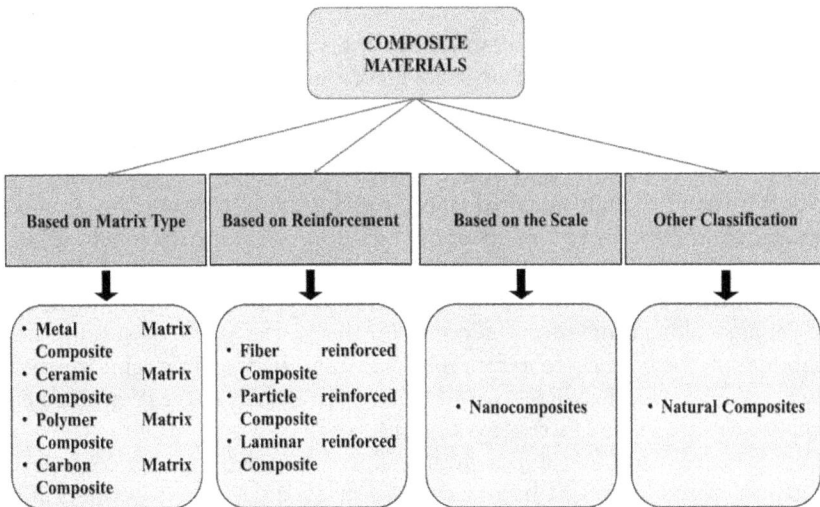

Figure 2.1 Broad classification of composites.

2.2.1 Classification Based on the Matrix Type

2.2.1.1 Metal Matrix Composites (MMCs)

The MMCs have a metallic matrix in the prepared composite. The MMCs are quite popular, but not as much as PMCs. The mechanical characteristics of MMCs are higher compared to PMCs in terms of stiffness, strength, better resistance to corrosion and elevated temperatures, as well as better fracture toughness. Due to higher melting points of metals, the MMCs require reinforcement that are stable/non-reactive under extreme thermal conditions. Hence, it is essential to select reinforcements with a higher modulus [11]. Many metals as well as metal alloys are ideal as matrix materials, but only a few metals with less density are suitable for applications related to lower temperatures. Some commonly used metal matrices are magnesium, aluminium, and titanium, especially for applications pertaining to aircrafts. For elevated temperatures, cobalt and alloys of cobalt-nickel alloy are given preference. For MMCs to have enhanced strength, reinforcements of high elastic modulus need to be used, which further enhance the strength-to-weight ratios of the formed composites, making it more than that of conventional alloys. It is the mechanical characteristics as well as the melting point that for diverse temperatures influence the composite's service temperature. Hence, the majority of compounds of metals as well as ceramics may be employed for alloys possessing lower melting points [12]. With an increase of metal's melting point, there is an increase of complication in the selection of reinforcement for a MMC.

2.2.1.2 Ceramic Matrix Composites (CMCs)

The CMCs comprise of a matrix made of ceramics with dispersed ceramic phases of carbides/oxides. Ceramics are solid materials with an extremely tough, the usual ionic bond or a covalent bond. The CMCs have exceptional resistance to corrosion, outstanding compressive strength, better resistance to thermal shocks, greater melting points, better quality of dynamic load bearing capability, and stable nature at elevated temperatures (over 1500°C) make them a good prospect for applications involving elevated temperatures [13]. The attempts of the addition of reinforcements to CMCs for enhancing strength have failed due to the combination of lower tensile strain and higher elastic modulus of ceramics. The reason behind this is inadequate elongation of the ceramic matrix at the point of rupture and inability of the CMCs to effectively transmit load to the reinforcement particles. This can be avoided with sufficient quantities of the reinforcement because elevated tensile strengths of reinforcement would raise the capacity of load bearing in CMCs. But adding fibres of great strength to a weak matrix of ceramics hasn't been successful at all times due to formation of weaker composites. Monofilament fibres of the reinforcements such as SiC, titanium boride or zirconium oxides, alumina, etc. that have elevated elastic moduli have been added to CMCs to enhance strength. To a certain extent, the issues are resolved. But if there is a scenario in which the matrix's coefficient of thermal expansion is higher than that of the added reinforcement in a CMC, then the obtained composite would not have grander characteristics such as the strength because of the formation of micro-cracks inside the CMCs while it is cooling. This micro-cracking between the consecutive fibres would completely lower the tensile strength of the CMCs produced [14].

2.2.1.3 Polymer Matrix Composites (PMCs)

Polymers are best suited for lightweight applications as well as to obtain good mechanical characteristics due to their lower densities and good strength to weight ratios. The fabrication of PMCs is relatively easier in comparison to other composites. Hence, in aeronautics, wide use of high-temperature resins is seen. The polymers are further classified as thermoplastics and thermosets.

The structure of thermoplastic molecules is either 1D or 2D; due to this, the thermoplastics have the tendency to soften at higher temperatures and exhibit low melting points. This facilitates retention of properties at the time of cooling and endorses the typical methods of compression for moulding the components. The main idea behind fabrication of fibre-reinforced thermoplastic resins is to enhance the polymer characteristics and obtain enhanced properties such that these polymer composites can substitute metals in the processes like die casting. Nucleation as well as morphology of a thermoplastic in crystalline form is enhanced by a reinforcement. The resins in a thermoplastic (amorphous/crystalline) are capable of changing their creep over a wide temperature range. Irrespective of a range in temperature, there is a constraint to use resins and the reinforcements in that PMCs system could enhance the load of failure, including the resistance against creep enhancement. Some of the commonly used resins of thermoplastics are polystyrene, polyethylene, polyamides, nylon, and polypropylene [15]. The use of reinforcements in thermoplastic PMCs can also impact the shape retention as well as shrinkage of the polymer depending upon the nature of reinforcement. There is no involvement of chemical reactions in thermoplastic PMCs which is advantageous as compared to thermosetting PMCs because chemical reactions would lead to liberation of heat/gases. The entire process of production of PMCs is the cumulative duration of heating, shaping as well as cooling of the entire polymer structure. The resins of thermoplastic are moulded components and ideal for reinforcing with fibres that can result in isotropic composite properties due to the randomness in the fibre dispersion. But it is feasible for directional alignment of fibres during moulding. The thermoplastic PMCs forego their strength at high temperatures, but adding fillers as reinforcements can enhance the resistance to heat. It is the positive characteristics of thermoplastics such as toughness, rigidity, and their ability to renounce creep that make them significant in the fabrication of PMCs. The PMCs' applications include control panels in automobiles, encasements of electronic components, etc. There has been an increment in the applications of thermoplastic-reinforced resins with a development in technology. In modern times, thermoplastics are seen as large reinforced sheets that can be moulded as desired through heating and sampling, which simplifies the fabrication methods of bulky PMCs by eliminating cumbersome components of moulding [16].

Thermosets, post-curing, possess a well-bonded 3-D molecular structure; hence, only decomposition occurs in thermosets unlike in thermoplastics, where melting/hardening occurs. Altering the resin composition would alter the characteristics or curing conditions of a thermoset polymer. The properties of thermosets could be retained under the conditions of partial curing for longer durations, making thermosets extremely flexible in manufacturing. Therefore, thermosets are ideal for fabricating advanced PMCs of short, premixed fibre reinforcements and epoxy, phenolic polyamides as the matrix materials. If not for thermosets, there wouldn't have been such a huge research development in the field of structural engineering. Some commonly used thermosets are epoxy, polyester, and phenolic polyamides and the applications of these thermosets are seen in printed circuit boards, automotive and aerospace components, systems of defence, etc. Thermosets are prepared via direct condensation polymerization in which water is the by-product of this reaction [17]. This presence of water results in the formation of composites with voids that would hinder the strength as well as dielectric characteristics of the composite. Epoxy as well as polyester phenolic are the major kinds of thermosetting resins.

Epoxies are most commonly employed in composites prepared via a filament winding process to prepare prepresses. Epoxies have satisfactory stability against chemicals, exceptional adherence, and also shrink slowly at the time of curing without emitting any volatile gases. But still, application of these resins is expensive and also they wouldn't last at 140°C temperatures. Hence, these cannot be used for higher-temperature applications [18].

The resins of polyester are inexpensive and can be accessed in a very easy manner, which make them popular in diverse fields. It is possible to store liquid polyester at ambient temperatures for many months or years and simply; when a catalyst is added in an extremely short duration, the polymer matrices can be cured. Their usage is in structure-related applications and automobiles. A polyester when cured becomes transparent but sometimes is rigid and at times flexible but becomes resistant to chemicals as well as environmental changes. Polyester are extremely compatible with glass fibres and, based on the resin type and the nature of applications, they can be used in temperatures greater than 75°C [19]. Aromatic polyamides are extensively used in the fabrication of advanced PMCs in structure-based applications in temperatures of 200–250°C for longer durations.

2.2.1.4 Carbon-Carbon Composites (CCCs)

The use of carbon is generally seen at high temperatures in atmospheres that are inert. Carbon is brittle and anisotropic in nature, its sensitivity is flawed, and it has variable characteristics. So, the application of carbon to fabricate CCCs for structures as well as huge complex bodies is restricted. The CCCs comprise of carbon in a matrix as well as reinforcements that possess the beneficial characteristics like refractoriness, toughness, enhanced thermal shock resistance, high-speed frictional characteristics, and greater strength of both the ceramics as well as fibre-reinforced composites [20]. Since the CCCs retain their mechanical characteristics at elevated temperature conditions, they are widely used in aircraft brakes, rocket nozzles, and heat shields in space vehicles. Moreover, CCCs are an ambassador for the carbon science development, are biocompatible, and chemically inert, which makes them significant in the fields of chemicals and medicine. The CCCs' mechanical properties are defined by the orientation of reinforcements (randomly/strongly oriented) and the method of manufacturing.

2.2.2 Classification Based on Reinforcements

The reinforcements are fibrous materials that are inert, strong, woven or nonwoven, and integral to the matrix within a composite, together strive in the enhancement of a composite's physical characteristics. The most significant variation among a reinforcement and a filler is that unlike in a filler, reinforcements significantly enhance the flexural strength as well as tensile strengths within a composite and reinforcements strongly bond with the matrix. Some of the used reinforcements are boron, asbestos, ceramic and synthetic fibres, carbon, graphite and metal glass, flock, sisal, chopped paper, jute, and macerated fabrics. Diverse reinforcement characteristics provide diverse composite characteristics. The nature of reinforcements within a composite can be in the form of particles, which have no orientation or shape; whiskers, with a definite shape, smaller lengths, and diameters; or fibres that are much longer than whiskers, have one long axis, and a circular/oval shapes at the cross-sections. The various reinforcements used in composites are represented in Fig. 2.2.

Reinforcements should be stronger, stiffer, and must enhance the composite's resistance to failure as the main intention of using reinforcements is to make the composite stronger. Apart from strength, reinforcements also provide resistance against heat, corrosion, and enhanced rigidity or conduction. These properties are influenced by the reinforcement's characteristics and can be tailored as desired [21]. Therefore, reinforcements should have no or extremely low ductility and must be very brittle. In certain composites, fibres or bundles of fibres are only employed when fabrication methods like filament winding is used. In other cases, fibres are arranged

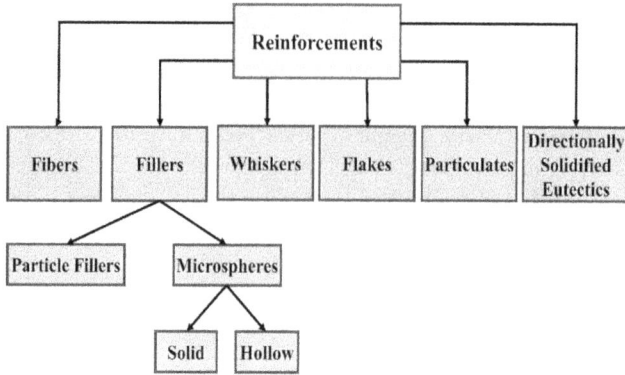

Figure 2.2 Classification of reinforcements.

like sheets (fabrics) for better handling and it is the diverse orientation of fibre arrangement within the sheets that ensures diverse properties.

2.2.2.1 Fibre-Reinforced Composites (FRCs)

Fibres are important reinforcements, which transfer the strength to matrix as desired, thereby enhancing the composite's characteristics. Fibres of glass were the first to be used as reinforcements, followed by ceramics and metals to fabricate heat-resistant and stiffer composites.

The performance of fibres is influenced by several parameters such as fibre length, geometry, orientation, fibre composition, as well as matrix's mechanical characteristics. A fibre orientation determines a composite's greatest strength, which is along the fibre length [22]. This will not ascertain that longitudinal fibres withstand loads irrespective of the applied direction. It's just that for an applied load along the length, longitudinal fibres exhibit optimum strength that can significantly decrease for a small change in the load's angle. Only certain applications have loads in one direction; therefore, it is wise to have diverse fibre orientations for heavier loads.

Monolayer tapes have continuous or discontinuous fibres of unidirectional orientation and stacked as plies of filament layers also in the same orientation. Complex orientations can also be done through computers for analysing the variations as per desired needs. As such, it is possible to modify the strength of unidirectional fibre planar composites that possess isotropic characteristics. In the case of non-quasi-isotropic, angle-plied composites, plies in number as well as plies in orientation determine the characteristics having constant ratios of the variables of composite with a weaker matrix in comparison to reinforcement fibres [23]. Therefore, having assumed that the percentage of volume is equal along all of the three axes, it can be said that a fibre's strength would be 1/3rd times the unidirectional fibre composite. Also, short fibres can have random orientations which can be 3-D in nature. This is possible through, sprinkling fibres in a plane, adding liquid/solid matrix prior or post the deposition of the fibres. Among the several methods for random orientations of fibre, 2-D composites would possess a strength that is 1/3 times that of unidirectional fibre-stressed composite, whereas a 3-D composite will have 1/5th the strength. There isn't any observation of strength in composites with extremely strong matrix but strength can be applied in those matrices with diverse orientations. The strength along the length may be obtained by assuming effective strength reduction in fibres via an approximate value inside composites with a non-longitudinal orientation of fibres and strong matrix.

Fibre composites comprise of short or continuous fibres (filaments) and it has been proven that the orientation of filaments is better irrespective of their performance [22]. Filaments are extremely long in comparison to their diameters and hence they are fabricated via continuous production. Filaments are ideal for diverse matrix types as they exhibit twisting, winding, knitting, and weaving which are similar to a fabric's properties. Due to their exceptional strength as well as low dense nature, filaments significantly impact a composite's mechanical characteristics or their response to fabrication methods. Composites comprising of ceramic and glass as well as multi-purpose short fibres with ideal orientation exhibit greater strength than composites using continuous fibres. Fibres that are short are known to exhibit strength [24]. The continuous fibre within composites are fabricated via the method of filament winding, where fibres impregnated by a matrix are wound around a mandrel to obtain the end product under suitable fibre orientations. In the case of composites, where fibres of short length are fabricated via the method of open or close moulds, the efficiency of the fibres is less in spite of lower costs of the input in comparison to filament winding.

In present times, the majority of the used fibres are solids of circular cross-section that can be easily handled or fabricated and in which some have non-conventional geometries. It is the hollow fibres that are capable of enhancing a composite's mechanical properties [25]. Within a fibre composite, the length of the fibre as well as the diameter are the criteria to determine a composite's properties and it is the small diameters that produce exceptional strength because of the eliminated surface defects. The development of filaments that are flat and thin gave rise to the use of rectangular cross-sectioned fibres for structural applications requiring high strength as the rectangular geometry facilitate seamless packing of the fibres. It is the hollow fibres within a composite that exhibit superior characteristics in terms of better compressive strengths as well as stiffness. The compressive strength within a composite of hollow fibre is lower along the transverse direction in comparison to the composites of solid fibre when the hollowness within the fibre is greater than half the total diameter of the fibre. But it is tough to fabricate or handle such fibre composites [26].

In general, composite reinforcement is carried out by fibres that organic and inorganic. Organic fibres are elastic and flexible with lower density, whereas in comparison to organic fibres, inorganic fibres are highly rigid, extremely stable at high temperatures, and possess high elastic modulus. But it is the organic fibres that give composites multiple advantages. Some of the common fibres in use are made of glass (E-glass, R glass, and A-glass) [27], quartz, high silica, silicon carbide, graphite, boron, metal wires, alumina, aramid, and fibres of multiphase. Organic fibres occupy a major market share among which boron, silicon carbide, and graphite are utilized in fabricating advanced composites.

2.2.2.2 Particulate-Reinforced Composites (PRCs)

Composites of ceramics as well as metals exhibit microstructures comprising of diverse geometrical particles such as circle, triangle or square in a phase combined with another dispersed phase but all of similar dimensions are termed as PRCs. It is the dispersion concentration of size as well as volume that is segregated from that of the dispersion materials that are hardened. Particulate composites have a dispersed size in microns and a volume concentration over 28%.

Hence, there isn't much difference observed among particulate composites as well as dispersion strengthened composites. Even the strengthening mechanism for each of them is different. The matrix strength is based on reinforcement dispersion as dislocations motion is arrested which requires greater fracture force to counter the dispersion restriction. The particulates provide strength to the composite via hardness and hydrostatic coercion among the fillers w.r.t the matrix [28]. Isotropy is attained within a 3-D reinforced composites due to the three-dimensional orthogonal planes. Due to heterogeneity, the characteristics of the materials are sensitive to the characteristics of the individual constituents and interfacial characteristics as well as the array geometries. The strength of a composite is influenced by particle diameters, spacing between adjacent particles, as well the reinforcement's volume fraction [29]. Also, the characteristics of the matrix impact the particle-reinforced composite's behaviour.

2.2.2.3 Laminar-Reinforced Composites (LRCs)

The number of LRC combinations are equal to the quantity of available materials. LRCs are basically materials with diverse material layers bonded together with the layers being two or greater than that arranged in an alternate or a specific manner as desired by the application for which they are used. There are multiple areas within laminates of clad as well as sandwich, which is a must, irrespective of their adherence to rule of mixture from the perspective of strength and modulus. Diverse methods of powder metallurgy such as hot pressing, brazing, roll, and diffusion bonding can be utilized in fabricating diverse foils, sheet alloys, and sprayed or powder materials. LRCs cannot have high strength as in FRCs but in comparison to FRCs, material isotropy is easily achieved in foils and sheets along two dimensions [30]. In LRCs, there can be over 92% of sheets or foils, an extremely high percentage but FRCs cannot have such a high fibre percentage, which is limited to just above 75%. It is the cost effective, single-layered metal-metal laminates that provide LRCs some special characteristics and they are fabricated via cladding or pre-coating. In pre-coating, a thin, continuous metal layer is deposited over a substrate obtained via hot dipping or chemical or electroplating. But metal cladding is for intense conditions with dense surfaces.

There exist multiple sheet-foil combinations that act as adhesives in low-temperature conditions. These types of metal or plastic material can be combined with another constituent. The pre-finished or pre-painted metal-organic laminate is considered to be the best, since it eliminates the secondary finishing process [31]. As many as 95% of metal-plastic laminate combinations such as vinyl-metal and organic films-metals are currently in use, which are fabricated via the method of adhesive bonding.

2.2.2.4 Other Fibre Types Used in Composites

2.2.2.4.1 Whiskers

Whiskers are discontinuous, short, single crystal fibres almost without any defects and have diverse cross-sections. Whiskers are generally 3–55 nm long and made of silicon carbide, copper, iron, and graphite and unlike particles, whiskers have the ratio of length to width greater than 1. Whiskers exhibit exceptional strength around 7000 MPa [32]. Though nature comprises of certain geological structures that can be termed "whiskers," the initial development of whiskers occurred in a laboratory. Considering to be a by-product of another structure, whiskers were disregarded initially. But later on they became popular due to research on crystal structures and defects influencing strength in metals. The whiskers-reinforced composites were fabricated via slip-casting and powder metallurgy. It has been proven in laboratories that metal-whisker composites tend to provide strength at elevated temperatures, but due to the fine and small size of whiskers, it is difficult to handle them and use them to achieve composites with enhanced characteristics. The strength of the whiskers has been proven to have an inverse variation with its effective diameter. The 2 to 10 µm range has been proven to be the ideal diameter range for achieving composites with good characteristics.

Ceramic whiskers exhibit high modulus, valuable strengths, as well as small densities. Ceramic whiskers are ideal for low-weight structural composites since they possess exceptional specific modulus and strength [33]. Ceramic whiskers are extremely resistant to oxidation, mechanical damage, and high temperatures. They also are highly responsive in comparison to metal whiskers. But handling them is a huge issue, since they get easily damaged. Hence, ceramic whiskers haven't been used widely for commercial applications.

2.2.2.4.2 Flakes

Flakes often substitute fibres due to their ability to be packed densely. Flakes of metals existing within a polymer matrix are electrically as well as thermally conductive [34], whereas glass or mica flakes act as insulators. In comparison to fibres, production of flakes is cheaper. But they have demerits, such as lack of control in geometries and sizes as well as presence of defects. For example, flakes of glass tend to crack at the edges or have notches that hamper the composite's strength. Flakes can't have parallel orientations in a matrix, which results in uneven strength. Flakes are held together with the help of a binder inside a matrix and, as per the desired component, flakes either exist in smaller quantities or exist in the entire composite. Flakes are advantageous for structure-based applications in comparison to fibres. For instance, flakes within a matrix provide greater strength over fibres. Moreover, angle plying isn't an issue in the case of flakes. The theoretical elastic modulus of flakes is higher than fibres and flakes are cheaper to fabricate along with easier handling when in small quantities.

2.2.2.4.3 Fillers

Fillers are added to polymer matrices to replace a certain matrix portion and modify or improve the composite's characteristics such as better strength at lesser weights. Fillers can also be added to a composite in the form of a second phase. In such cases, the composites will have a porous, honeycomb structure. The infiltrating filler may be completely independent from the matrix and bind it similar to fibres or powders or a filler could just fill the voids. Powdered fillers are also considered to be particulate composites. The porous/spongy nature of composites is due to the method of processing and by making use of metal impregnates, tolerance as well as strength of such composite's matrices can be enhanced. Such filled composites include graphite, metal castings, ceramics, and components fabricated via powder metallurgy. There is a specific design and a predetermined geometry behind a honeycomb-structured matrix and not a natural phenomenon. Foam/resin impregnated hexagonal shaped sheets are employed as core material for sandwich composites.

Fillers within a composite are either the main constituent or additionally used. Fillers may possess irregular structures, or they might possess precise geometries such as spheres, polyhedrons, or short fibres. Though the idea behind addition of fillers to composites is not to have any visual enhancements, in certain cases fillers might add opacity or colour to a composite. Fillers are inert and therefore can modify the properties of any resin across diverse directions and [35] also help in overcoming the limitations as well within a composite. It is the geometry, filler particle size, and blend along with surface treatment and size dispersion that influence the composite's characteristics in the end.

Plastic fillers have the tendency to behave as if they are two diverse constituents. Neither do they alloy nor do they accept bonding; instead they mutually develop. They abstain from having chemical interactions with each other which is essential to avoid destruction of each other's characteristics. In certain filled composites, the matrix just becomes a housing and all the characteristics desired are achieved via the fillers. Though a matrix is majorly included in the composite, the high quantities of the fillers make them the primary constituents. Fillers provide enhanced strength, stiffness, good resistance to abrasion and temperatures, porosity stability, and approving coefficient of thermal expansion [36]. But fabrication of fillers is restricted as they tend to reduce the longevity of certain resins and also weaken the final composites in some cases. There are also some inhibitions pertaining to the curing of resins.

2.2.2.4.4 Microspheres

Microspheres are one of the most utilized fillers. Their ability to alter the composites without impacting the output characteristics along with their own features such as strength, specific gravity, stable size of the particles, and controlled density are primary reasons for the popularity of microspheres. Microspheres, manufactured out of solid glass, are ideal for plastics [37]. These have a coating of binding agent which bonds itself and also the surface of the microsphere to resin within a composite, which enhances the strength of bonding. The liquid absorption within the separations around adjacent spheres is also eliminated. Due to the relative lower densities, solid microspheres impact the weight as well as the commerciality of composites. As per research, strength is inherited to the final composite comprising solid microspheres.

Hollow microspheres, comparatively larger than solid spheres of glass, are silicate based and fabricated via controlling the material's specific gravity. They have a wider particle size range and are utilized in polymers. The hollow microspheres of silica commercially come in diverse compositions comprising of compounds. It is because of this modification, the microspheres are considered to be less sensitive against moisture leading to reduced particle attraction. This is extremely essential in polymer composites that are highly filled by liquid and in which increased viscosity restricts the quantum of loaded fillers. Previously, hollow microspheres were clubbed with thermosetting resin composites [38]. There has been a change in the scenario with the emergence of multiple novel, strong spheres, which, in comparison to hollow microspheres, exhibit a five times higher static crush strength and four times longer shear strength. The use of ceramic, alumino, silicate microspheres in thermoplastic composites have enhanced the strength and abrasion resistance as compared to siliceous microspheres and made them ideal for applications related to high pressures. It has been possible to make use of hollow microspheres to fabricate light compounds dominated by resins due to their specific gravity being lower than that of pure resin. Hence, the hollow microspheres are widely used in lightweight applications of aerospace and automotive for lower energy consumption. The low crush resistance of hollow microspheres refrains them from being used in composites needing moulding at high pressures or elevated shear mixing. Hollow microspheres, upon calculative usage, are capable of removing crazing at the bends of PVC pipes where bending stresses are predominantly seen [37].

Microspheres (hollow/solid), exhibit characteristics that are in direct relation to their spherical geometry and allows them to have a behaviour similar to a miniature ball bearing. Thus, they enhance flow characteristics and also facilitate uniform stress dispersion along the resin matrix. The freely oriented and edge-free microspheres within PMCs produce smooth surfaces since spheres comprise the smallest surface area to volume ratios.

2.2.2.4.5 Directionally Solidified Eutectics or Composite Solidification

Composites are directionally solidified to generate in-situ fibres, which are integral to the melt precipitation during solidification of the alloys. This results in the formation of eutectics in which degeneration of the molten material occurs, resulting in multiple phases at steady temperature. The occurrence of the reaction post the phase solidification results in solidified eutectics [39]. At the time of solidification, nucleation of the crystals occurs from a cooler portion, which forms a structure having multiple grains or particles of crystals which grow into each other. For unidirectional solidification, random coalescence isn't allowed to take place.

2.2.3 Classification Based on Scale

2.2.3.1 Nanocomposites

Nanocomposites are the materials with two or more independent constituents, of diverse physical as well as chemical characteristics that are always distinctively segregated at a microscopic level, whereas they, together, consist of one phase in a dimension lower than 100 nm [40]. In other words, nanocomposites integrate multiple phases of nanodimension into one continuous macrophase to synergise the physical/chemical characteristics of the composite that is different as well as superior to that of the independent constituents. The formed nanocomposite comprises of a continuous matrix phase that engulfs another phase comprising nanofiller (reinforcement) that have their own distinctive phases that have multifunctionality structurally as well as in terms of material characteristics. The scientific development in 20th century resulted in the innovation of nanotechnology, which could produce multifunctional nanomaterials that are sustainable and exhibit greater levels of performance. With the development of novel materials and tools for characterization, nanotechnology paved the way to the next-generation nanocomposite materials, which can be controlled with ease and also comprise of many functionalities of engineering. The classification of nanocomposites is similar to that of the composites where in depending on the nature of matrix, they are classified as polymer nanocomposites (PNCs), metal nanocomposites (MNCs), ceramic nanocomposites (CNCs), and carbon nanocomposites.

Polymers are macromolecules comprised of repeated structural units, connected through covalent chemical bonds. These are used as matrixes for nanocomposites as they comprise of properties like light weight, easy processing, cheaper manufacturing, as well as decent substrate adhesion [41]. PNCs have polymer matrices with nanostructured reinforcements based on which the characteristics of the nanocomposite can be determined. Adding diverse organic/inorganic nano-reinforcements to polymer matrices results in enhanced characteristics of PNCs than the conventional polymers.

MNCs comprise of metal/alloy matrices that are ductile and reinforced by nanoscaled materials. Typical matrix materials employed in MNCs are iron, aluminium, copper, titanium, nickel, etc. Typical materials used as reinforcements in MNCs are carbides (B_4C/SiC), oxides (SiO_2, Al_2O_3), nitrides (AlN/Si_3N_4), and few other elements [42,43]. MNCs exhibit exceptional mechanical characteristics such as superior strength, specific modulus, as well as stability at elevated temperatures.

Ceramics are basically brittle, but have decent resistance against wear; they are exceptionally stable against chemicals as well as elevated temperatures. To negate this brittleness, ceramic matrix materials are being clubbed with nano-reinforcements in the form of particles, fibres, or whiskers to significantly improve the mechanical characteristics and achieve CNCs. The fabrication method of CNCs is essential in determining multifunctional characteristics of the nanocomposite as desired by the advanced application.

The nanocomposites are being extensively used across diverse fields of engineering like structures, safety, transportation, sensing and actuators, environmental remediation, energy storage, defense systems, biomedicine, novel catalysts, and EM absorption [44]. The reason behind this is the multifunctional properties like electrical, mechanical, optical, magnetic, catalytic, and biological characteristics being imbibed into various types of nanocomposites.

2.2.4 Other Classifications of Composites

2.2.4.1 Natural Composites

These are the composites comprising reinforcements or sometimes the matrix obtained from renewable resources or the nature such as plants or wood. Natural composites are integral to both plants as well as animals. An example of natural fibre composites is integration of plant-derived fibres with a polymeric matrix. Some of the natural fibres are wood, bamboo, sisal, coconut, hemp, jute, cotton, flax, kenaf, abaca, wheat straw, banana leaf fibres, etc. The natural fibre composites exhibit properties such as light weight, low production of energy, and CO_2 sequestration or decreasing the greenhouse effect [45].

Wood is a natural composite of long cellulose fibres (reinforcements) held together by a weaker lignin (matrix). Cotton also has cellulose but has no lignin to bind it hence it is weak. The combination of these two weak substances (lignin and cellulose) forms a stronger composite.

The bone is also a composite made from the hard, brittle, calcium phosphate, hydroxyapatite surrounded by a soft protein, termed "collagen." Even the fingernails and hair comprise of collagen. But collagen alone wouldn't have been of any use in the skeleton system. But a combination of these two form a strong, natural composite that is capable of supporting the entire body. Some other illustrations of natural composites are teeth, pearls, and other related structures of a shell.

REFERENCES

[1] Jones RM. *Mechanics of composite materials*. CRC Press; 2014.
[2] Clyne TW, Hull D. *An introduction to composite materials*. Cambridge University Press; 2019 Jul 11.
[3] Chawla KK. *Composite materials: Science and engineering*. Springer Science & Business Media; 2012 Sep 26.
[4] Kumlutaş D, Tavman IH, Çoban MT. Thermal conductivity of particle filled polyethylene composite materials. *Composites Science and Technology*. 2003 Jan 1; 63(1): 113–117.
[5] Tsai SW, Hahn HT. *Introduction to composite materials*. Routledge; 2018 May 2.
[6] Yashas Gowda TG, Sanjay MR, Subrahmanya Bhat K, Madhu P, Senthamaraikannan P, Yogesha B. Polymer matrix-natural fiber composites: An overview. *Cogent Engineering*. 2018 Jan 1; 5(1): 1446667.
[7] Christensen RM. *Mechanics of composite materials*. Courier Corporation; 2012 Mar 20.
[8] Altenbach H, Altenbach J, Kissing W. Classification of composite materials. In *Mechanics of composite structural elements* (pp. 1–14). Berlin, Heidelberg: Springer; 2004.

[9] Rajak DK, Pagar DD, Kumar R, Pruncu CI. Recent progress of reinforcement materials: A comprehensive overview of composite materials. *Journal of Materials Research and Technology*. 2019 Nov 1; 8(6): 6354–6374.

[10] Saravana Bavan D, Mohan Kumar GC. Potential use of natural fiber composite materials in India. *Journal of Reinforced Plastics and Composites*. 2010 Dec; 29(24): 3600–3613.

[11] Kainer KU, editor. Basics of metal matrix composites. *Metal Matrix Composites*. 2006 Aug 21: 1–54.

[12] Chawla KK. Metal matrix composites. In *Composite materials* (pp. 197–248). New York, NY: Springer; 2012.

[13] Low IM, editor. *Advances in ceramic matrix composites*. Woodhead Publishing; 2018 Jan 20.

[14] Skinner T, Rai A, Chattopadhyay A. Multiscale ceramic matrix composite thermomechanical damage model with fracture mechanics and internal state variables. *Composite Structures*. 2020 Mar 15; 236: 111847.

[15] Shalin RE, editor. *Polymer matrix composites*. Springer Science & Business Media; 2012 Dec 6.

[16] Dang ZM, Yuan JK, Zha JW, Zhou T, Li ST, Hu GH. Fundamentals, processes and applications of high-permittivity polymer–matrix composites. *Progress in Materials Science*. 2012 May 1; 57(4): 660–723.

[17] Thakur VK, Thakur MK. Processing and characterization of natural cellulose fibers/thermoset polymer composites. *Carbohydrate Polymers*. 2014 Aug 30; 109: 102–117.

[18] Downey MA, Drzal LT. Toughening of carbon fiber-reinforced epoxy polymer composites utilizing fiber surface treatment and sizing. *Composites Part A: Applied Science and Manufacturing*. 2016 Nov 1; 90: 687–698.

[19] Li Z, Haigh A, Soutis C, Gibson A, Wang P. A review of microwave testing of glass fibre-reinforced polymer composites. *Nondestructive Testing and Evaluation*. 2019 Apr 29; 34(4): 429–458.

[20] Savage E. *Carbon-carbon composites*. Springer Science & Business Media; 2012 Dec 6.

[21] Balasubramanian M. *Composite materials and processing*. Boca Raton: CRC Press; 2014.

[22] Bunsell AR, Joannès S, Thionnet A. *Fundamentals of fiber reinforced composite materials*. CRC Press; 2021 Mar 28.

[23] Rahmani H, Najafi SH, Saffarzadeh-Matin S, Ashori A. Mechanical properties of carbon fiber/epoxy composites: Effects of number of plies, fiber contents, and angle-ply layers. *Polymer Engineering & Science*. 2014 Nov; 54(11): 2676–2682.

[24] Harris M, Potgieter J, Archer R, Arif KM. Effect of material and process specific factors on the strength of printed parts in fused filament fabrication: A review of recent developments. *Materials*. 2019 Jan; 12(10): 1664.

[25] Hang Y, Liu G, Huang K, Jin W. Mechanical properties and interfacial adhesion of composite membranes probed by in-situ nano-indentation/scratch technique. *Journal of Membrane Science*. 2015 Nov 15; 494: 205–215.

[26] Korikov AP, Kosaraju PB, Sirkar KK. Interfacially polymerized hydrophilic microporous thin film composite membranes on porous polypropylene hollow fibers and flat films. *Journal of Membrane Science*. 2006 Aug 1; 279(1–2): 588–600.

[27] Wallenberger FT, Hicks RJ, Simcic PN, Bierhals AT. New environmentally and energy friendly fibreglass compositions (E-glass, ECR-glass, C-glass and A-glass)–advances since 1998. *Glass Technology-European Journal of Glass Science and Technology Part A*. 2007 Dec 1; 48(6): 305–315.

[28] Hunt WH, Herling DR. Metal matrix composites. *Advanced Materials & Processes*. 2004 Feb 1; 162(2): 39–42.

[29] Wu C, Ma K, Wu J, Fang P, Luo G, Chen F, Shen Q, Zhang L, Schoenung JM, Lavernia EJ. Influence of particle size and spatial distribution of B4C reinforcement on the microstructure and mechanical behavior of precipitation strengthened Al alloy matrix composites. *Materials Science and Engineering: A*. 2016 Oct 15; 675: 421–430.

[30] Staab G. *Laminar composites*. Butterworth-Heinemann; 2015 Aug 11.

[31] Singh J, Bhunia H, Saini JS. Mechanical properties of clay/TiO$_2$ epoxy hybrid nanocomposites (Doctoral dissertation).

[32] Pan D, Zhang X, Hou X, Han Y, Chu M, Chen B, Jia L, Kondoh K, Li S. TiB nano-whiskers reinforced titanium matrix composites with novel nano-reticulated microstructure and high performance via composite powder by selective laser melting. *Materials Science and Engineering: A*. 2021 Jan 2; 799: 140137.

[33] Chen X, Li T, Ren Q, Wu X, Li H, Dang A, Zhao T, Shang Y, Zhang Y. Mullite whisker network reinforced ceramic with high strength and lightweight. *Journal of Alloys and Compounds*. 2017 Apr 5; 700: 37–42.

[34] Luo J, Cheng Z, Li C, Wang L, Yu C, Zhao Y, Chen M, Li Q, Yao Y. Electrically conductive adhesives based on thermoplastic polyurethane filled with silver flakes and carbon nanotubes. *Composites Science and Technology*. 2016 Jun 6; 129: 191–197.

[35] Buchalla W, Attin T, Hilgers RD, Hellwig E. The effect of water storage and light exposure on the color and translucency of a hybrid and a micro-filled composite. *Journal of Prosthetic Dentistry*. 2002 Mar 1; 87(3): 264–270.

[36] Friedrich K, Zhang Z, Schlarb AK. Effects of various fillers on the sliding wear of polymer composites. *Composites Science and Technology*. 2005 Dec 1; 65(15–16): 2329–2343.

[37] Amos SE, Yalcin B. *Hollow glass microspheres for plastics, elastomers, and adhesives compounds*. Elsevier; 2015 Apr 30.

[38] Crotts G, Park TG. Preparation of porous and nonporous biodegradable polymeric hollow microspheres. *Journal of Controlled Release*. 1995 Aug 1; 35(2–3): 91–105.

[39] LLorca J, Orera VM. Directionally solidified eutectic ceramic oxides. *Progress in Materials Science*. 2006 Aug 1; 51(6): 711–809.

[40] He Q, Yuan T, Wang Y, Guleria A, Wei S, Zhang G, Sun L, Liu J, Yu J, Young DP, Lin H. Manipulating the dimensional assembly pattern and crystalline structures of iron oxide nanostructures with a functional polyolefin. *Nanoscale*. 2016; 8(4): 1915–1920.

[41] Ajayan PM, Schadler LS, Braun PV. *Nanocomposite science and technology*. John Wiley & Sons; 2006 Mar 6.

[42] Mohit H, Selvan VA. Physical and thermomechanical characterization of the novel aluminum silicon carbide-reinforced polymer nanocomposites. *Iranian Polymer Journal*. 2019 Oct; 28(10): 823–837.

[43] Hu Z, Tong G, Nian Q, Xu R, Saei M, Chen F, Chen C, Zhang M, Guo H, Xu J. Laser sintered single layer graphene oxide reinforced titanium matrix nanocomposites. *Composites Part B: Engineering*. 2016 May 15; 93: 352–359.

[44] Reddy B, editor. *Advances in nanocomposites: Synthesis, characterization and industrial applications*. BoD – Books on Demand; 2011 Apr 19.

[45] Daniel IM, Ishai O, Daniel IM, Daniel I. *Engineering mechanics of composite materials*. New York: Oxford University Press; 2006.

Chapter 3

Different Techniques for Designing and Fabrication of Advanced Composite Materials

Subhash Singh

Department of Mechanical and Automation Engineering, Indira Gandhi Delhi Technical University for Women, New Delhi, India

Rama Kanti

Centre for Nanotechnology, Indian Institute of Technology Roorkee, Uttrakhand, India

Vikas Kumar

Department of Automotive Technology, Mechanical Department Division, Federal TVET Institute, Addis Ababa, Ethiopia

CONTENTS

DOI: 10.1201/9781003327370-3

3.1 INTRODUCTION

Advanced materials are an important part of the engineering field. Composites are the result of advancements in the field of materials and are used in diverse applications of engineering. The composites comprise of two distinctive phases called the matrix and reinforcement phases in which the matrix phase engulfs the reinforcements is comparatively weaker than the reinforcement phase. Due to their superior performance as well as exceptional properties such as exceptional strength, toughness, light weight, decent resistance to corrosion, prominent stiffness, ability to absorb high energy, and inferior levels of thermal expansion, the composites have substituted the regular ceramics, metals, and polymers [1]. Many composite applications include structural applications in the fields of aerospace, automotive, biodegradable implants in orthopaedics, biomedical stents and drug delivery systems, super capacitors in electronics, etc. [2,3]. Fabrication technique plays a significant role in determining the physical and mechanical characteristics of the composite. Therefore, it needs to be given a prime importance while looking into various composite applications in the engineering field [4]. The selection of the manufacturing technique of the composite should preferably in concurrence to the type of material and its structural design. In other words, this method of selecting the nature of manufacturing technique must be integral and interactive to the material of the composite selected since the fabrication technique depends of the matrix in a composite [5]. Sometimes, during manufacturing, there may be certain issues that could arise such as porosity, voids, dry spots, and degradation of the polymer in the case of polymer matrix composites. Hence, the fabrication technique should aim at certain aspects such as reinforcement orientation/location control, control over the thickness of the ply, fibre volume ratio, voids, as well as residual stresses need to be looked into and the final dimensions should also be taken care of [6]. There shouldn't be any deviations from the pre-set values of the temperature during manufacturing. Dispersion of temperature as well as curing should be uniform throughout and should be carried out as quickly as possible. The best fabrication process should produce high-quality composites at an optimum cost. In general, good quality manufacturing methods are expensive, particularly in the areas of tooling as well as production-related costs. The methods of fabrication slightly vary for polymer matrix composites and the metal/ceramic matrix composites. The subsequent sections in this chapter broadly discuss the various manufacturing methods for all kinds of composites classified on the basis of the matrix type.

3.2 MANUFACTURING METHODS FOR FIBRE-REINFORCED/ POLYMER MATRIX COMPOSITES

The classification of fibre-reinforced composites or polymer composites may be done as composites having chopped fibre, continuous fibre, woven fibre, and hybrid fibres. The fibre type within a composite determines the type of manufacturing method to be employed. The fabrication of these composites can be done through various methods such as hand lay-up, spray-up, autoclave moulding and pultrusion, resin transfer moulding (RTM), filament winding, and automated tape laying.

3.2.1 Hand Lay-up Process

The process of hand or the wet lay-up is the oldest and the most basic method used in the fabrication of composites with the least required infrastructure. It is an open mold process in which the steps involved in the fabrication are completely manual and quite simple. Initially, a layer of release gel is applied onto the surface of the mold. This release gel doesn't all the polymer sheets to stick to the mold surface. Thin sheets of plastic are placed at the bottom of the mold surface to obtain good quality surface. A woven/chopped strand mat of the reinforcement is cut according to the dimensions of the mold cavity and placed over the mold surface after the plastic sheet [7]. A liquid comprising a thorough and proportionally mixed thermosetting polymer and a hardener (curing agent) is poured over the reinforcement fibre mat placed in the mold. Using a brush, the liquid is spread uniformly along the reinforcement mat and then over the spread mixture, and another reinforcement mat is placed. A hand roller applying pressure is then moved along the Fibre mat-polymer surface to eliminate the trapped air (if any) as well as the excessive polymer liquid. The process is repeated till the desired layer quantities are stacked and the desired thickness of the composite is achieved. Finally plastic sheet is placed on top of the polymer layers and release gel is again applied on the inner surface of the upper mold plate, which is placed over the stacked polymer layers applying pressure [8]. The entire setup is left for a certain duration for the curing process to take place at room or any pre-determined temperature. Once the mold is opened, the obtained final product is a polymer matrix composite that further undergoes secondary operations as per the requirements. Fig. 3.1 represents a schematic of a hand lay-up process.

The curing time of the composite is determined by the nature of the polymer employed in the prepared composite. For a conventional epoxy system, the duration of curing varies in the range of 24–48 hours at room temperature. This is ideal for composites comprising a thermosetting polymer matrix. Some of the commonly employed polymers in this process are unsaturated polyester, polyvinyl ester, polyester, epoxy, polyurethane, and phenolic resins. Fibres like aramid, carbon, glass, and naturally

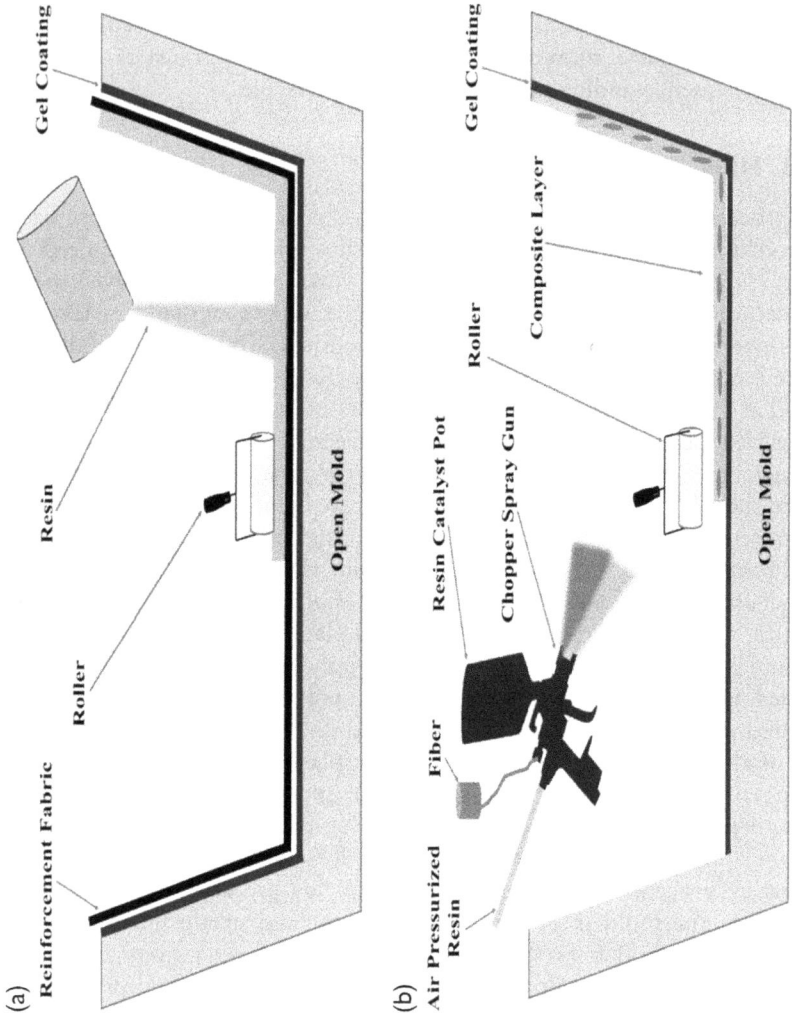

Figure 3.1 **(a)** Hand lay-up process; **(b)** spray lay-up process [6,9].

obtained Fibres are the commonly employed as reinforcements of diverse orientations. The hand lay-up process requires less infrastructure as well a capital in comparison to the other techniques. It is a challenging task in this process to achieve high productivity as well as high-volume fraction of the reinforcements within the composites. Therefore, the achieved composite characteristics wouldn't be exceptionally good to be used in extreme conditions and so its applications are limited to certain areas of components in aircrafts, automobiles, and hulls of boats.

3.2.2 Spray Lay-up Process

The spray lay-up technique is similar to that of the hand lay-up process except that a hand gun is utilized in this process to spray a combination of the resin and chopped Fibre, as shown in Fig. 3.1b. An inexpensive mold is employed for fabricating large products. The resin-Fibre mixture is sprayed into the mold and allowed to completely cure at ambient temperature, following which the final composite is obtained. Prior to curing, the resin-Fibre mixture is rolled using the rollers to get rid of the voids as well as bubbles [9]. Rollers and brushes are employed to clean the wet Fibre as well as the air. The mechanical characteristics of the final composite are influenced by the orientations as well as the constraints of the Fibres.

3.2.3 Autoclave Moulding Process

The method of autoclave moulding is employed to fabricate advanced composites that are capable of high performance in applications of aerospace, marine, military, infrastructure, and transportation. The restraints of this process are limited to geometry and size but the obtained composites have good tolerances in terms of dimensions. This method is expensive due to its low volume of production and huge labour requirement. Thermoplastic as well as thermosetting resins comprising glass, aramid, carbon, fabrics, and fibres are generally fabricated using this method. The setup of a typical autoclave process is shown in Fig. 3.2a [10].

The material an autoclave uses is a partially cured fabric reinforcement, pre-impregnated with a polymer resin called prepeg. The sheets of prepeg are systematically oriented, cut as per the desired dimensions and stacked as layers called lay-up. The autoclave has a bleeder, breather system which is inclusive of a fabric made or glass fibre to facilitate the absorption of excessive resin and the escapade of volatile gases at the time of curing. The prepeg lay-up and auxiliary material assembly is sealed over a tool plate with the help of a vacuum bag. Inside the autoclave chamber, a temperature-pressure-vacuum-time cycle undergoes to facilitate curing [11]. The cycle starts with an initial heating of the lay-up at 2–4°C/min rate up to 110–125°C temperature under full vacuum conditions for removal of volatile gases and complete melting of the resin without excessive bleeding

(a)

(b)

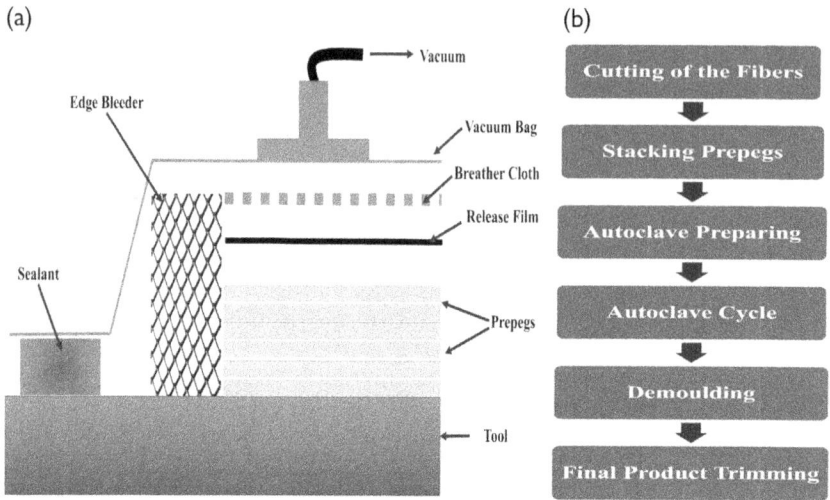

Figure 3.2 (a) Autoclave setup; (b) steps in autoclave process [10].

taking place in the resin. Vacuum is maintained for 1 hour dwell period at 110–125°C, after which 550–690 kPa of pressure is applied following up with the vacuum removal in the end. At last, temperature within the chamber is elevated by 2–4°C/min rate to a maximum of 175°C and held for 2 hours. The setup is gradually cooled to avoid micro cracks and to reduce the residual stresses. The various steps involved in an autoclave process are represented in Fig. 3.2b.

3.2.4 Vacuum Bag Moulding

The process of vacuum bag moulding strives to create a vacuum to get rid of the captured air, volatile gases, and the excessive resin. After completing the lay-up process of the reinforcement fabric as well as the resin, a non-adhering film of PVA or nylon is placed on top of the lay-up and sealed. The PVA/nylon film acts as a bag over a mould within which a vacuum is created and maintained till the curing of the entire composite takes place at ambient or a desired specific temperature [12]. The atmospheric pressure is utilized for sucking the air from under the vacuum bag, thereby compacting the layers of the composite, resulting in a laminate of superior quality.

3.2.5 Pressure Bag Moulding

This process is similar to that of the vacuum bag moulding except for the air pressure. In this, the entrapped gases as well as the excessive resin under the PVA/nylon film (bag) are removed through the application of air pressure.

This bag encloses the Fibre lay-up as well as the resin. In some cases, steam under pressure is employed in place of air, which serves both the purposes of curing the composite as well as removing the extra air [13].

3.2.6 Filament Winding

In filament winding process, preimpregnated or resin coated reinforcement loop is wound under tension to a rotating mandrel. This process is ideal for products with surfaces of revolution such as tubes, pipes, pressure vessels, rocket motor cases, conical structures, etc. The tension in the fibre results in compaction. Products having uniform as well as controlled orientations of Fibre and high specific strength are produced via filament winding. The process of filament winding is seen in Fig. 3.3 [14].

The strands could be preimpregnated with partially cured resins (dry winding) or may be allowed to pass via resin bath to be wetted (wet winding). Considering the various methods of winding, for planar winding process, the feeding arm of the fibre rotates along the length of the stationery mandrel. The helical winding process, a fibre shuffling back and forth, is fed at a restrained velocity to a rotating mandrel to obtain the required helical angle. Even the products without surfaces of revolution can be fabricated via this method. The curing of wound mandrels can also be carried out within an autoclave under controlled conditions of curing cycle [15]. Certain components such as rotor blades or a helicopter or hat part can be fabricated via a filament winding process.

3.2.7 Resin Transfer Moulding (RTM)

The RTM process, employs a closed mould in which the resin is transferred to a pre-existing reinforcement (woven/strand mat) placed over a lower half of a mould. This mould surface is coated with a release gel for a composite to be easily removed and then the mould is closed using a clamp. The

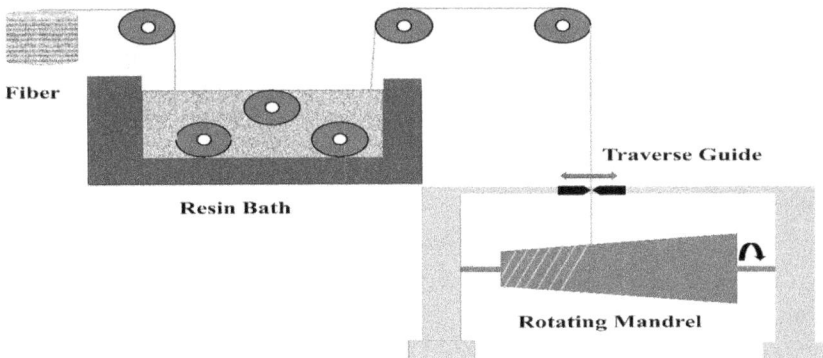

Figure 3.3 Filament winding process [14].

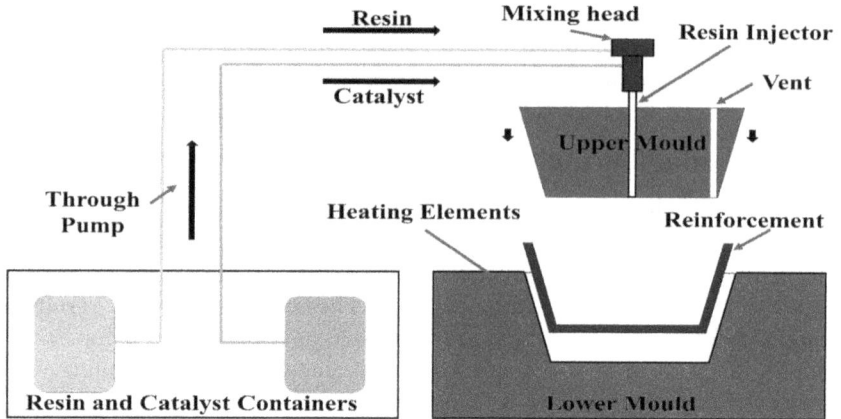

Figure 3.4 Resin transfer moulding process [16].

mechanism of clamping employed is either press or a perimeter clamping. The uniform pumping of the resin occurs via mould ports seeking the support of a catalyst as well as vacuum, resulting in air displacements via vents. Once cured, the composite is removed by opening the mould. The resin as well as the catalyst are store in diverse containers (Fig. 3.4) in which resin is stored in a larger one [16]. Each container has its own outlet that pumps the respective contents into a mixing chamber where the resin as well as the catalyst are mixed thoroughly. Resin is then injected via an injector, into a mould cavity comprising two halves (upper/lower) with an integrated heating system. The unwanted gases escape via the vents at the time of clamping.

The mould in resin transfer moulding can be hard/soft as per the run's time length. Thermosets such as polyester/epoxy are employed in a soft mould, whereas aluminium/steel are employed in a hard mould, which further determines the cost as well as the longevity of the moulds. The cycle time of RTM can be lowed via automation in the case of complex geometries, by making use of preformed Fibre reinforcement. The resin's viscosity is essential for RTM since it influences the resin's injection time [17]. Highly viscous resins need higher injection pressures that can displace the Fibres, resulting in a phenomenon called Fibre wash.

3.2.8 Pultrusion

Pultrusion is a closed moulding, composite preparation process that is continuous and automated in which pulling of the material occurs through dies. The continuous Fibre mats/roving of the reinforcement is unrolled from a creel and allowed to pass via an open resin tank, dipping and wetting them completely. These resin-impregnated fibres are allowed to pass through a die with a heating arrangement to get the desired geometry,

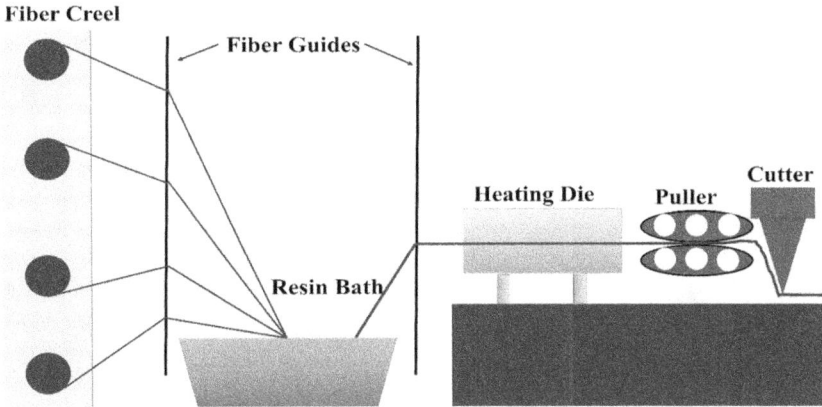

Figure 3.5 Pultrusion process [18].

which also facilitates curing. The cured composite fibres coming out from the hot die are then pulled by grippers. At last, an integrated cutter cuts the composites into the required dimensions, as shown in Fig. 3.5 [18].

In some cases, fillers are also added within the resin bath, which combine with the passing fibre reel [19]. In most cases, excessive resin is eliminated in the hot die because of excessive pressure but certain pultrusion systems make use of a pre-former placed after the resin tank and prior to the die. This pre-former is essential for pultrusion as it removes the excessive polymer, leading to the formation of uncured composites, which are sequentially aligned and passed along the hot die. A dysfunction of the pre-former could affect the quality of the composite. Pultrusion is employed for continuous production of thermosetting polymer of a uniform cross section. The fibre alignment, dispersion, as well as the resin impregnation are exceptional because of this uniform cross section of the final composite. The pultrusion production rate is high, but cross-sectional variability isn't possible in pultrusion, which has a lower investment cost in comparison to other processes.

3.2.9 Automated Tape Laying (ATL)

The ATL is a well-established automated production method for the composites. Wide tapes of Fibre in a single direction are laid over a mould through a roller system that is loaded with varied degrees of articulation as per the composite's complexity that is being fabricated. The automated tape deposition is done at elevated velocities with great control and over large parts in the ATL process. Though the ATL isn't a method for placing Fibres, there is complete control over the cutting as well as orientation of the Fibres, facilitating complex processes instead of simple laying of plies to form a laminate, as shown in Fig. 3.6 [20]. There are devices in ATL that

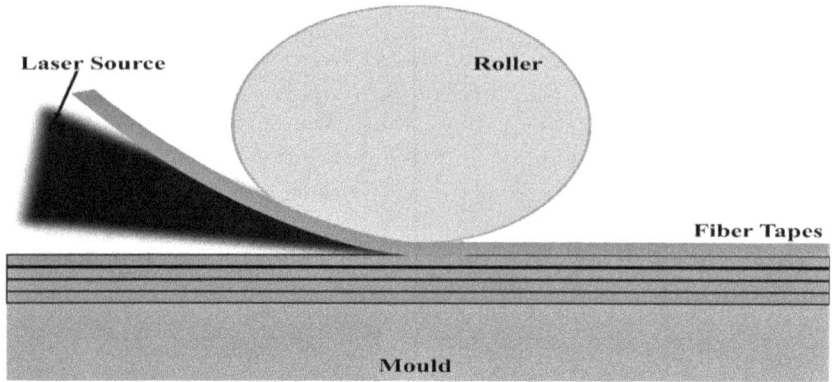

Figure 3.6 Automated tape laying process [20].

are dedicated to place the Fibre prepegs over contoured tools and also along a specified path as per the composite structure's requirements. The sharpness of a turn within the geometry of the composite is influenced by the tape's width along a specific direction. Gradual turns are seen for wider tapes, whereas sharper turns are observed in narrow tapes as per the geometry of complex-shaped composites.

3.3 MANUFACTURING OF METAL/CERAMIC MATRIX COMPOSITES

3.3.1 Solid State Processing

The fabrication of composites via solid state involves mutual diffusion and bonding among the matrix-reinforcement phases in solid state under the influence of elevated pressures as well as temperatures. Some commonly employed solid state fabrication methods of composites are powder metallurgy (PM), diffusion bonding, and spark plasma sintering (SPS).

3.3.1.1 Powder Metallurgy

Powder metallurgy (PM) is among the oldest known techniques in composite fabrication, which employs blend press sintering where the reinforcement is in the form of particulates/whiskers. Initially, blending of the matrix as well as the reinforcement powders is done, which are then compacted in the influence of elevated pressures inside a mould. This follows sintering of the compacted powder mixture under the matrix's melting point in a controlled atmosphere to facilitate diffusion bonding between the matrix-reinforcement phases existing in a solid, powdered form and also to eliminate oxidation. Sintering lessens porosity as well as mechanical integration. Sintering is influenced by heating rate and time as well as surrounding conditions. The secondary processes then follow the sintering process [21,22]. The composite of desired characteristics is finally obtained post secondary operations which then can be utilized for any of the desired applications. The steps involved in PM process are represented in Fig. 3.7a [22].

The recent advancements resulted in the development of a quick, hybrid microwave sintering, which uses both methods of heating and a microwave. In conventional sintering, heat is transmitted to the composite material via conduction or radiation within the furnace. In the case of sintering by a hybrid microwave, electromagnetic field energy transmission to the composite material which facilitates quick molecular interactions due to volumetric heating, resulting in lower sintering temperature and time. Therefore, reheating can be eliminated. Enhanced material density has been observed in the hybrid microwave sintering technique [23].

Another advancement in the PM process is the semi-powder metallurgy, which has been employed in the fabrication of Mg composites, as shown in Fig. 3.7b. Mg powder has been combined with ethanol and also reinforcements combined separately with ethanol. The matrix reinforcement combination is done via magnetic stirrer [24]. To achieve homogeneity mixture of Mg ethanol was combined to mixture of reinforcement–ethanol in which the latter was added as droplets. The combination was transformed to powder through vacuum drying and filtering as a variation to the conventional PM and subsequently the steps involved in PM were employed. Despite the dense composites and enhanced mechanical characteristics, semi-powder metallurgy isn't suitable for complex composite geometries. The binder, which holds the matrix and the reinforcement during sintering, needs to be removed later and in this case it is complicated.

3.3.1.2 Diffusion Bonding

Diffusion bonding is another popular composite fabrication technique which involves solid matrix and reinforcements phases. The matrix, a thin 100 µ foil and the reinforcement, a long fibre, are systematically arranged one over the other in a die or a ceramic preform at high temperatures and pressed, as shown in Fig. 3.8. The obtained composite is a multi-layered laminate. This process is employed for components with simple geometries like tubes/plates [25]. Wettability is the prime factor that influences in the process because applying pressure leads to molecular diffusion among the matrix-reinforcement layers which bonds them together. Energy exchange occurs under the applied pressure as well as the temperature and the composite lamina tries to obtain a shape that is further endorsed by a die to obtain definite geometry. Diffusion bonding can also facilitate integration between two different parts [26].

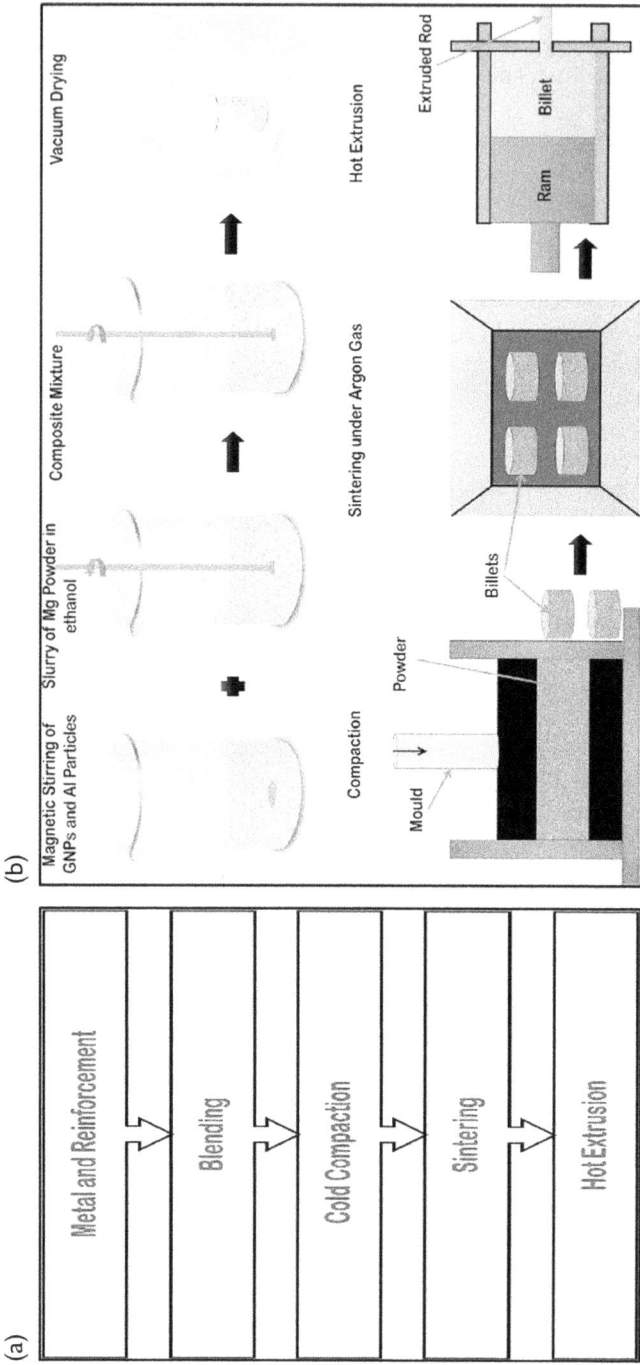

Figure 3.7 **(a)** Powder metallurgy process; **(b)** semi-powder metallurgy process [22].

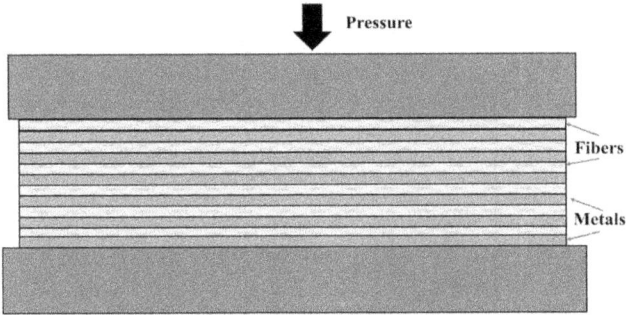

Figure 3.8 Diffusion bonding [25].

3.3.1.3 Spark Plasma Sintering (SPS)

The SPS is an ideal method for solid state fabrication of thermally conductive composites. This process involves sintering by applied pressure when a DC charge electrical pulse passes via a powder compact that is conductive in nature [27,28]. This is also feasible for materials that are non-conductive but that application is extremely limited [26]. The heat in case of conventional sintering is externally supplied but in SPS, heat among the particles of powder is internally generated. The SPS employs extremely short sintering times (5–20 minutes approx.). Due to its conductivity, inclusion of fibres of carbon within a metal powder will result in joule heating as well as the discharge of spark, thereby melting the metal particles' surfaces. This will bond the metal-carbon powder well and reduce the thermal resistance among the matrix and the reinforcements.

3.3.2 Liquid State Processing

This classification involves manufacturing methods in which reinforcements in particulate or powdered form are added to the metal matrix existing in a molten state. Generally reinforcements of ceramics are employed in this process. Wettability is a major factor in the fabrication of composites through this process.

3.3.2.1 Stir Casting

Stir casting is ideal and widely employed for large-scale fabrication of composites comprising up to 30% volume fraction of reinforcement. In this process, a metal matrix is the molten form to which reinforcement is solid, particulate form is added by continuous stirring of the molten matrix. Vigorous stirring of the matrix gets rid of particle agglomeration. In the present scenario, two-stage mixing of the stir casting method is being carried out in which the metal is heated over its liquidus and solidus points

(a)   (b)

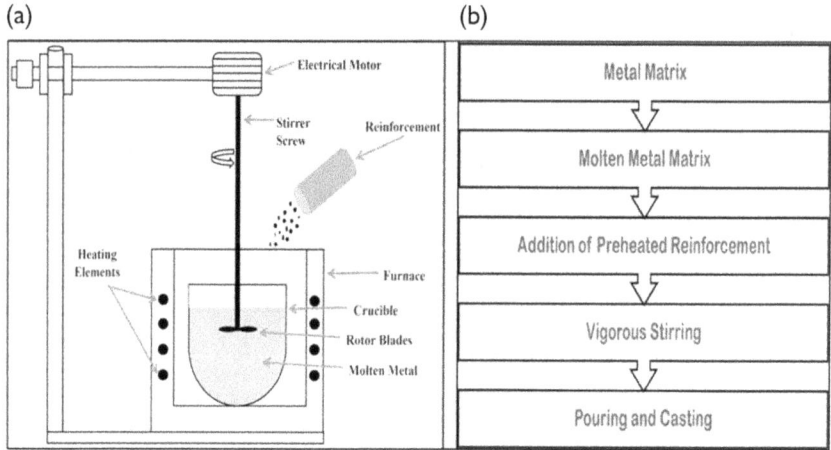

Figure 3.9 **(a)** Setup of stir casting process; **(b)** various steps involved in stir casting [22].

to attain a semi-solid form within a furnace. Subsequently, pre-heated reinforcement particles, ready to be poured, are added to the semi-solid matrix melt through continuous stirring, as shown in Fig. 3.9a, to facilitate uniform reinforcement dispersion and high mechanical characteristics of the casting [22]. Non-uniform particle dispersion will eventually result in casting failure. The highly viscous matrix melt gets rid of the gas layer over the surface of the particle and facilitates easy mixing of the secondary molten form. Typical methods are utilized in cooling the obtained composite post-mixing. It is the secondary operations post-casting that reduce porosity [29]. Various steps involved in the stir casting method are listed in Fig. 3.9b.

Stir casting has certain disadvantages like non-uniform reinforcement dispersion, agglomeration, and sedimentation of the reinforcement particles at the time of pouring. Multiple factors such as wettability, mixing strength, and solidification rate impact the dispersion of reinforcement [30]. Therefore, it is essential to select optimum parameters of processing to achieve acceptable dispersion in the reinforcement particles.

3.3.2.2 Squeeze Casting

Squeeze casting, carried out in a closed mould, was developed in 1931 and it can be categorized as direct and indirect squeeze casting methods. To fabricate composites, a preform of the reinforcement is placed within a closed die and a high-pressured (100 MPa) molten metal is injected into the die. This molten metal passes through the pores of the placed preform and then is left to cool, which upon solidification, forms the composite.

3.3.2.2.1 Direct Squeeze casting

Direct squeeze casting is also called liquid metal forging. As seen in Fig. 3.10a, the lower half of the die is pre-heated and has a preform placed in it through which the poured molten metal will pass in favorable conditions of pressure. This will produce the desired composite. The reinforcement preform is fabricated via vacuum/press forming methods. It is essential for the metal melt to possess sufficient wettability and be flowable enough to infiltrate through the preform [31]. Forging is then carried out on in conditions of elevated temperature and pressure. The pressure is

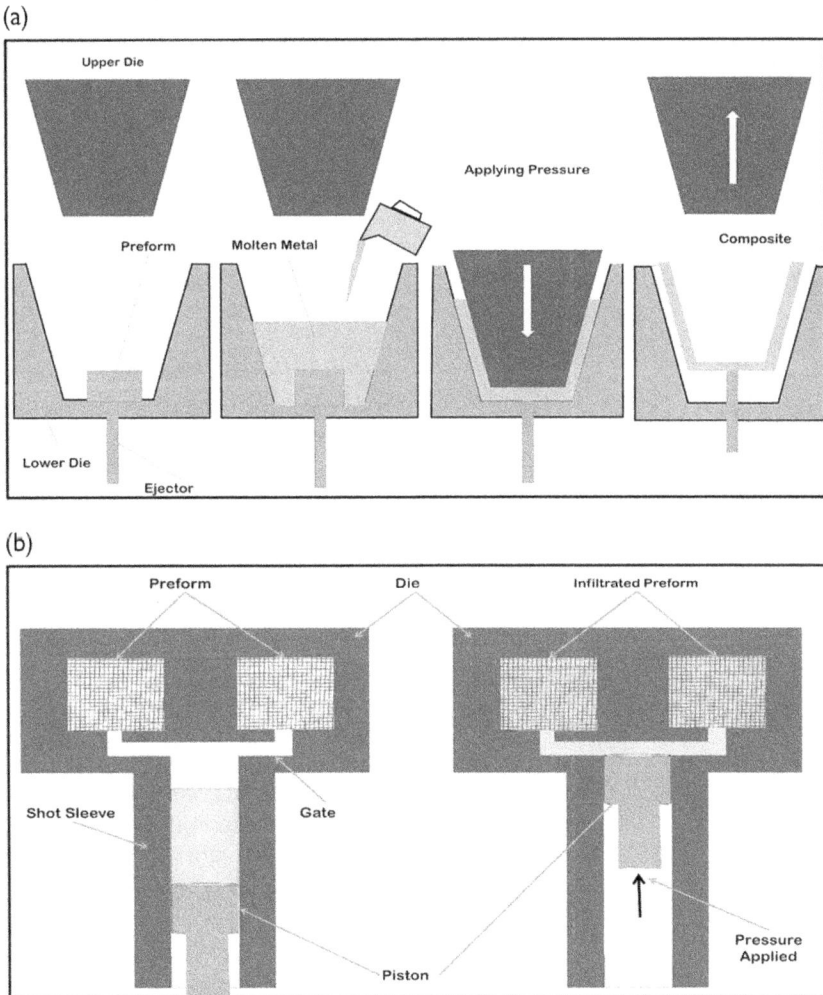

(a)

(b)

Figure 3.10 (a) Direct squeeze casting; (b) indirect squeeze casting [22].

sustained until the entire casting is solidified upon which the composite is taken out. This method is ideal for lower melting point material composites having reinforcement volumes lower than 50%, like Mg, because casting having small grains and low shrinkage and porosity are produced, which gives exceptional mechanical characteristics. Pressure applied in squeeze casting should be perfectly judged, especially while using Mg since it can completely oxidize the composite casting if there are any gases trapped inside because of turbulence of the metal melt. These gases can also damage the reinforcement, leading to low-grade casting. Therefore, care needs to be taken while choosing the pressure range. Mass production of composites isn't feasible with squeeze casting due to the dimensional and geometrical checks in this process [32]. The present scenario witnesses squeeze casting of Mg composites in two steps, which include low pressured infiltration of the material and high-pressured solidification to overcome the above pressure-related limitations.

3.3.2.2.2 Indirect Squeeze Casting

This method is just like the high-pressured processes of die casting with vertical/horizontal machines. The melt of the metal is injected via a closed tube through a shot sleeve comprising multiple large-sized gates and into the cavity of a die at speeds as low as 0.5 m/sec (Fig. 3.10b). It is the lower speeds of injection and deprived turbulence of the melt that facilitate uniform die filling and eliminate the chances of gases being trapped and avoid porous castings. Shrinkage porosity is also lowered due to low superheat of melt [33]. This is because fluidity isn't the reason for filling of the die; instead, it is the high pressure that is employed. When the castings are large and heavy, shrinkage porosity can be high because of extreme pressures; in that case, squirting of melt is done from hot spots to shrinkage pores that emerge eventually eliminating the pores. Molten metal moves in composites with large freezing ranges to produce castings of good quality at moderate pressures (Fig. 3.11).

3.3.2.3 Spray Deposition

Spray deposition is a technique for producing composites by spraying metal over a substrate. The reinforcement is added to a melt as in stir casting or pooled by injecting the reinforcement onto the stream of metal spray spraying or the metal-reinforcement are sprayed simultaneously as a matrix over a substrate. The methods of spray deposition are segregated according to the stream of droplet obtained through melt bath and cold metal's continuous feeding to rapid heat injection area. Injecting ceramic powder to a spray for producing composites has been widely discovered commercially having restrained success. The speed of the droplets is around 20 to 40 m/s, which results in the formation of a thin semi-solid layer over the substrate

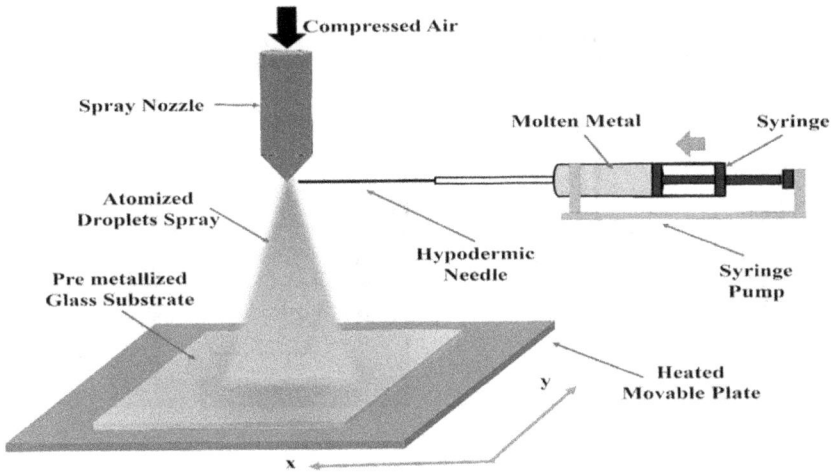

Figure 3.11 Spray deposition method [34].

[34]. The composite thus fabricated exhibits dispersed reinforcement (ceramic) particles. The as-sprayed nature exhibits 5 to 10% of porosity. In case of thermal spraying speeds are much higher with lower porosities. But the issues lie in fabrication of metal matrix composites of acceptable porosity and also to achieve uniform dispersion of the Fibres.

3.3.3 Vapour State Processing

Generation of a film that is thin over a material's surface (substrate) through a substance deposition in the vapour phase is feasible and the resulting coating of one material over another is similar to a composite. The application of this process is limited only to thin film coating due to longer durations in deposition of thick coatings of films. The vapour methods are classified as chemical vapour deposition (CVD) and physical vapour deposition (PVD). There are many processes that have been developed for the production of thin films. Those processes are roughly classified into CVD and PVD processes.

3.3.3.1 Chemical vapour Deposition (CVD)

In CVD, a reaction between flowing materials in gaseous form react with a solid, hot material's surface, depositing the by-products of the reaction over the substrate as a thin layer. Three types of CVD methods are employed: a) Reducing the metal halides via hydrogen in the presence of a catalyst, b) decomposing the phases of gas using thermal energy and reacting a substrate with phases of gas. The substrate's temperature determines the rate of crystallization/deposition. Hence, even the deposited

(a)

(b)

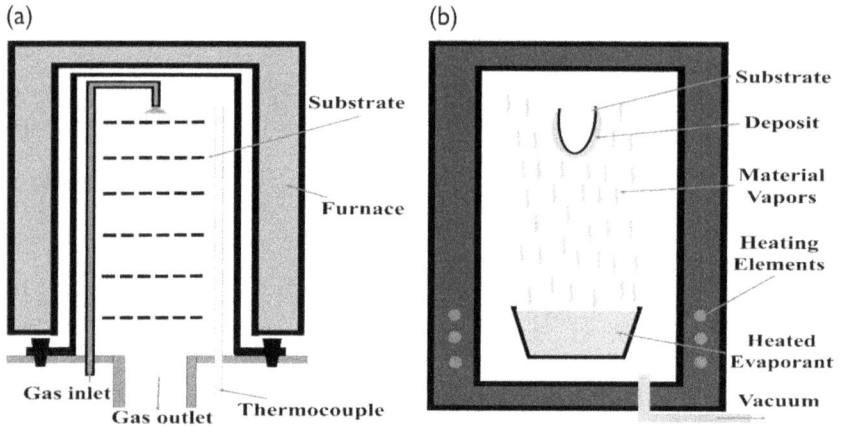

Figure 3.12 (a) CVD process; (b) PVD process [26].

layer's microstructure is determined by the substrate temperature [35]. Usually, the growth of diverse single crystals is observed at elevated temperatures, whereas phases that are amorphous/polycrystalline grow at low temperatures. For coating, it is essential that polycrystalline deposits are fine, exhibit homogeneity and dense mechanical characteristics. A CVD setup comprising a heating system near the internal walls of the furnace is represented in Fig. 3.12a and is a "hot/cold wall type" that can be used for coating multiple substrates simultaneously [26]. When induction heating frequency is high, the same furnace wall turns cold. Post-CVD, vast variations in the coefficient of thermal expansion among the substrate and coating materials will result in the crack formations at the interface of coating layer and the substrate. This crack formation can be avoided when thinner coatings are used and certain coating material characteristics are compromised. Another way is to make use of a diffusion layer or an intermediate phase comprising both materials at the coating-substrate interface.

3.3.3.2 Physical vapour Deposition (PVD)

PVD comprises a series of methods for depositing thin material layers ranging from nanometres (nm) to micrometers (μm). Fibres are passed on a continuous basis via an area having metals of high partial pressure, which need to be deposited. vapour is generated and included within the process which post condensing, gets deposited at 5 to 10 μm in 1 minute, over a Fibre substrate to generate a coating. Finally, Fibre consolidation is done through hot pressing. PVD methods are eco-friendly and conducted in a vacuum conditions. PVD comprises the following basic stages. Initially, a solid material is vapourized with the support of heat, vacuum, and gas plasma. The generated vapour is carried to the substrate material in conditions of partial or complete vacuum. Finally, the substrate condenses over the substrate's surface and thin films are produced.

PVD methods are classified as thermal evapouration and sputtering. The sputtering process is further classified as conventional sputtering as well as ion beam sputtering. Thermal evapouration focusses on vapourizing the solid evapourant through heat in a vacuum medium. On the other hand, conventional sputtering vapourizes the solid evapourant with the help of plasma by bombarding the source target. Accelerated gaseous ions of argon, nitrogen or oxygen, in a pressure range of 1 to 10 Pascal are liberated from the negative charge of the evapourant which form the plasma source [26]. A deposition rate of 0.1 μm is achieved in sputtering technique. In this, temperature of the substrate becomes extremely high due to it being exposed to electrons having high velocities. The mechanism of ion beam sputtering is similar to the conventional sputtering except that vapourization is carried out by irradiating the source at 0.1 to 1 Pascal though a separate chamber of ion source [36,37]. The irradiation releases atoms that get deposited over the substrate. Another difference is that, in ion beam sputtering, substrate temperatures are extremely low in comparison to the conventional sputtering as there is no exposure of the substrate to high velocity electrons. Hence deposition can be better controlled. Irrespective of the method to vapourize the source evapourant, all of the above PVD techniques rely on substrate coating post the vapour condensation.

The typical PVD process (thermal evapouration) is represented in Fig. 3.12b. Speaking in terms of fabricating composites via PVD, the matrix in the vapour form is deposited over a reinforcement added in the form of a substrate. The PVD methods of coating are relatively slow but vapourization is much quicker due to the existence of vacuum and lack of mechanical discrepancies over the area of interface [26]. The final composite is obtained once all the coated Fibres are stacked and hot pressed. Around 80% of homogeneous Fibre composition is retained of extreme uniformity in their dispersion. Also, volume fractions of the Fibre are precisely monitored as per the coating thickness.

3.4 CONCLUSION

This book chapter discusses the various fabrication techniques of composite materials. It has been proven that manufacturing method influences the characteristics of the finally obtained composite. Hence, it is essential to focus and enhance upon the various fabrication methods to achieve exceptional composite quality. Many considerations have to be done to select the manufacturing method such as the type of Fibre, its orientation, size used in a composite, or the nature of matrix material, etc. In theoretical aspects it is to be understood that a manufacturing method generating composites of higher volumes of Fibre will be best suited for that particular composite and even the composite obtained via that process exhibits superior mechanical characteristics.

REFERENCES

[1] Daniel IM, Ishai O, Daniel IM, Daniel I. *Engineering mechanics of composite materials*. New York: Oxford University Press; 2006.

[2] Iqbal MM, Kumar A, Singh S. Biodegradable composite materials for orthopedic implant: A review. In *AIP Conference Proceedings* 2021 May 13 (Vol. 2341, No. 1, p. 040031). AIP Publishing LLC.

[3] Pradeep AV, Prasad SS, Suryam LV, Kumari PP. A comprehensive review on contemporary materials used for blades of wind turbine. *Materials Today: Proceedings*. 2019 Jan 1; 19: 556–559.

[4] Noor, MF, Hasan, F, Bhardwaj, S, Hasan, S. Reverse engineering in customization of products: Review and case study. In *Ergonomics for improved productivity 2021* (pp. 587–592).Springer, Singapore.

[5] Bains PS, Sidhu SS, Payal HS. Fabrication and machining of metal matrix composites: A review. *Materials and Manufacturing Processes*. 2016 Apr 3; 31(5): 553–573.

[6] Divya HV, LL Naik, B Yogesha. Processing techniques of polymer matrix composites–A review. *International Journal of Engineering Research and General Science*. 2016 May; 4(3): 357–362.

[7] Elkington M, Bloom D, Ward C, Chatzimichali A, Potter K. Hand lay-up: Understanding the manual process. *Advanced Manufacturing: Polymer & Composites Science*. 2015 Jul 3; 1(3): 138–151.

[8] Fong TC, Saba N, Liew CK, De Silva R, Hoque ME, Goh KL. Yarn flax fibres for polymer-coated sutures and hand lay-up polymer composite laminates. In *Manufacturing of natural fibre-reinforced polymer composites* (pp. 155–175). Springer, Cham; 2015.

[9] Cripps D, Searle TJ, Summerscales J. *Open mold techniques for thermoset composites. Comprehensive Composite Materials*. 2000; 2: 737–761.

[10] Varma I, Gupta VB, Sini N. *Thermosetting Resin—Properties*. 2018; 2: 401–468.

[11] Hodgkinson JM, editor. *Mechanical testing of advanced fibre composites*. Woodhead Publishing; 2000 Nov 10.

[12] Moeller Jr M, Shane C, Jha R. Fabrication of composite laminates using a vacuum bag process. *Annual Symposium on Undergraduate Research Experiences*. 2007 Aug 2; 1: 214.

[13] Marsh G. Prepregs—Raw material for high-performance composites. *Reinforced Plastics*. 2002 Oct 1; 46(10): 24–28.

[14] Mack J, Schledjewski R. Filament winding process in thermoplastics. In *Manufacturing techniques for polymer matrix composites (PMCs)* (pp. 182–208). Woodhead Publishing; 2012 Jan 1.

[15] He HW, Gao F. Effect of Fibre volume fraction on the flexural properties of unidirectional carbon Fibre/epoxy composites. *International Journal of Polymer Analysis and Characterization*. 2015 Feb 17; 20(2): 180–189.

[16] Marques AT. Fibrous materials reinforced composites production techniques. In *Fibrous and composite materials for civil engineering applications* (pp. 191–215). Woodhead Publishing; 2011 Jan 1.

[17] Cheng QF, Wang JP, Wen JJ, Liu CH, Jiang KL, Li QQ, Fan SS. Carbon nanotube/epoxy composites fabricated by resin transfer molding. *Carbon.* 2010 Jan 1; 48(1): 260–266.

[18] Balasubramanian K, Sultan MT, Rajeswari N. Manufacturing techniques of composites for aerospace applications. In *Sustainable composites for aerospace applications* (pp. 55–67). Woodhead Publishing; 2018 Jan 1.

[19] Peng X, Fan M, Hartley J, Al-Zubaidy M. Properties of natural Fibre composites made by pultrusion process. *Journal of Composite Materials.* 2012 Jan; 46(2): 237–246.

[20] Baley C, Kervoëlen A, Lan M, Cartié D, Le Duigou A, Bourmaud A, Davies P. Flax/PP manufacture by automated fibre placement (AFP). *Materials & Design.* 2016 Mar 15; 94: 207–213.

[21] Tun KS, Gupta M, Srivatsan TS. Investigating influence of hybrid (yttria+ copper) nanoparticulate reinforcements on microstructural development and tensile response of magnesium. *Materials Science and Technology.* 2010 Jan 1; 26(1): 87–94.

[22] Prasad SS, Prasad SB, Verma K, Mishra RK, Kumar V, Singh S. The role and significance of Magnesium in modern day research – A review. *Journal of Magnesium and Alloys.* 2021 Jun 24; 10: 1–61.

[23] Leong Eugene WW, Gupta M. Characteristics of aluminum and magnesium based nanocomposites processed using hybrid microwave sintering. *Journal of Microwave Power and Electromagnetic Energy.* 2016; 44: 14–27.

[24] Rashad M, Pan F, Tang A, Asif M, Hussain S, Gou J, Mao J. Improved strength and ductility of magnesium with addition of aluminum and graphene nanoplatelets (Al+ GNPs) using semi powder metallurgy method. *Journal of Industrial and Engineering Chemistry.* 2015 Mar 25; 23: 243–250.

[25] Kopeliovich D; SubsTech2012. Solid state fabrication of metal matrix composites. www.substech.com

[26] Nishida Y. Introduction to metal matrix composites: Fabrication and recycling. Springer Science & Business Media; 2013 Jan 13.

[27] Imanishi T, Sasaki K, Katagiri K, Kakitsuji A. Thermal and mechanical properties of VGCF-containing aluminum. *Transactions of the Japanese Society of Mechanical Engineers.* 2008; 74(741): 655–661.

[28] Imanishi T, Sasaki K, Katagiri K, Kakitsuji A. Effect of CNT addition on thermal properties of VGCF/aluminum composites. *Transactions of the Japanese Society of Mechanical Engineers.* 2009; 75(749): 27–33.

[29] Saravanan RA, Surappa MK. Fabrication and characterisation of pure magnesium-30 vol. % SiCP particle composite. *Materials Science and Engineering: A.* 2000 Jan 15; 276(1–2): 108–116.

[30] Chen L, Yao Y. Processing, microstructures, and mechanical properties of magnesium matrix composites: a review. *Acta Metallurgica Sinica* (English Letters). 2014 Oct; 27(5): 762–774.

[31] Thandalam SK, Ramanathan S, Sundarrajan S. Synthesis, microstructural and mechanical properties of ex situ zircon particles (ZrSiO4) reinforced Metal Matrix Composites (MMCs): A review. *Journal of Materials Research and Technology.* 2015 Jul 1; 4(3): 333–347.

[32] Ye HZ, Liu XY. Review of recent studies in magnesium matrix composites. *Journal of Materials Science*. 2004 Oct; 39(20): 6153–6171.

[33] Etemadi R. Effect of processing parameters and matrix shrinkage on porosity formation during synthesis of metal matrix composites with dual-scale Fibre reinforcements using pressure infiltration process (Doctoral dissertation, The University of Wisconsin-Milwaukee).

[34] Champagne VK. *The cold spray materials deposition process.* Elsevier Science; 2007.

[35] Park JH, Sudarshan TS, editors. *Chemical vapour deposition.* ASM international; 2001.

[36] Savale PA. Physical vapour deposition (PVD) methods for synthesis of thin films: A comparative study. *Archives of Applied Science Research* 2016; 8(5): 1–8.

[37] Mahan JE, editor. *Physical vapour deposition of thin films.* Wiley Publications; 2000.

Chapter 4

Ultrasonic Vibration-Assisted Machining of Advanced Materials

Mamta Kumari, Ashok Kumar Jha, and Subhash Singh

Department of Mechanical and Automation Engineering, Indira Gandhi Delhi Technical University for Women, New Delhi, India

CONTENTS

4.1 INTRODUCTION

Some materials are specially engineered to exhibit novel or enhanced properties that give super performance relative to conventional materials; these are known as advanced materials. Advanced materials fulfil efficient and satisfactory expectations in material removal processes and give best performance results. Therefore, they are becoming famous worldwide in industries with a high range of applications. In previous years, the advanced materials were not as widely used as nowadays, because of the lack of technologies, techniques, facilities, and limited areas of application. Formation of advanced materials has exposed the range of application not only in automobile industries but also in medical fields. A wide span of utilization has a great impact in the fields of manufacturing, and this can be justified by extended demands of advanced materials starting from tools used in operating a human body to materials used in different parts of automobiles and various segments in aerospace industries. In almost every field of fabrication and development, advanced materials have given their contribution. Adaptations of new technologies have not only helped in the formation of advanced materials but also opened the opportunities in enormous areas of research and development.

The fundamental advantages of using advanced materials over 40 years are the properties on different domains which present their greater physical strength, good ratios of strength density, superior hardness and thermal quality, electrical, magnetic, optical, mechanical, and chemical properties. Its effects on different range of temperature or sustainability include some other advantages like corrosion resistance, weight savings, good stiffness behaviour, and excellent surface. Efficient production rate at lower energy with good quality products are possible. Some examples are presented in this particular work which are most important. Advanced materials with distinguished characteristics used in research and development areas are aluminum-lithium alloys and shape memory alloys, rapidly solidified and porous metals, structural ceramics like alumina, SiC, beryllium, titanium carbide, boron nitride, and thorium. Some engineering polymers like polyphenylene sulphide (PPS), polyetheretherketone (PEEK), polyacrylate, and different polyamide-imides are included. Advanced composites are made using metal, polymer, and ceramic matrix with reinforcements in the form of whiskers, particles, and fibres. The reinforcement materials can be aluminium, boron, carbon, silicates, and polymers. Electronic, optical, and magnetic materials like barium, indium, gallium, zirconium, and yttrium are also used. Materials for medical and dental like polylactic acid reinforced with carbon, glasses of calcium phosphate, and alumina are few other examples [1]. Implementation of various techniques has clearly identified the potential needs and rapid growth of advanced materials in applications involving distinguished features of composite materials, structural ceramics, polymers or matrix polymers, and glass fiber–reinforced plastics. Composites have been developed to use in car building and fabrication work on a frequent basis having the quality to replace most expensive materials with less in required properties [2]. Some materials like titanium alloys, silicon carbides, zirconia-based ceramics, carbon fibres, reinforced materials, and semiconductor materials are now widely utilized in aerospace industries and in the field of bio-medicines for various purposes [3]. In this context, one more advanced material is in the queue, i.e. magnesium alloys in biomedical applications because of their biodegradable nature, non-toxicity, low density and elastic structural stability, biocompatibility, and elastic modulus capacity compared with natural bones. Hence, it is widely used in medical practice [4]. Advanced materials are comparatively considered stronger, lighter, having more durability, and can replace the conventional materials in various functional areas. This is possible with less added cost and increased rate of profitability which ultimately give the good results resulting in customer satisfaction. One of the major reasons behind using advanced materials frequently nowadays is because of reliability, maintainability, and reparability, including adequate properties, quality, and cost after machining process. Ultimately advanced materials have impacted and influenced the market for its better product performance without sacrificing the quality [5]. In recent years, the manufacturing of advanced materials has opened various ways of development in materials science and technology which have naturally decreased the failure rates of tool materials which were larger in traditional machining operations like turning, milling, and grinding. The scenario has completely changed after using non-conventional machining operations for processing advanced materials because a variety of technologies are being involved to get the best results of end products with better properties than conventional processes. The credit goes to advanced materials and non-conventional machining processes, which have become so dynamic in the field of preparation and fabrication of different materials. The basic reason of using advanced materials in more elaborated from industrial uses, medical fields, and defence for making different types of tools because it can machine difficult-to-machine materials. Intricate holes, blind holes, and complex cavities are not often possible on the materials processed through conventional operations.

4.2 APPLICATION OF ADVANCED MATERIALS

There are various other ranges of applications apart from those given in the introduction part covering industrial, medical, and manufacturing areas, etc. and are listed below in Table 4.1.

4.3 MACHINING OF ADVANCED MATERIALS

Increasing demands for precision work on material has come to that point of development where there is no excuse of forming bad-quality product components. Technologies has so advanced that there are different types of machining processes available in the field of machines & manufacturing of advanced material into finished products for different useful work. Earlier conventional machining process was only the option to meet this requirement with some limitation of not having the capability of machining brittle materials, machining of micro-components, some hard and advanced materials. Further, there is necessity of eliminating or reducing cutting fluids usage during machining process due to various disadvantages like transmission of heat from tool to work piece, mechanical properties of tool due to regular use of cutting fluid, ecological hazards to tool and work pieces, and to achieve better surface finish, high accuracy, and high precision at low cost. To facilitate high precision work and machining operations, it is also very important to introduce non-conventional machining process which has the capability of machining hard to machine materials and almost every type of advanced materials with accuracy [12].

Table 4.1 Applications of advanced materials

S. No.	Material	Properties	Applications	References
1	Inconel718	High-temperature strength, thermal stability, corrosion resistance	Micro-milling, turbines for aero-engines and impellers	[5]
2	Nickel-Titanium (NiTi) alloys	Shape memory effect, pseudo elasticity, good damping capacity, good corrosion resistance, excellent biomechanical compatibility	Automobile, aerospace, medical devices	[6]
3	In-situ TiB$_2$/7050-Al metal matrix composites (MMCs)	Improved strength, low density, increased wear resistance	Aero-engine	[7]
4	Sapphire (αAl$_2$O$_3$)	Multi-purpose mono crystalline material. Excellent optical, chemical, thermal, and physical properties.	LED substrates, windows fields, optical fibre sensors, laser amplifiers	[8]
5	Ti-6Al-4 V	Specific strength, outstanding corrosion resistance, and excellent low temperature resistance	Aerospace components	[9]
6	Magnesium alloys	Biodegradable nature, low density and elastic structural stability, biocompatibility, non-toxicity, biodegradability	Medical practice	[4]
7	AlSiMg0.75 alloys	Low density, high heat conductivity for heat dissipation, and excellent mechanical strength	Nano metric, surface roughness for reflective mirrors used in aerospace work	[10]
8	Ceramics-Whisker reinforced Al$_2$O$_3$, SiAl$_x$N	High hardness and high wear resistance	High-speed machining, tools preparation	[11]
9	Inconel 718 with coated carbide tools	High-temperature strength, superior creep resistance, and elevated corrosion resistance	Nuclear and marine industries	[11]
10	Titanium alloy	Thermal conductivity is low, good chemical reactivity, Young's modulus is low	Production of aviation impellers, aerospace, and biomedical structural components	[3]
11	Si$_3$N$_4$	Greater strength over broad range of temperature, good resistance to wear, good thermal shock, and chemical resistance	Automobile industry, turbine blades and weld positioners	[3]
12	Mono-crystal silicon	Exceptional mechanical as well as electrical properties, semiconductor material	Electronics	[3]

In this particular work, a brief information regarding conventional, non-traditional manufacturing or non-conventional machining operations as well as advanced machining process for advanced materials are presented because of their significance and related aspect of discussion before going into deeper level of explanation. Manufacturing processes mainly or broadly can be classified into two groups: primary and secondary manufacturing processes. The former gives basic size/shape to the material as desired as forming, casting, powder metallurgy, etc. The secondary manufacturing processes mainly take part in the final shape, size, controlling the dimensional accuracy, surface finish, and some other aspects. Material removal processes are mainly the secondary manufacturing processes, and this again can be divided into conventional and non-conventional machining processes based on different aspects of application and the machining process. Material removal occurs in chip form with the application of mechanical force on the work piece with the help of cutting tools which is harder than the work piece material under some machining conditions. Application of forces on the work piece and tool induce the plastic deformation inside the work, which leads to shearing along the shear plane and chips are generated. In the era, with advanced engineering materials like ceramics or composites having superior properties (thermal, chemical, mechanical), it has become tough to machine using conventional abrasion and cutting based machining processes which leads to higher machining expenses while forming complex shapes. To achieve desired surface finish, quality, and dimensional accuracy of the machined parts seems quite difficult and fails to acquire the required goals in metal machining with addition of some unwanted material from the parent materials. Mechanical energy is provided by cutting tool, which in continuous contact with the work, causes shear deformation along the shear plane, resulting in chip formation with various limitations of not forming the desired shapes, complex and intricate details on the advanced materials, and putting conventional processes in pressure for its limited capacity. These disadvantages have led to the development and deployment of a novel, non- traditional machining (NTM) processes. In this, material removal from the work piece is done through various energy forms, like electrical, thermal, chemical, mechanical, or their combination. In NTM, there isn't any direct contact between cutting tool and the work piece [13]. The classification of NTM processes is shown in Fig. 4.1.

All non-traditional manufacturing processes as stated in Fig. 4.1 have their distinguished process principles and capacity of operating on different range of materials. The process principle of Abrasive Jet Micro Machining (AJMM) is an advanced version of abrasive jet machining (AJM), both of which are responsible for material removal due to erosion as air is mixed with abrasives are accelerated and directed towards the work piece or on the targeted zone with huge pressure used for micro-machining and micro-fluidic capillary electrophoresis devices. The major disadvantage of using

Classification of NTM Process

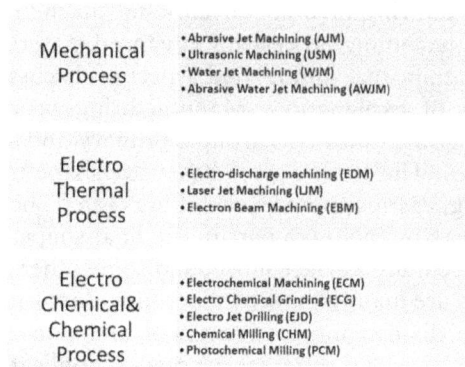

Mechanical Process	• Abrasive Jet Machining (AJM) • Ultrasonic Machining (USM) • Water Jet Machining (WJM) • Abrasive Water Jet Machining (AWJM)
Electro Thermal Process	• Electro-discharge machining (EDM) • Laser Jet Machining (LJM) • Electron Beam Machining (EBM)
Electro Chemical& Chemical Process	• Electrochemical Machining (ECM) • Electro Chemical Grinding (ECG) • Electro Jet Drilling (EJD) • Chemical Milling (CHM) • Photochemical Milling (PCM)

Figure 4.1 Classification of NTM processes.

this process is when compressed jet is used to propel the erodent abrasive particles, they diverge after the nozzle exit which is simultaneously responsible for increase in the size of blast zone which affects the target zone and nozzle life [14]. The process principle of Electrical Discharge Machining (EDM) is completely different from other NTM processes. It is used for machining on difficult-to-machine materials for various purposes like complex geometries in small batches or even on job-shop basis. The major disadvantage of using the EDM process for operating on advanced materials is that materials should be electrically conductive in which repetitive spark discharges through electric pulse generators having dielectric fluid supply between tool and work piece where no mechanical cutting takes part. So, there is no contact between the work piece and tool as this is an electro-thermal non-traditional machining process, where electrical energy is used to generate electrical spark which is in very much precise and controlled manner. Material removal is due to the spark's thermal energy between the closest points of electrode act as tool and work piece within a dielectric fluid, mainly by erosion of electrically conductive materials by series of high-frequency, spatially discreet electrical discharges (sparks) between a tool and work piece. Material is removed by the spark from the electrode as well as the work piece, thereby increasing the sparking gap at that point. This helps the subsequent spark to produce the next-closest points in between the work and electrode as material is removed by heat. So, the spark melts small quantities of material from the electrodes and work piece. This is considered as a major drawback for the life of the tool [15]. Continuing with this context, electro-chemical machining process (ECM) also majorly contributes to material removal of the advanced materials. In this non-traditional machining operation, D.C. voltage of 10–25 volts is supplied to the pre-shaped cathode's inter-electrode gap, which acts a tool and work piece is anode. Material removal is mainly by anodic dissolution mechanism at the time of electrolysis process. The electrolytic solution, e.g. NaCl aqueous, is made to flow at high speed of 10–60 m/s within an inter-electrode gap (0.1–0.6 mm). 20 to 200 A/cm^2 current density is usually passed through the gap. The process is basically based on Faraday's laws of electrolysis having the capacity to generate an approximate tool's mirror image over the work piece. The ECM is more suitable for mass production [16]. There are some other NTM processes available that have features useful for advanced materials, but ultrasonic vibration assisted process has much dominant advantages over other NTM processes. This is more suitable and advantageous in terms of ease of working and the process parameters are considered friendly with both metallic and non-metallic composites, ceramics, reinforced materials, etc. The aim of this chapter is to focus on ultrasonic vibration-assisted machining, its advancement in recent years, and working on advanced materials, which is huge as far as research and development is concerned. It has been absorbed after analyzing various experimental examination of NTM processes of hard and brittle material which are both electrically conductive and non-conductive. For example, glasses, plastics reinforced by carbon fibre, ceramics, and metal matrix composites are characterized by their superior hardness, high resistance to wear, and low toughness against fracture having various applications in automotive industries, aerospace, computer and mobile screens, optics, and micro-fluidic devices because of their superior mechanical characteristics. It is really difficult to machine such types of materials by conventional processes. To overcome this limitation in advanced material manufacturing, ultrasonic vibration-assisted machining was first used in the year 1927 [17] and with some modification further used in 1945 by L. Balamuth. This process was used for surface finishing in EDM because it was found that USM reduced both cutting forces and torque and improved surface roughness. Ultrasonic means a vibratory wave with greater frequency than the human ear hearing upper frequency limit (usually more than 16 kc/s) [17]. This can cultivate in diverse mediums as per its essential elastic densities as well as properties [21].

4.3.1 Process Principle of Ultrasonic Vibration-Assisted Machining

Material removal is mainly done due to indentation process as cracks generate on hard, brittle surfaces. As tool is made up of strong, tough, and ductile materials such as steel, stainless steel, and other metallic alloys, it is vibrated by a transducer (piezoelectric ceramics/magneto-strictive) to convert high-frequency electrical energy into corresponding mechanical vibrations. This is applied in machining and material is removed by mechanical effect of, indentation, cavitation, and brittle fracture of a work piece [3, 18, 19, 20]. The ultrasonic wave generation can be affected by multiple factors like amplitude, ultrasonic frequency, etc. Ultrasonic frequency is the number of times a complete oscillation occurs in a second and amplitude is the maximum value of ultrasonic intensity, which is the quantity of power passing through a unit cross-sectional area. A particular connection between an ultrasonic generator and metal plate should be there to provide an ultrasonic vibration in the presence of a medium like horn, which acts as a tool holder for initialization of vibration followed by erosion. Ultrasonic vibration has much higher frequency than natural frequency which can maintain and improve the stability in the complete manufacturing system without disturbing the low frequency vibration [21].

4.3.2 Working Procedure

The given diagram of USM in Fig. 4.2 explains the working principle. Transducers are used to provide ultrasonic vibration which is driven through appropriate signal producer tailed by amplifier. At the tip of the tool, material removal occurs from the work piece. Fine, irregular-shaped abrasives particles are mixed with water-based slurry are sent as a stream near the machining zone between the work piece and tool tip through a nozzle. When the tool vibrates, abrasive particles act as indenters and etch both work piece and tool tip due to initiate crack formation.

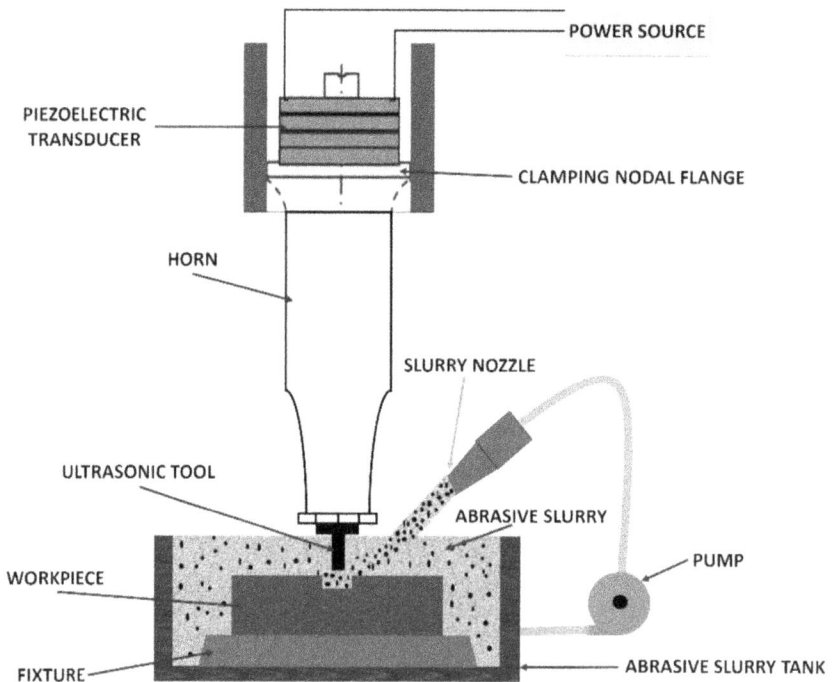

Figure 4.2 Ultrasonic vibration-assisted machine.

4.3.3 Improvement in the Ultrasonic Vibration-Assisted Process to Perform on Advanced Materials

This present book chapter aims only to deliver a broad area of employment and application of advanced material on ultrasonic vibration-assisted manufacturing process. However, it is framed that utilization of advanced materials is a huge and wide topic to discuss in full frame that cannot be completely covered in one chapter so here the limited scope is shown that plays a very pivotal role in enhancing the product quality and also give the great aspects of research work i.e. some significant process parameters and surface roughness and its characterization.

4.3.3.1 Process Parameters

After going through various experimental works on process parameter-based review papers some crucial parameters have come out with remarkable impact on surface quality on the work material as well as on the tool using the UVAM process. In most of the processes, the fundamental and foremost aim is to get good, rather than better material removal rate (MRR) than the previous one. So, MRR is considered to be one of the significant aspects towards evaluation of the process parameters. Fig. 4.3 shows the clear picture of some important process parameters.

- **Grain size** – Rajathesh et al. (2019) and Ramulu et al. (2005) [22] have explained the impact of grain or abrasive size on the material removal rate using UVAM process on some advanced materials like ceramic silicon carbide, titanium diboride, and boron carbide.

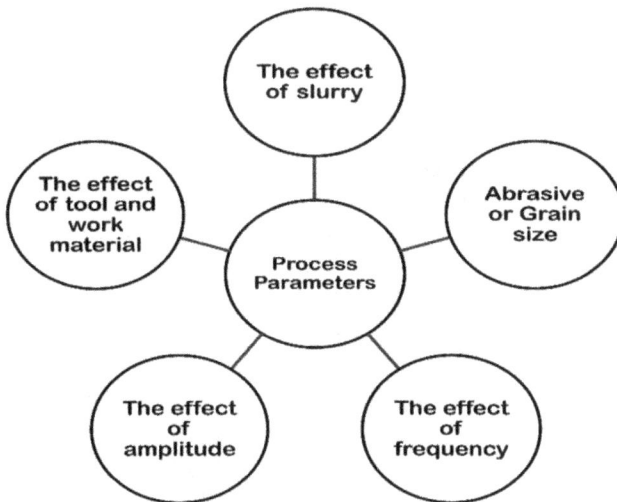

Figure 4.3 Important process parameters.

Grits were used to define and explain the analysis regarding micro hardness, MRR, and surface integrity. It has been stated that there is a strong relation between grain size and MRR that is directly responsible for occurrence of erosion from the work material. Usually, it has been observed that when machining SiC, the size of the abrasive grains, influenced the MRR rate. Initially MRR increased, but after some interval, the rate of material removal decreased. This happened because, the continuous use of abrasive grit particles blunted the edges with passage of time. In this regard, the response of different advanced materials shows different picture. As in the case of titanium diboride/silicon carbide the graph of MRR vs diameter, the pattern of MRR was not the same as found on SiC work piece. The MRR decreased with increase of grit size. On the contrary, for boron carbide, the MRR decreased with a decrease in grit size. For this, instance surface roughness was measured and found to be dependent on grit size. The smoother the grit or abrasive grain size, the finer is the surface finish. In this regard, some more experimental work has been studied and included in this present work. The high carbon steel reflects better results and remarkable improvement in MRR because carbon steel has higher hardness than titanium, it has found to be increasing with the increase in the size of the grain, i.e. coarseness of the grain particles. The MRR was observed to be decreasing by a small rate when the size of the grain particle is increased by 320 to 500. The reason behind this is, the coarser the grain, better will be the ability to impinge on the work with high rate of energy. This tends to decrease the surface roughness. In the experimental experience with coarse boron carbide particles, improved MRR has been observed when compared with Al or SiC as revealed by Kararia et al. (2010) [22].

- **The effect of slurry** – The slurry used in the process of UVA machining is the mixture of abrasive particles, water or oil or some other substances that can easily form a paste like structure in order to fulfil the required purpose of erosion on the work. Some abrasives like boron carbide, aluminium oxide, and silicon carbide are generally used in the process, as Rajathesh et al. (2019) and Sumit et al. (2009) have stated in their studies. Boron carbide is the fastest among other two abrasives having better cutting capability and comparatively be used for a longer period of time. Aluminium oxide is not commonly used in machining process because of its wear property than other stated materials. As far as silicon carbide is concerned, the advantage of using this material is, abundance availability, lowest cost and better MRR. This results in good flow rate and machining capability with 30–60% concentration of slurry, i.e. mixture of abrasive and other binding substances like water or oil. Change in the concentration of slurry can affect the performance of the machining by slowing down the machining due to the dust settling on the worktable. So, in some intervals of time, the concentration can be monitored and maintained to enhance the performance and output results. Again, temperature of the slurry also influences the machining and should be maintained between 5–6°C by using the refrigerator cooling system whenever required. It has been observed that machining gives the best results when slurry concentration is kept at 30%. Hence, a statement came into existence that MRR decreases with the increase in concentration of slurry percentage due to increase in viscosity. The reason behind this is flow rate of slurry at work tool interface will face difficulty to work with flow characteristics.

- **The effect of frequency on MRR** – Rajathesh et al. (2019) and Sumit et al. (2009) have explained that MRR is also dependent on variation of frequency provided to the machine. It gives different responses on varying the frequency. Hence, frequency is one of the prime parameters to influence the material removal rate. Theoretically, frequency is directly proportional to MRR but in practical work, little variation has been observed. It varies from 15,000 Hz to 25,0000 Hz lower limit to upper limit, respectively. To get adequate frequency level, it is necessary to provide required amount of amplitude to the tip tool interface to get the profound results followed by provision of resonant frequency maintenance, which maximizes the utilization of the machining.

- **The effect of amplitude on MRR** – Amplitude is produced by giving the vibration to the tool, generally not sufficient for cutting operation so tool should relate to the transducer through a concentrator so that it can convert wave to obtain the required amplitude. This can be provided at the end of the tool for machining and increased amplitude is fed to the transducer connected with concentrator in variable design and cross-sectional form to transmit adequate amount of vibration. MRR is directly proportional to amplitude followed by vibration, as magneto-strictive material reduces the strength of the transducer amplitude, while the vibration amplitude varies in between 0.01 to 0.06 mm, as stated by Rajathesh et al. (2019) and Sumit et al. (2009).

- **Influence of tool and work materials on machining** – As the tool always plays an important role in cutting and forming operations, it should be very strong and hard to protect it from quick wear and tear. The life of the tool depends on the hardness of it and some parameters during machining operation. The harder the tool, the faster will be the failure rate. So, selection of optimum parameters is required as too much vibration to the tool leads to unacceptable metal removal rates and surface finish will also be affected. Materials having tough, malleable properties i.e. alloys of steel and stainless steel can give satisfactory results in machining as compared to aluminium and its alloys. This is because of its mechanical properties and tool life. In the USM process, the mass of the tool has a significant role as it absorbs ultrasonic energy used in operating the work piece. Reduction in the machine efficiency takes place with longer tool because there will be over stressing of the tool due to energy concentration near the tool tip. Considered length is usually less than 25 millimetres and slenderness ratio also must be less than 20. The tool under-sizing is directly in proportion to the abrasive grain size i.e. sufficient if

Tool size = (Hole size) – 2(Abrasive size) [22].

Some very important investigation has been done on work material, according to input variables, i.e. different work thickness, work homogeneity including some other factors. This is because parameters are essential for composites (i.e. WC-Co, Al/SiC, WC-Ni, etc.), where the concentration of binder phase materials is essential to vary material properties of the work [17].

4.4 CONCLUSIONS

- In ultrasonic wave, the frequency is higher than the upper frequency of human ear hearing limit (usually more than 16 kc/s), which can cultivate in diverse mediums based on their essential densities and elastic properties. It is generally used for machining difficult-to-machine materials.
- A major advantage of using UVAM on advanced materials is that this operation can be performed on electrically conductive and non-conductive materials and having more than 40HRC hardness.
- High quality of soldering and brazing can be performed on horn used in UVAM with great mechanical strength, good acoustic characteristics, and corrosion resistance. Horn and tool design has significant role in enhancing the resonance properties ultimately related to MRR. For good tool wear resistance, elastic, fatigue strength, toughness, and hardness can be given prime importance.
- Some advanced materials like ceramic silicon carbide, titanium diboride, and boron carbide grits are used to explain the analysis regarding microhardness, MRR, and surface integrity. When the size of SiC grains is increased, the MRR rate increases initially and later, rate of MRR decreases. In the case of titanium diboride/silicon carbide, the pattern of MRR is different from SiC. For boron carbide, MRR decreases with a decrease in grit size, whereas in silicon carbide, MRR increases.
- Surface roughness also is dependent on grit size. The smoother the grit or abrasive grain size, the finer the obtained surface finish.
- Among them, aluminium oxide, boron carbide and SiC, boron carbide has better cutting ability over others. Aluminium oxide is not commonly used in machining process because of its wearing property in comparison to other materials. Silicon carbide is abundantly available

and cheaper with better MRR, resulting in good flow rate to the machining capability with concentration of slurry around 30–60%.

- Temperature of the slurry also influences the machining and should be maintained between 5–6°C by using the refrigerator cooling system whenever required.
- Machining gives the best result when the slurry concentration is kept at 30%. Hence, it can be said that MRR decreases with the increase in slurry viscosity concentration. Adequate size of the abrasive with high slurry concentration increases MRR.
- Frequency and amplitude are directly proportional to MRR, where frequency varies from 15,000 Hz to 25,0000 Hz lower limit to upper limit, respectively. Amplitude followed by vibration, as magneto-strictive material, reduces the strength of the transducer vibration which varies from 0.01 mm to 0.06 mm.
- Considering the length of the tool and slenderness ratio, they should be kept 25 millimetre and less than 20, respectively, for optimum results.
- Materials with tough malleable properties, i.e. alloys of steel, can provide satisfactory machining results as compared with aluminium and its alloys because of their mechanical properties and tool life.
- In the USM process, the mass of the tool has a significant role as it absorbs ultrasonic energy used in operating the work piece. Machine efficiency reduces because of a longer tool due to over stressing. This is due to energy concentration near the tool tip.

REFERENCES

[1] T. R. Curlee, S. Das, R. Lee, D. Trumble, Advanced materials: Information and analysis needs, energy division information and analysis, September 1990, pp. 1–29.

[2] A. Modrea, S. Vlase, H. Teodorescu-Draghicescu, M. Mihalcica, M. R. Calin, C. Astalos, Properties of advanced new materials used in automotive engineering, *Optoelectronics and Advanced Materials – Rapid Communications* 7(5–6) (May–June 2013), pp. 452–455.

[3] Z. Yang, L. Zhu, G. Zhang, C. Ni, B. Lin, Review of ultrasonic vibration-assisted machining in advanced materials, *International Journal of Machine Tools and Manufacturing* 156 (2020) 103594.

[4] E.D. Iorioa, R. Bertolinia, S. Bruschia, A. Ghiottia, Design and development of an ultrasonic vibration assisted turning system for machining bio absorbable magnesium alloys, *Procedia CIRP* 77 (2018) 324–327.

[5] B. Fang, Z. Yuan, D. Li, L. Gao, Effect of ultrasonic vibration on finished quality in ultrasonic vibration assisted micro milling of Inconel718, *Chinese Journal of Aeronautics* (June 2021); Volume 34 (6); pp. 209–219.

[6] D. Zhang, Y. Li, H. Wang, W. Cong, Ultrasonic vibration-assisted laser directed energy deposition in-situ synthesis of NiTi alloys: Effects on

microstructure and mechanical properties, *Journal of Manufacturing Processes* 60 (December 2020) 328–339.

[7] X. Liu, W. Wang, R. Jiang, Y. Xiong, K. Lin, J. Li, C. Shan, Analytical model of cutting force in axial ultrasonic vibration-assisted milling in-situ TiB_2/ 7050Al PRMMCs, *Chinese Journal of Aeronautics* (20 August 2020) 34(4); 160–173.

[8] Y. Wang, Z. Liang, W. Zhao, X. Wang, H. Wang, Effect of ultrasonic elliptical vibration assistance on the surface layer defect of M-plane sapphire in microcutting, *Materials and Design*, 192 (2020) 108755.

[9] L. Zhu, C. Ni, Z. Yang, C. Liu, Investigations of micro-textured surface generation mechanism and tribological properties in ultrasonic vibration-assisted milling of Ti–6Al–4V, *Precision Engineering Journal* 2019; May 1; 57; 229–243.

[10] Y. Bai, Z. Shi, Y. J. Lee, H. Wang, Optical surface generation on additively manufactured AlSiMg0.75 alloys with ultrasonic vibration-assisted machining, *Journal of Materials Processing Technology* 280 (2020) 116597, www.elsevier.com/locate/jmatprotec

[11] Z. Peng, X. Zhang, D. Zhang, Performance evaluation of high-speed ultrasonic vibration cutting for improving machinability of Inconel 718 with coated carbide tools Tribology International journal homepage: http://www. elsevier.com/locate/triboint

[12] M. N. Kumar, S. K. Subbu, P. V. Krishna, A. Venugopal, Vibration assisted conventional and advanced machining: A review, 12th Global Congress on Manufacturing and Management, *GCMM 2014 Procedia Engineering* 97 (2014) 1577–1586.

[13] S. Chakraborty, A case-based reasoning approach for non-traditional machining processes selection from *Advances in Production Engineering & Management* (ISSN 1854–6250) 11 December 2016) 311–323.

[14] R.H.M. Jafar, H. Nouraei, M. Emamifar, M. Papini, J.K. Spelt, Erosion modeling in abrasive slurry jet micro-machining of brittle materials, *Journal of Manufacturing Processes* 17 (2015) 127–140.

[15] M. P. Jahan, M. Rahman, Y.S. Wong, A review on the conventional and micro-electro discharge machining of tungsten carbide, *International Journal of Machine Tools & Manufacture* 51 (2011) 837–858.

[16] K. P. Rajurkara, M. M. Sundaramb, A. P. Malshe, Review of electrochemical and electro discharge machining, CIRP conference on electro physical and chemical machining (ISEM) procedia CIRP 6 (2013) 13–26.

[17] R. P. Singh and S. Singhal, Rotary Ultrasonic Machining: A Review. Materials and Manufacturing Processes, 31 (2016) 1795–1824, Copyright # Taylor & Francis Group, LLC ISSN: 1042-6914 print=1532-2475 online DOI: 10.1080/10426914.2016.1140188

[18] S. Boral, S. Chakraborty, A case-based reasoning approach for non-traditional machining processes selection, *Advances in Production Engineering & Management* ISSN 1854–6250, 11 (December 2016) 311–323.

[19] T. B. Thoe, D. K. Aspinwall, M. L. H. Wise, Review on ultrasonic machining, *International Journal of Machine Tools and Manufacture*. Elsevier Science Ltd, 38(4) (1998) 239–255.

[20] H. El-Hofy, E book by *Fundamentals of Machining Process: Conventional and Non-conventional Process*, 3rd edition from CRC Press/Taylor & Francis Group 2018.

[21] F. Ninga, W. Cong, Ultrasonic vibration-assisted (UV-A) manufacturing processes: State of the art and future perspectives, *Journal of Manufacturing Processes* 51 (2020) 174–190.

[22] B. C. Rajathesh, G. Vinay, S. C. Akki, Dr T. S. Nanjundeswaraswamy, Process parameters of ultrasonic machining, *International Journal of Recent Technology and Engineering (IJRTE)* ISSN: 2277–3878, 8(4) (November 2019).

Chapter 5

3-D Printing Processes for Biomedical Applications

S. V. Satya Prasad

Department of Production and Industrial Engineering, National Institute of
Technology Jamshedpur, Jharkhand, India

Ravi Verma

Institute of Electronics, Microelectronics and Nanotechnology,
Villeneuve-d'Ascq, France

S. B. Prasad

Department of Production and Industrial Engineering, National Institute of
Technology Jamshedpur, Jharkhand, India

Subhash Singh

Department of Mechanical and Automation Engineering, Indira Gandhi Delhi
Technical University for Women, New Delhi, India

CONTENTS

DOI: 10.1201/9781003327370-5   77

5.1 INTRODUCTION

The typical manufacturing has many stages and expensive infrastructure before obtaining the finished product. As a result, it becomes difficult to produce the products on time. Moreover, through conventional manufacturing methods, it isn't possible to achieve intricate geometries necessary for biomedical application [1]. Conventional manufacturing techniques are also expensive pertaining to the factors such as resources, effort, and time. Therefore, this has paved the way to 3-D printing, which is ideal for effective manufacturing of complicated designs in terms of time and cost. This is especially suitable for biomedical applications. The contemporary 3-D printing concept initially came into existence in 1986, which was followed by stereolithography (SLA) in the subsequent year [2]. The SLA made use of digital data and it is after the invention of SLA, other rapid prototyping techniques such as fused deposition modelling (FDM), bed fusion, inkjet printing, etc. came into existence. The idea behind 3-D printing process includes sequential layers being deposited over one another in a 3-D space to fabricate different as well as complicated geometries [3]. This process is also termed "additive manufacturing (AM)," a kind of rapid prototyping (solid freeform fabrication). The control on 3-D printers is through computers which scan through a digital, computer aided design (CAD) model, developed using CAD software [4]. The principal methods in 3-D printing are solid/liquid/powder based, which include diverse procedures for fabricating 3-D objects using ceramics or polymers [5].

3-D printing is reputed for processing material for a long time now. The emergence of many novel 3-D printing technologies has elevated its significance in healthcare sectors like pharmaceuticals, dentistry, and biomedicine. The applications of 3-D printing include implant designs for various body parts, therapeutic delivery [6], collagen manipulation through biomaterials in tissue engineering for generating the bone and surgical planning through printed surgical models [7,8]. Some other applications include printing of prosthetic trabecular bones, skin tissues, cartilages and artificial stents. More specifically, 3-D printing has been utilized in the customization of patient related products (personalized medicine/dosage), which has enhanced the treatment process [9]. Some applications of 3-D printing are shown in Fig. 5.1. The concept of personalized medicine refers to integration of patients' genetic analysis for patient specific therapy [10]. In the recent past, bioprinting has seen a rapid growth using which, cell seeding

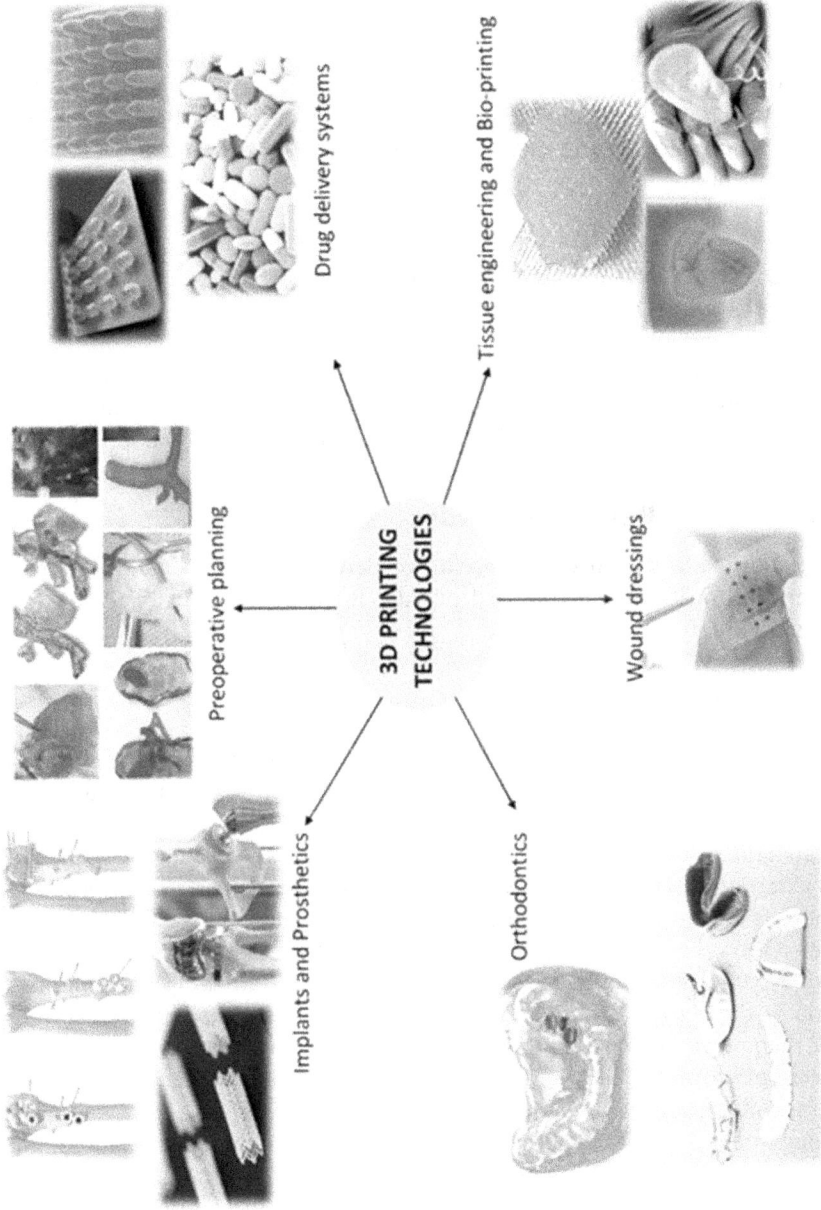

Figure 5.1 Applications of 3-D printing methods [10].

can be done within a three-dimensional space in a definite manner. The commercial grade 3-D printers cost over $250,000. But the development in the biological substrates such as BioBots, MakerBots, and RepRap which has brought down the cost of bioprinting. Though these printers have limitations, they triggered the growth of new materials and new applications. Therefore, the researchers could implement newer techniques in 3-D printing and address the various issues pertaining to the biomedical applications which was not possible through the conventional fabrication techniques. So, in this present chapter, the various 3-D printing processes in relation to biomedical applications are discussed.

5.2 3-D PRINTING METHODS FOR BIOMEDICAL APPLICATIONS

According to the ISO/ASTM 52900 standard, the number of established additive manufacturing (3-D printing) methods are seven, which are discussed in detail. The overview of all 3-D printing methods is represented in Fig. 5.2.

5.2.1 Printing Based on Powder

The techniques mentioned here use a powder bed as the primary material for their fabrication. But the used powder is modified in diverse ways to obtain a cohesive structure.

5.2.1.1 Binder Jetting Method

In this technique, a jet of binder in liquid form is sprayed onto a definite are on the powder bed. Upon completion, another layer of powder is spread and on top of that, a subsequent binder layer is jetted [3]. This way, the process is repeated till the desired component geometry is achieved. There is no usage of any external heat source. As seen in Fig. 5.3a, the binder jetting process comprises a printing nozzle that is programmed to move in X-Y directions, which sprays the binder. The droplets of the binder are deposited over loose powder particles over the bed, which leads to local hardening and solidification of the layers. This is due to wetting of the particles. The fabrication plate then moves upwards and the next layer of the powder from the powder delivery platform is deposited over the previous solidified layer [11]. A thin layer is rolled out flat, as the roller passes over the powder bed. This way a 3-D component is made, which later is hardened by the process of sintering. Generally metal powders are used in this process. It has a resolution of 50 to 400 μm. This process is most commonly applied in the fabrication of degradable implants (Fe alloys) [12]. The process is cost effective, can be used for large components, color printing is possible, and there is no additional support structure is needed. But the main disadvantage is the fabricated

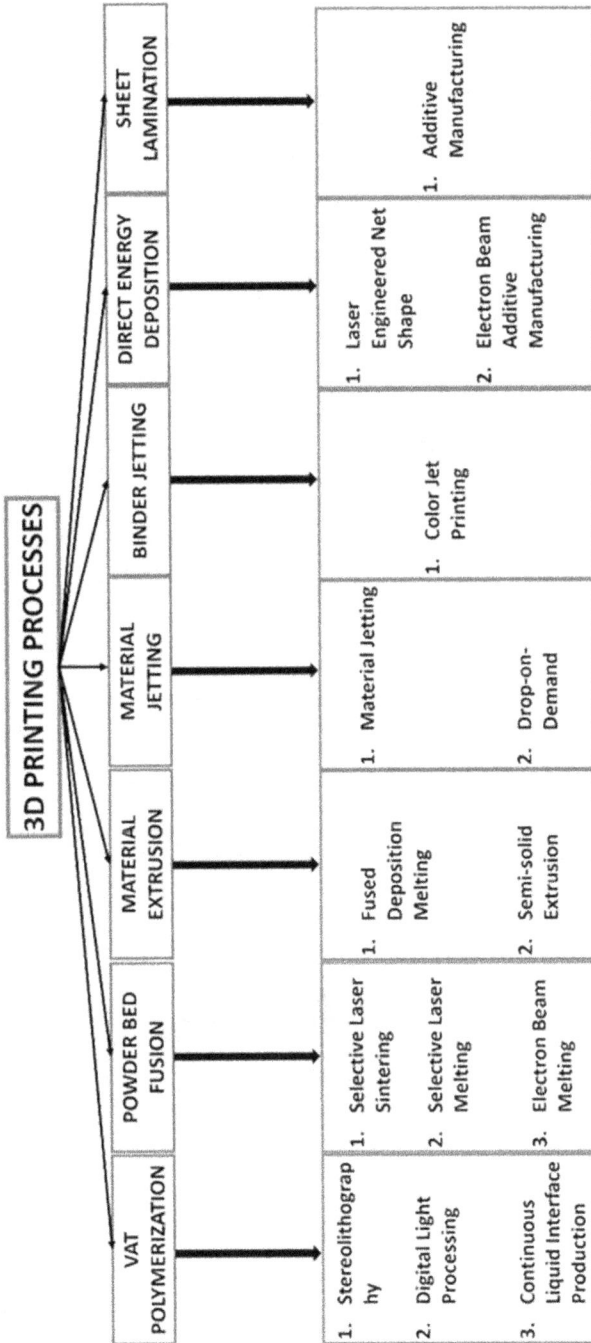

Figure 5.2 Classification of 3-D printing methods [10].

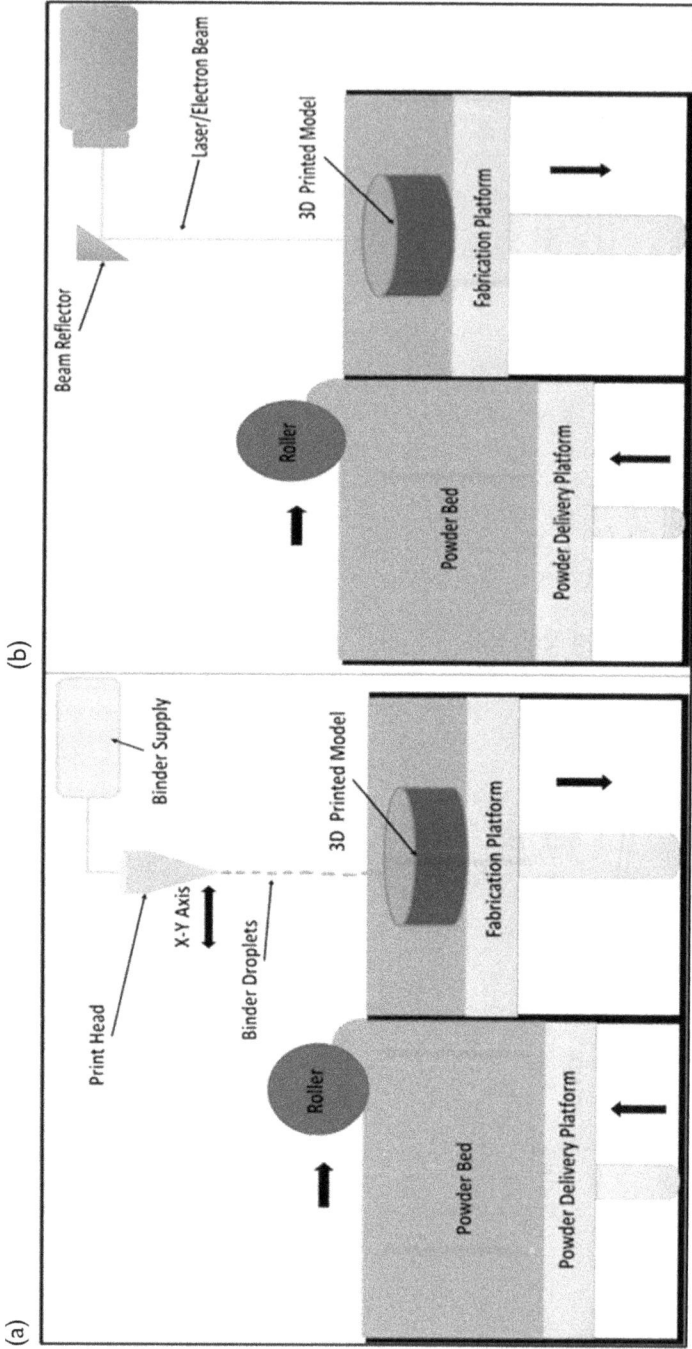

Figure 5.3 **(a)** Binder jetting process; **(b)** powder fusion process [11,13].

objects have low strength and also post-processing is required, which adds to the manufacturing costs. The powders can be hazardous for respiration.

5.2.1.2 Powder Bed Fusion (PBF)

The difference between binder jetting and PBF lies in the fact that fusion of the powder particulates takes place due to the use of lasers. This method is classified into electron beam melting (EBM), selective laser sintering (SLS), selective heat sintering (SHS), selective laser melting (SLM), and direct metal laser sintering (DMLS). The SLS technique is widely used. In DMLS, 3-D components are fabricated from metals or conducting materials in powdered form. The other PBF techniques make use of thermoplastics or ceramics as the base powder. The setup, working principle and mechanism of all the PBF processes are almost similar with an exception where the process of powder fusion varies. The DMLS, SLS, and SLM processes make use of high-powered lasers. But in the case of SHS, a thermal print head comes in contact with the base powder to melt it. EBM makes use of an electron beam; therefore, the process is carried out inside a vacuum chamber. In this process (Fig. 5.3b), there are two chambers, one in which the base powder is stored and the other in which the 3-D model is fabricated. The base powder is melted by the help of laser/electron beams or thermal print head, allowing the particle fusion [13]. The powder (metal, thermoplastics/ceramic) used in the process is fed into the building area with the help of a roller and flattened out. The laser/electron beam head can traverse along the horizontal direction. After a powder layer melts, the print bed goes down to a depth equal to the desired height of the subsequent layer which is deposited over the previous, solid layer. The general height of the layers is around 0.1 mm. This process is repeated till the 3-D model of desired dimensions is fabricated. The fabricated model is allowed to cool for a certain time till it gains its hardness. The printing resolution in the process is decent at 100 to 200 µm and can be used in orthopedic/dental applications, metallic or degradable rigid implants [14,15]. The PBF process is quicker compared to binder jetting and provides good strength to the fabricated products. It doesn't require any additional solvents. The process however is expensive and the printing resolution is decent. Post-processing is needed in some cases to achieve good quality of the final product.

5.2.2 Printing Based on Material Deposition

In this method, a nozzle is used to deposit layers of printing material. The different types of material deposition based printing methods are as follows.

5.2.2.1 Directed Energy Deposition (DED)

In DED, there is a head for supplying wire which moves in multi axial direction. It is also termed 3-D laser cladding or direct metal deposition.

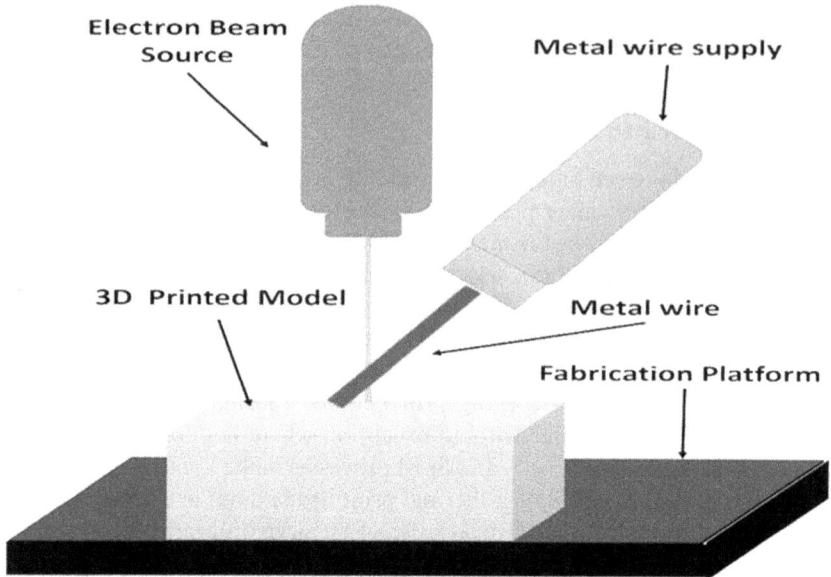

Figure 5.4 Direct energy deposition process [16].

In DED, a constant metal stream is directly extruded through a nozzle situated on top of a multi-axis arm and deposited over a print surface at any angle. This is due to the presence of multi-axis arm which facilitates movement along four to five axes. A laser or a thermal energy source is utilized to constantly melt the flowing metal layer wise (Fig. 5.4) instantaneously, once the material is deposited over the print surface. Hence, the metal deposition and the fusion take place in a coherent manner. The 3-D printing is done from the bottom to top direction. The shape of the final component is structured by placing a control over the angle and the metal feed rate [16].

This method generally has a resolution of 250 to 400 μm and has limited medical use. DED is advantageous due to its high accuracy. It facilitates good control over the grain structures of 3-D printed models. The DED technique has a specific use of 3-D printing metal/metal alloy components. This is primarily used in repairing of defective, 3-D printed metal/alloy components as the method has fast operating speeds. But it is expensive and has a low resolution. It is also utilized for attaching extensions for the already fabricated 3-D metallic models. The components should also be post-processed after fabrication in some cases.

5.2.2.2 Material Extrusion

The process of material extrusion is controlled by a computer. Generally, thermoplastics/ceramics/bio inks are used as the printing materials in this process which are passed through a movable nozzle. Fig. 5.5a shows

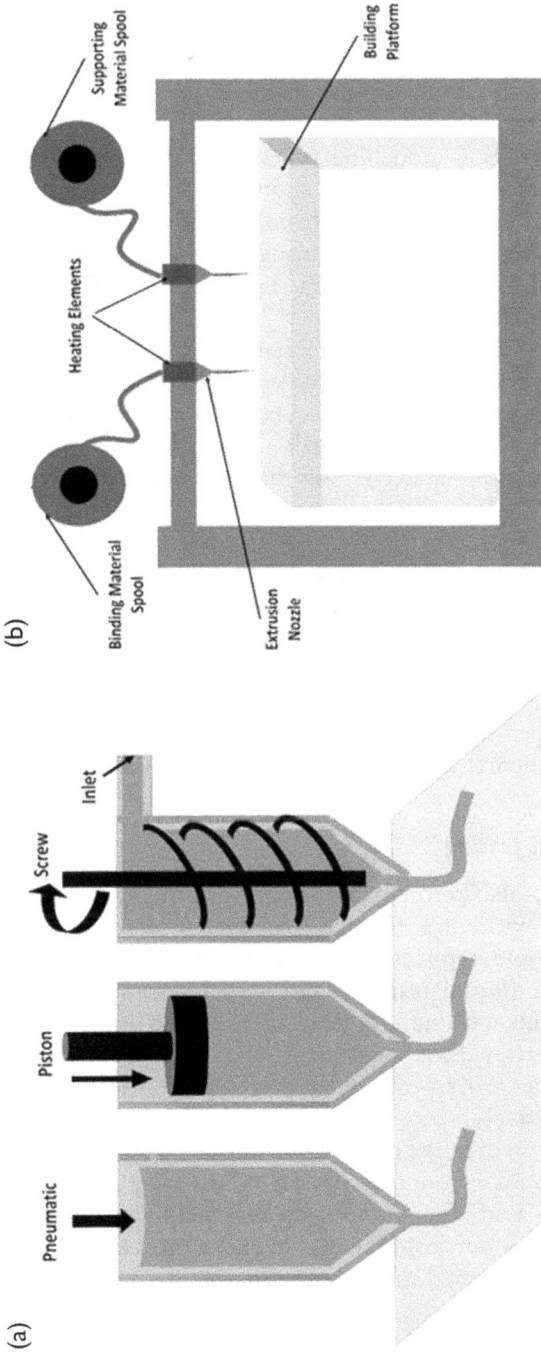

Figure 5.5 **(a)** Working principle of extrusion-based 3-D printing; **(b)** fusion deposition modelling process [14, 17].

various ejection mechanisms of inks in metal extrusion process. These materials are melted into a molten/semi-liquid state and continuously deposited as layers over a printing bed. The semi-liquid nature of the layers helps in binding, resulting in fusion prior to curing at room temperature. The metal extrusion process can be classified as

- Fusion deposition modelling (FDM): FDM comprises of a thermoplastic filament wound over spools of 1.75 or 3 mm diameters. This filament is unwounded from spools and melted by supplying heat. The filament passing through a nozzle, is extruded and deposited as layers on a platform. The process of FDM is depicted in Fig. 5.5b [17].
- 3-D plotting: This is a versatile material extrusion process. The technique comprises a movable nozzle (extrusion print head) and a cartridge that is capable of traversing both horizontally (X, Y axes) as well as vertically (Z axis). The nozzle movement is computer controlled, whereas the print material flow has pneumatic control (varying air pressure).

This method can be utilized along with hydrogels as well as pastes to print along with living cells at a resolution of 100 to 200 µm. This method can also be utilized in printing soft tissues in organs and cell scaffolds for culture [18]. Rigid or soft models of anatomy are also printed which can be used for planning surgeries. In this process, color printing is possible at low costs with open source designs. The process is slow, has low resolution, and the nozzle imparts large shear forces over the cells.

5.2.2.3 Material Jetting

Material jetting, also termed the "inkjet printing process," is a reprographic method without a contact. The data in digital form is obtained from computer and replicated over a substrate. In the process, layers of photopolymer (liquid resin) or bio inks are sprayed through nozzles like jets or droplets with simultaneous layer curing by UV light. After this, the process repeats with a subsequent layer. The schematics of working principle and the process of inkjet printing are seen in Fig. 5.6a and Fig. 5.6b, respectively.

The classification of inkjet printing is continuous inkjet printing (CIJ) and drop-on-demand inkjet printing (DOD). The timing of ink droplet being generated creates a difference among the two methods. In CIJ, a continuous jet of the ink comes out from the nozzle under pressure, whereas in DOD, droplets of ink are formed through electric signals. An actuator producing electric pulses determines the timing and volume of the droplet coming from piezoelectric heads, as seen in Fig. 5.6a. The material jetting process is the only 3-D printing process which gives the best Z-direction resolutions.

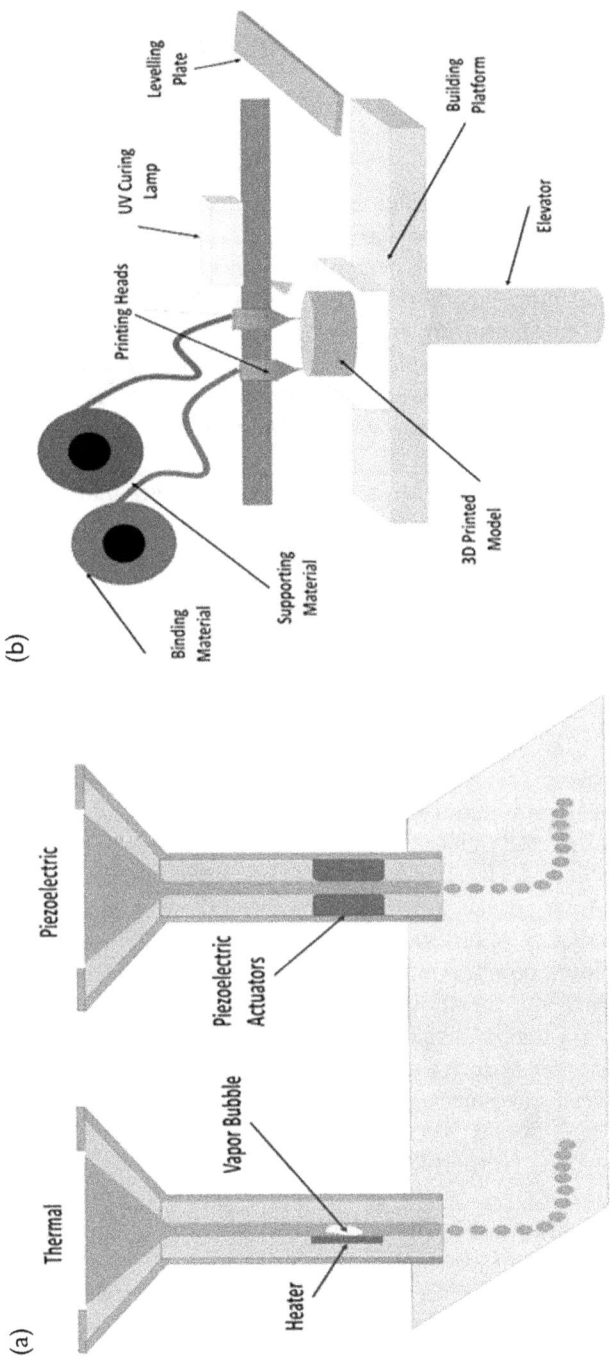

Figure 5.6 **(a)** Principle of inkjet printing; **(b)** inkjet printing process [14, 17].

Therefore, this method with resolution of 20 to 100 μm is ideal in printing soft tissues in organs and complex cell scaffolds for culture [18]. Moreover, the printed cells are viable. However, the process is slow and there is a possibility of material wastage.

5.2.2.4 Bioprinting

This isn't a specific 3-D printing technique, instead, diverse 3-D printing methods are included in bioprinting. The principle behind this includes suspended cells within a bioink being deposited through nozzle/extrusion or laser-based bioprinting methods. The laser-based bioprinting methods are appropriate for 3-D printing of hydrogels except SLS. Laser-based bioprinting works on the principle of light deposition in a definite pattern therefore in printing of cross-linked hydrogels, photo-crosslinkable pre-polymers are needed. Some renowned techniques are

- Two-Photon Polymerization: A photosensitive material is polymerized by the application of a near-infrared laser pulse (titanium-sapphire) at a wavelength of 800 nm within a focal volume. Inside a photosensitive resin, this beam is capable of absorbing two, 800 nm wavelength photons at once, which act as a single photon of 400 nm wavelength (UV light) source. This will initiate a chemical reaction among monomers as well as photoinitiator molecules [19]. This nonlinear excitation will ensure photopolymerization only at the focus of near-infrared laser pulse as shown in Fig. 5.7a. This laser focus movement results in construction of a 3-D model within the photosensitive material's volume. This is possible at lower resolution than the directed light's diffraction limit.
- Laser-induced forward transfer: This system comprises of three parts: a pulsed laser source, a donor substrate, and an acceptor substrate as shown in Fig. 5.7b [20]. A laser beam is focused on membranes coated with a bioink comprising of cells. The coating layer absorbs the laser pulse. Once the threshold is achieved by incident energy, these cell droplets are then transferred towards the acceptor substrate. Due to this, shear stress decreases over the cells because of the nonexistent orifice. This movement results in the formation of high resolution structures (2-D and 3-D) from the droplets. Microscopic resolutions are achieved (~10 to 100 μm) [21].

Drop-on-demand printing or inkjet is another method in which picolitres of cell droplets consisting of bio inks are dropped with extreme precision to be integrated with the fibers. Cross-linking of these fibers is done prior to deposition of subsequent layers, which gives rise to a three-dimensional structure [23]. Extremely small droplet volumes result in resolutions up to sub-100 μm.

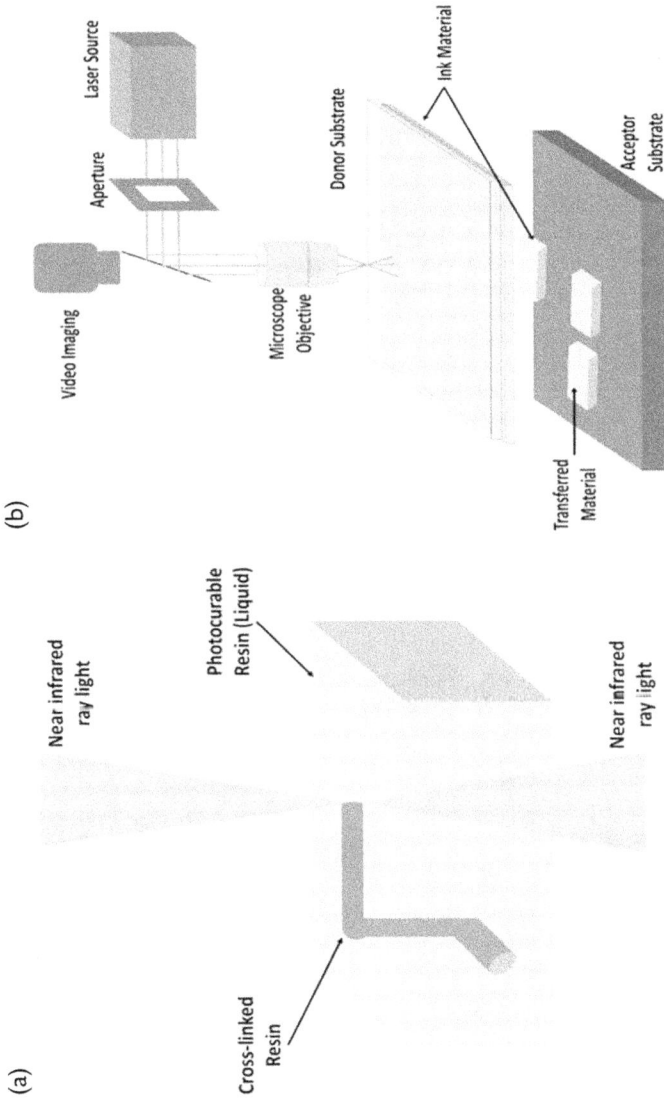

Figure 5.7 **(a)** Two-photon polymerization; **(b)** laser-induced forward transfer process [20,22].

5.2.3 Liquid Reservoir

5.2.3.1 Stereolithography (SL) Method

Stereolithography is the oldest and the first commercial 3-D printing method developed in 1986. It is a VAT polymerization method. 3-D digital parts are designed in CAD software and converted into STL files. These STL files are further transferred to a 3-D printing system. The SL setup comprises of a container with photopolymer liquid and a laser source (UV light). The UV light moves in the horizontal plane whereas the fabrication bed moves in the vertical plane. In this process, the printing process takes place one layer at a time. Fig. 5.8 shows the important parts of SL setup and how the fabrication is carried out.

The process of curing is carried out through concentration of visible/UV light over the resin for solidifying it. The resin, upon absorbing the photon, hardens to a depth more than the bed's height [24]. This results in the polymerization between unreacted functional groups and the subsequent layers. Then bed is then moved down by a distance of one layer inside a VAT liquid comprising of bioresin/photopolymer. The layer post polymerization is raised out of the resin. The process is then repeated for each subsequent layer one at a time. High resolutions of 1.2 to 200 µm are achieved therefore, this process can be utilized in printing both hard and soft tissues for organs, cell scaffolds for culture [25]. But due to the

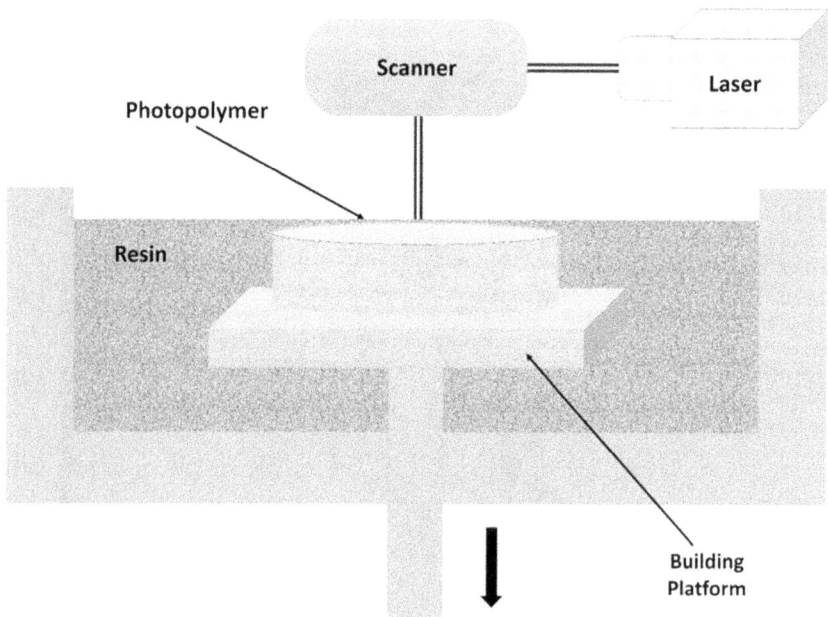

Figure 5.8 Configuration of stereolithography technique [19,24].

existence of pending reactive functional groups, it is essential to wash the resin off or cure the 3-D component with UV light after fabrication.

5.2.4 Material Sheets

5.2.4.1 Sheet Lamination

Another name for this process is laminated object modelling (LOM). The method involves cutting of material sheets (metal, plastic, ceramic, or paper) using blade or laser where every material sheet represents a CAD model slice. Post slicing, all the sheets are stacked and bounded together with a binder. The LOM process is represented in Fig. 5.9. The process involves solid state bonding; hence, it isn't necessary for the material to attain melting point for occurrence of bonding [17]. In some cases, heat is also given to join the layers together. A 3-D inner design is revealed post-removal of the cut sections. Resolutions up to 1 mm are achieved and the process is used in fabrication of macroscopic model of anatomy. The process is cost effective and needs no additional support structures. Nevertheless, the process is slow and there is a lot of material wastage. There is also a possibility of delamination of the printed 3-D object.

5.2.5 Nanofabrication

The process in which structures ideal for medicine as well as electronics and with sizes lower than 100 nm are constructed is called nanofabrication.

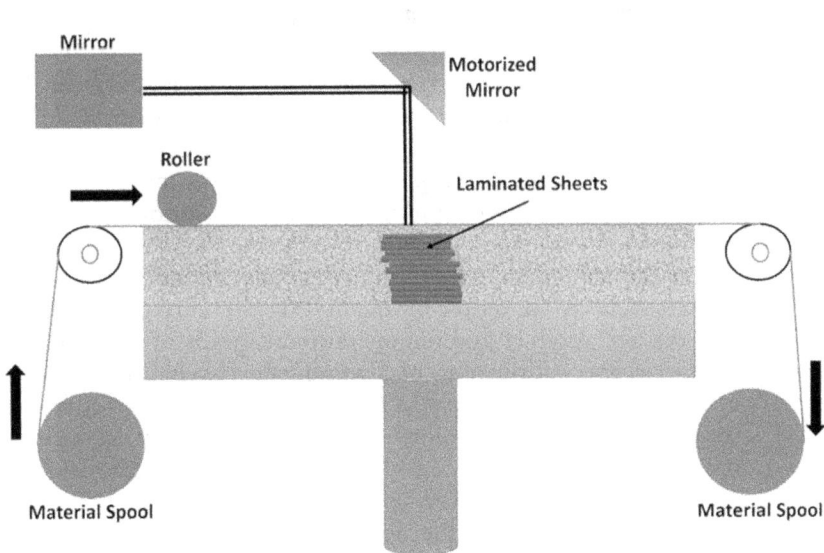

Figure 5.9 Laminated object modelling process [17].

This isn't considered as a typical 3-D printing technique, but this employs similar principles. There are two types of nanofabrication methods.

5.2.5.1 Top-down Approach

In this, a large material is deconstructed so as to achieve the required nanostructure. It is similar to the conventional machining process where material removal takes place. The steps involved in this technique are complex and expensive with limited modification flexibility. Moreover, the variable nature of manufacturing process can't be controlled as a result of which nanostructures produced through this method have limited reproducibility [26].

5.2.5.2 Bottom-up Approach

This type of nanofabrication involves building blocks through self-assembly of atoms/molecules or through printing a nanostructure of atomic resolutions. This approach has additive nature and similar to 3-D printing methods. The applications are numerous in tissue engineering as well as biomedicine. Some examples are fabrication of biomaterials comprising of distinctive biological characteristics, delivery and sequestration of drugs, circulation of waste binders, and toxins. The applications also involves vaccines where immunogenicity of compounds is preserved and transplant rejection is minimized by immuno-isolation [27].

5.3 NOVEL METHODS OF 3-D PRINTING

Some of the current advancements pertaining to novel methods of 3-D printing in the field of biomedical engineering have been discussed. The research related to interdisciplinary fields is the bridge between advanced materials and the methods of 3-D printing. The objects made according to traditional methods of 3-D printing are due to addition of multiple layers. But, these conventional methods, form a step like structure at the edges which is termed "stair-step effect." The conventional 3-D printing has certain restrictions in terms of quality of the surface generated, geometry of the end product and the operational speeds of the methods. Another issue is the fabrication of components having both multi-functionality as well as complexity. The purpose of these novel techniques is to enhance the formability and the printability of various biomaterials without compromising on their original characteristics at the time of 3-D printing. Moreover, construction of multifunctional and complex 3-D models is the unique ability of these novel 3-D printing processes.

5.3.1 Electrohydrodynamic Printing

Electrohydrodynamic (EHD) printing, a non-conventional method in 3-D printing, has exhibited potential in the biomedical field. It has an exceptional resolution in printing with liquid inks obtained as droplets/jets when an electric field exists (Fig. 5.10) [28].

Since its development, 15 years ago, for application in the biomedical field, EHD hasn't been much used due to its complex and specific nature. But EHD is ideal for thermolabile drugs and exerts better control in the printing process with higher precision [29,30]. One system with two methods was designed by Ahmad et al. which can shift from electrohydrodynamic spraying to EHD and vice versa [31]. The idea is for dispensing nanosuspensions in biomedical field. It was observed that EHD is simple for using and ideal for a range of biomaterials. Another finding suggested using EHD method for printing of composites as well as polymers [32]. Scaffolds with higher precision of polyurethane and PMSQ were 3-D printed using the EHD method to be applied in tissue engineering. The EHD method was employed for fabricating graphene-based PCL and PEO dual core matrices for nerve regeneration/restoration applications. Graphene was employed for the growth of neural stem cell and the addition of dopamine hydrochloride was done to enhance graphene matrix's biocompatibility [33]. The integration of these matrices was by the EHD method was successful. The obtained results are encouraging for restoring outer nerves. EHD printing was used to produce a film of flexible composite drug having two simultaneous drugs. Such layers can be rolled and stocked inside commercial

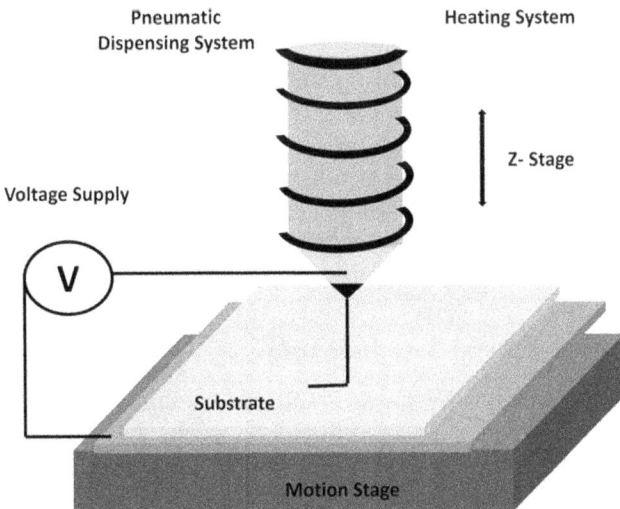

Figure 5.10 Electrohydrodynamic printing technique [28].

oral dosage capsules. This folding/unfolding ability of the drug enhances drug release through enhanced surface area as well as retention duration. Therefore the dosage can be reduced. In spite of promising signs, EHD 3-D printing has limited biomedical uses. Major disadvantage of the EHD 3-D printing process is the low rate of manufacturing and also complex nature of the process in comparison to other 3-D printing methods, which give similar results.

5.3.2 Novel SLA

A unique 3-D printing methods was developed by Kelly et al. which makes use of computed tomography known as computed axial lithography (CAL) [34]. The CAL technique was described through the method of image reconstruction of computed tomography, a technique broadly applied in non-destructive testing as well as medical imaging. Current computed tomography advancements for cancer treatment applications give a technique termed "intensity modulated radiation therapy (IMRT)." This IMRT enables the target areas of tumors within the body of the patient to bare crucial dose of radiation in 3-D. Replacing the patient is a material that is photoresponsive which is exposed to CT scans. This gives out complex 3-D models that are flexible, smooth and free from stair-steps. A polymer-based viscous liquid having photo-curable grafts was utilized by the research enthusiasts. Oxygen molecules were dissolved and materials were created to have a reaction with a particular threshold patterned light to achieve solidification. Any 3-D object with desirable geometry can be fabricated through the projection of light over a rotating cylinder.

With this technique, a geometry with a centimeter scale may be completed under a minute's duration. A wide range of shapes can be manufactured in 30–300 s of time having lateral sizes up to 55 mm. There is also a possibility of adding new components to the existing component which is not convenient in the traditional methods of 3-D printing. It isn't necessary for the printing materials to be transparent, opaque objects can also be fabricated by allowing the visible light to be absorbed by a dye molecule. The wavelength of the light should be different from that of the curing wavelength [35].

5.3.3 Multi-material 3-D Printing

An attempt was made by Kang et al. to overcome the issues in 3-D printing of complex vascular cellular networks by employing multi-material printing technique. A system termed an "integrated tissue-organ printer (ITOP)" was developed to facilitate the 3-D printing of a human scaled tissue having any desired geometry [36]. This was possible due to multi-dispensing systems for pattering and extruding of multi cell-laden hydrogels within a single system. Poly (Ɛ-caprolactone) and pluronic F-127 hydrogel acted as supporting and sacrificing layers. Many methods as well as materials were developed. Also, a carrier material of optimum features and with a capability to position cells in liquid state at specified locations within three dimensional structure was also developed. A complex nozzle with a low resolution of 2 μm for biomaterials as well as a 50 μm resolution for cells was developed. Lastly, photo cross-linkable cell laden hydrogels having photocurable characteristics post the passage of cell was also developed [37].

There was a simultaneous printing of outer mold for sacrificial hydrogel which served as the supporting layer. The oxygen as well as the nutrients were able to diffuse inside the printed tissue due to the microchannels lattice structure. The ITOP was successful in creating diverse 3-D models having many kinds of cells as well as biomaterials. It exhibited prospects for manufacturing diverse kinds of vascularized tissues. Another instance is creating a unique organs-on-chip devices with 3-D bioprinting systems [38]. Many biocompatible, functional inks having soft materials were generated. The properties of piezoresistivity and large conductance of the inks enabled self-assembly of them within the physio-mimetic laminar cardiac tissues. The microphysiological for the heart were 3-D printed in one step and it was used for observing responses of the drugs as well as contractile mechanism of the laminar tissues (cardiac). Complicated 3-D components can be printed in less time (fractions of seconds) using extrusion-based methods. But multi-material 3-D printing, centimeter-scaled 3-D objects (soft robots of locomotive and foldable structures of origami) consisting of two different kinds of silicon inks or epoxy, with different stiffs can be printed in minutes at 10 to 40 mm per second speeds.

5.3.4 Embedded 3-D Printing

This method can be utilized for printing models like complex tissues. Initially, this technique was employed for printing a three-dimensional network having interconnected channels inside a matrix comprising of acellular hydrogel as well as silicone with the help of a viscoelastic sacrificial ink [39]. This was possible post the process of curing and removal of sacrificial ink. In the process of embedded three-dimensional printing, viscoelastic ink needs to be extruded inside a reservoir having low yield stress, large shear elastic modulus as well as ability to photo-crosslink. In order to meet the above requirements, Pluronic F127 tri-block with one part of hydrophobic polypropylene oxide, two parts of hydrophilic polyethylene oxide was created as a reservoir through chemical modifications of the hydroxyl compounds of hydrophilic parts with diacrylate groups. Following this lead, another embedded printing method, was proposed on the principle of supramolecular assembly of hydrogels (shear thinning) by guest-host complexes, two supramolecular hydrogels to be mixed.

In this, HA modified by adamantine was the guest and HA modified by β-cyclodextrin became the host. This was injected inside the supporting hydrogel for 3-D structure formations like channel/spiral. The intermolecular bond between the guest and the host resulted in quick generation of supramolecular assemblies. This method was successfully implemented in drug delivery applications.

5.3.5 4-D Printing

The initial idea of creating complex 3-D models can react to external environmental stimulus or the 4-D printing was given by Tibbit et al. A method was identified to create new design systems. The technique of 4-D printing makes use of stimuli-responsive or the smart materials rather than the traditional materials. Due to this, 3-D models that are capable of self-assembling/self-regulating are formed, which are capable of reacting to external stimuli within the environment by changing their shape [40]. In the current scenario, research is being done to manufacture 4-D printed models which are capable of modifying their geometry by twisting, bending, corrugating, and elongating w.r.t external stimuli such as light, heat, or humidity. As newer smart materials, novel printing methods, or mathematical modelling techniques of deformation develop, the 4-D printing will become more feasible [41]. The most researched smart materials for 4-D printing are material responsive to temperatures. The deformation mechanism in such materials depends on the shape memory effect [42]. Shape memory polymers (SMPs) are generally in use due to their ability to achieve their original shape post deformation under external stimulus. Their ease in printability is also a factor for their wide usage. In SMPs, value of T_g is larger than operating temperature. The shape programming can be done through heating or cooling ($>T_g$ or $<T_g$) treatments. For operating temperatures lower than T_g, they undergo a temporary deformation and when the operating temperatures are greater than T_g they regain their original shape. When SMP fibers were integrated with elastomeric matrix to form a hinge structure, the hinge deformed at a maximum angle of value of 20°. The deformation value is dependent on T_g. Similarly, a shape changing model created by write printing of UV photo crosslinkable poly (lactic acid) bioink was fabricated using SMPs and Shape memory nanocomposites. This exhibited an enhanced shape memory effect which gave rise to the configuration transformations of 3-D–1D–3-D, 3-D–2D–3-D, and 3-D–3-D–3-D [43].

Materials responsive to humidity deform by taking or releasing moisture used for 4-D printing. 3-D models printed from rigid polymer inks and humidity responsive materials was tested and by varying the moisture level, there was 200% of volume change in terms of unfolding and folding as compared to the original. But the final object became fragile after repeated folding/unfolding. Light responsive materials also facilitate stimulus responsive material printing methods. This is due to the fact that light focuses energy only at a desired area which enables rapid local control/switching of materials which respond to light. Temperature increases in the local area due to light absorption. Electric/magnetic fields are also useful for 4-D printing heat sources. Silicon rubber neodymium-iron-boron (NdFeB) was fabricated by Kim et al. by applying magnetic actuation during printing and exhibited shape change due to magnetic actuation [44].

Apart from physical stimuli, chemical as well as biological stimuli have also interested the researchers in 4-D printing and related materials. The 4-D printed models are capable of shape changing as well as functionality with the passage of time. This dependence on time for changing shape can generate different prospects for applications in biomedical actuators like self-tightening/self-bending valves, stents as well as staples [45]. Targeted therapy can be carried out through drug release by microbots and delivery, biosensors for diagnosis of issues in medical field with the application of external stimulus. Moreover, scaffolds can be fabricated with a capability of mimicking complex tissues in the human body which possess their own dynamic changes in tissue regeneration. 4-D printed tissues can respond to changes in the environment, change their geometry, and provide favorable microenvironment for regeneration of the tissue. This precise mimicking is not possible in case of the typical 3-D printed tissue models.

5.4 CONCLUSIONS

This current book chapter provides a detailed discussion of various 3-D printing processes and their advancements in biomedical applications. 3-D printing has proven to be advantageous compared to the conventional machining processes. There is growing interest on 3-D printing especially in biomedical applications. This is due to its unmatched abilities to create biological products easily. With further advancements, we can expect the costs of 3-D manufacturing to come down. In the biomedical field, there are challenges existing for 3-D printing in terms of fabrication speeds, resolutions, and large-scale fabrication with high biocompatibility. Bioprinting has exhibited success in fabricating 3-D models such as bone scaffolds, cartilage, heart, etc. But achieving large vascularized models with complex structures in biomedical applications is still a challenge. The problems such as longevity of bioink and microvascularization should be conquered to achieve full printing of organs and complex tissues. The focus should be advancements such as 4-D printing techniques, which can help achieve greater landmarks and eliminate the complications currently being faced in the biomedical field.

REFERENCES

[1] Zadpoor AA. Design for additive bio-manufacturing: From patient-specific medical devices to rationally designed meta-biomaterials. *International Journal of Molecular Sciences*. 2017 Aug; 18(8): 1607.

[2] Bourell DL. Perspectives on additive manufacturing. *Annual Review of Materials Research*. 2016 Jul 1; 46: 1–8.

[3] Ahangar P, Cooke ME, Weber MH, Rosenzweig DH. Current biomedical applications of 3-D printing and additive manufacturing. *Applied Sciences*. 2019 Jan; 9(8): 1713.

[4] Yeong WY, Chua CK, Leong KF, Chandrasekaran M, Lee MW. Indirect fabrication of collagen scaffold based on inkjet printing technique. *Rapid Prototyping Journal*. 2006 Aug 1; 12(4): 229–237.

[5] Rayate A, Jain PK. A review on 4-D printing material composites and their applications. *Materials Today: Proceedings*. 2018 Jan 1; 5(9): 20474–20484.

[6] Akoury E, Weber MH, Rosenzweig DH. 3d-printed nanoporous scaffolds impregnated with zoledronate for the treatment of spinal bone metastases. *MRS Advances*. 2019 Apr; 4(21): 1245–1251.

[7] Coelho G, Chaves TM, Goes AF, Del Massa EC, Moraes O, Yoshida M. Multimaterial 3-D printing preoperative planning for frontoethmoidal meningoencephalocele surgery. *Child's Nervous System*. 2018 Apr; 34(4): 749–756.

[8] Fairag R, Rosenzweig DH, Ramirez-Garcialuna JL, Weber MH, Haglund L. Three-dimensional printed polylactic acid scaffolds promote bone-like matrix deposition in vitro. *ACS Applied Materials & Interfaces*. 2019 Apr 11; 11(17): 15306–15315.

[9] Ngo TD, Kashani A, Imbalzano G, Nguyen KT, Hui D. Additive manufacturing (3-D printing): A review of materials, methods, applications and challenges. *Composites Part B: Engineering*. 2018 Jun 15; 143: 172–196.

[10] Al-Dulimi Z, Wallis M, Tan DK, Maniruzzaman M, Nokhodchi A. 3-D printing technology as innovative solutions for biomedical applications. *Drug Discovery Today*. 2020 Nov 16; 26(2): 360–383.

[11] Trenfield SJ, Madla CM, Basit AW, Gaisford S. Binder jet printing in pharmaceutical manufacturing. In *3-D Printing of Pharmaceuticals* 2018 (pp. 41–54). Springer, Cham.

[12] Hong D, Chou DT, Velikokhatnyi OI, Roy A, Lee B, Swink I, Issaev I, Kuhn HA, Kumta PN. Binder-jetting 3-D printing and alloy development of new biodegradable Fe-Mn-Ca/Mg alloys. *Acta Biomaterialia*. 2016 Nov 1; 45: 375–386.

[13] Badiru AB, Valencia VV, Liu D, editors. Additive Manufacturing Handbook: Product Development for the Defense Industry. CRC Press; 2017 May 19.

[14] Malda J, Visser J, Melchels FP, Jüngst T, Hennink WE, Dhert WJ, Groll J, Hutmacher DW. 25th anniversary article: Engineering hydrogels for biofabrication. *Advanced Materials*. 2013 Sep; 25(36): 5011–5028.

[15] Rahman Z, Ali SF, Ozkan T, Charoo NA, Reddy IK, Khan MA. Additive manufacturing with 3-D printing: Progress from bench to bedside. *The AAPS Journal*. 2018 Nov; 20(6): 1–4.

[16] Dass A, Moridi A. State of the art in directed energy deposition: from additive manufacturing to materials design. *Coatings*. 2019 Jul; 9(7): 418.

[17] Sireesha M, Lee J, Kiran AS, Babu VJ, Kee BB, Ramakrishna S. A review on additive manufacturing and its way into the oil and gas industry. *RSC Advances*. 2018; 8(40): 22460–22468.

[18] Lim KS, Levato R, Costa PF, Castilho MD, Alcala-Orozco CR, Van Dorenmalen KM, Melchels FP, Gawlitta D, Hooper GJ, Malda J, Woodfield TB. Bio-resin for high resolution lithography-based biofabrication of complex cell-laden constructs. *Biofabrication*. 2018 May 11; 10(3): 034101.

[19] Li J, Wu C, Chu PK, Gelinsky M. 3-D printing of hydrogels: Rational design strategies and emerging biomedical applications. *Materials Science and Engineering: R: Reports*. 2020 Apr 1; 140: 100543.

[20] Serra P, Piqué A. Laser-induced forward transfer: Fundamentals and applications. *Advanced Materials Technologies*. 2019 Jan; 4(1): 1800099.

[21] Ghidini T. Regenerative medicine and 3-D bioprinting for human space exploration and planet colonisation. *Journal of Thoracic Disease*. 2018 Jul; 10(Suppl 20): S2363.

[22] Billiet T, Vandenhaute M, Schelfhout J, Van Vlierberghe S, Dubruel P. A review of trends and limitations in hydrogel-rapid prototyping for tissue engineering. *Biomaterials*. 2012 Sep 1; 33(26): 6020–6041.

[23] Stringer J, Derby B. Formation and stability of lines produced by inkjet printing. *Langmuir*. 2010 Jun 15; 26(12): 10365–10372.

[24] Stansbury JW, Idacavage MJ. 3-D printing with polymers: Challenges among expanding options and opportunities. *Dental Materials*. 2016 Jan 1; 32(1): 54–64.

[25] Burton HE, Eisenstein NM, Lawless BM, Jamshidi P, Segarra MA, Addison O, Shepherd DE, Attallah MM, Grover LM, Cox SC. The design of additively manufactured lattices to increase the functionality of medical implants. *Materials Science and Engineering: C*. 2019 Jan 1; 94: 901–908.

[26] Biswas A, Bayer IS, Biris AS, Wang T, Dervishi E, Faupel F. Advances in top–down and bottom–up surface nanofabrication: Techniques, applications & future prospects. *Advances in Colloid and Interface Science*. 2012 Jan 15; 170(1–2): 2–7.

[27] Ruiz-Hitzky E, Aranda P, Darder M, Ogawa M. Hybrid and biohybrid silicate based materials: molecular vs. block-assembling bottom–up processes. *Chemical Society Reviews*. 2011; 40(2): 801–828.

[28] Han Y, Dong J. High-resolution electrohydrodynamic (EHD) direct printing of molten metal. *Procedia Manufacturing*. 2017 Jan 1; 10: 845–850.

[29] Wu S, Ahmad Z, Li JS, Chang MW. Fabrication of flexible composite drug films via foldable linkages using electrohydrodynamic printing. *Materials Science and Engineering: C*. 2020 Mar 1; 108: 110393.

[30] Fina F, Madla CM, Goyanes A, Zhang J, Gaisford S, Basit AW. Fabricating 3-D printed orally disintegrating printlets using selective laser sintering. *International Journal of Pharmaceutics*. 2018 Apr 25; 541 (1–2): 101–107.

[31] Ahmad Z, Thian ES, Huang J, Edirisinghe MJ, Best SM, Jayasinghe SN, Bonfield W, Brooks RA, Rushton N. Deposition of nano-hydroxyapatite particles utilising direct and transitional electrohydrodynamic processes. *Journal of Materials Science: Materials in Medicine*. 2008 Sep; 19(9): 3093–3104.

[32] Ahmad Z, Rasekh M, Edirisinghe M. Electrohydrodynamic direct writing of biomedical polymers and composites. *Macromolecular Materials and Engineering*. 2010 Apr 14; 295(4): 315–319.

[33] Wang B, Chen X, Ahmad Z, Huang J, Chang MW. 3-D electrohydrodynamic printing of highly aligned dual-core graphene composite matrices. *Carbon*. 2019 Nov 1; 153: 285–297.

[34] Kelly BE, Bhattacharya I, Heidari H, Shusteff M, Spadaccini CM, Taylor HK. Volumetric additive manufacturing via tomographic reconstruction. *Science*. 2019 Mar 8; 363(6431): 1075–1079.

[35] Tetsuka H, Shin SR. Materials and technical innovations in 3-D printing in biomedical applications. *Journal of Materials Chemistry B*. 2020; 8(15): 2930–2950.

[36] Kang HW, Lee SJ, Ko IK, Kengla C, Yoo JJ, Atala A. A 3-D bioprinting system to produce human-scale tissue constructs with structural integrity. *Nature Biotechnology*. 2016 Mar; 34(3): 312–319.

[37] Lind JU, Busbee TA, Valentine AD, Pasqualini FS, Yuan H, Yadid M, Park SJ, Kotikian A, Nesmith AP, Campbell PH, Vlassak JJ. Instrumented cardiac microphysiological devices via multimaterial three-dimensional printing. *Nature Materials*. 2017 Mar; 16(3): 303–308.

[38] Wu W, DeConinck A, Lewis JA. Omnidirectional printing of 3-D microvascular networks. *Advanced Materials*. 2011 Jun 24; 23(24): H178–H183.

[39] Highley CB, Song KH, Daly AC, Burdick JA. Jammed microgel inks for 3-D printing applications. *Advanced Science*. 2019 Jan; 6(1): 1801076.

[40] Tibbits S. 4-D printing: Multi-material shape change. *Architectural Design*. 2014 Jan; 84(1): 116–121.

[41] Zhang Z, Demir KG, Gu GX. Developments in 4-D-printing: A review on current smart materials, technologies, and applications. *International Journal of Smart and Nano Materials*. 2019 Jul 3; 10(3): 205–224.

[42] Momeni F, Sabzpoushan S, Valizadeh R, Morad MR, Liu X, Ni J. Plant leaf-mimetic smart wind turbine blades by 4-D printing. *Renewable Energy*. 2019 Jan 1; 130: 329–351.

[43] Wei H, Zhang Q, Yao Y, Liu L, Liu Y, Leng J. Direct-write fabrication of 4-D active shape-changing structures based on a shape memory polymer and its nanocomposite. *ACS Applied Materials & Interfaces*. 2017 Jan 11; 9(1): 876–883.

[44] Kim Y, Yuk H, Zhao R, Chester SA, Zhao X. Printing ferromagnetic domains for untethered fast-transforming soft materials. *Nature*. 2018 Jun; 558(7709): 274–279.

[45] Lui YS, Sow WT, Tan LP, Wu Y, Lai Y, Li H. 4-D printing and stimuli-responsive materials in biomedical aspects. *Acta Biomaterialia*. 2019 Jul 1; 92: 19–36.

Chapter 6

Additive Manufacturing for Fabrication of Composites

Kundan Kumar, Ashish Das, and Shashi Bhushan Prasad

Department of Production and Industrial Engineering, National Institute
of Technology, Jamshedpur, Jharkhand, India

CONTENTS

DOI: 10.1201/9781003327370-6

6.1 INTRODUCTION

Additive manufacturing (AM) has been recognized as a smart developed manufacturing technique where computer-aided design (CAD) models are used for making three-dimensional parts and based on the addition of materials directly layer by layer. The most important advantage of AM is complex parts are easily formed which cannot be formed by subtractive manufacturing processes. The design and fabrication of composite parts provide using AM which is direct and layer-by-layer fabrication as compared to conventional methods. 3-D printing used for AM is the manufacturing from CAD data to end-use parts directly. AM eliminates the necessity for complex intermediate tooling. It is considerable shortening lead times of the manufacturing. Apart from significant design freedom, AM produces more complex shapes. AM is referred to as the digital age of the industrial revolution and exact just-in-time production having advantages of the digital and tool-less technologies. It has taken the attention worldwide. The unique processing features of AM proposal other opportunities like reduction of waste, environmental conservation, and optimum designs for lean manufacturing [1]. Environmentally friendly product designs are another benefit of AM. In general, models assisting manufacturing are used in the product development process. The applications of AM have a wide range including automobile, aerospace, industrial, and biomedical [2].

AM processes are classified according to the raw materials state-input such as a solid layer, powder, molten, and liquid [2–4]. Several advantages of AM over traditional manufacturing methods (casting, machining, etc.) are as follows:

1. High degree of customization
2. High level of automation
3. Rapid product development process
4. Complex designs [4]

Different AM processes used for fabricating composite materials are Selective Laser Sintering (SLS), Fused Deposition Modeling (FDM), Direct Energy Deposition (DED), Selective Laser Melting (SLM), and Stereolithography (SLA) [5–7].

6.1.1 Composite

AM is the most original form of fabrication of composites. Composite materials have been used for lightweight components in many industries such as automotive, aerospace, marine, nuclear, and biomedical industries. They offer high performance and high mechanical strength for specific applications. A composite has at least two constituents. Major constituents (>50%) of composite are known as matrix i.e. metal, polymer, and ceramic. Manor constituent

(<50%) of composite is known as reinforcement i.e. fibers, sheets, particles, etc. AM is used for the fabrication of composites such as ceramic matrix, metal matrix, and polymer matrix. Furthermore, an effective application of AM in the fabrication of Functionally Graded Materials (FGM) owing to the ability to optimize the properties and control the composition of the built part. The missile nose cone is an example of FGM using AM.

This paper presents an overview of the development of AM processes for the fabrication of composite materials. In the present work, AM for composite materials has been studied for their objectives, applications, processing, results, and challenges.

6.2 AM USED FOR FABRICATION OF METAL MATRIX COMPOSITES (MMCS)

As compared to pure metals and alloys, unique features of MMCs such as high specific stiffness, high wear resistance, and greater specific strength to weight ratio increase their demands. They are used in several applications like aerospace, cutting tools, brake disks, turbine blades, high-temperature components, and biomedical applications [8,9]. Ductile metal is used as the matrix such as copper, magnesium, aluminum, titanium, etc. in MMCs. Hard-ceramic material is used as reinforcement in the form of fibers, particulates, and filaments. MMCs have enhanced mechanical properties such as strength, creep life, modulus, and wear resistance.

Traditional MMCs processes have many limitations that are overcome by AM and are considered as an alternative method. Three-dimensional (3-D) printing is one of the most prominent methods of fabrication of MMCs by AM. This method can be fabricated complex parts generally from powdered materials. Merits of this method are recycling of the used materials, design and printing of complex parts, waste elimination, the possibility of lightweight component printing, and gradient structures [10,11].

6.2.1 Fabrication Processes of MMCs by AM

The most common method of MMCs fabrication is powder-based AM that uses the processing of different powdered materials. The fusion-based AM process is also used in the melted matrix. Usually, "in-situ" and "ex-situ" are two main methods for the fabrication of MMCs. The fabricated composite remains the same composition as the starting powdered materials in the ex-situ approach. The reinforcement is externally synthesized in "ex-situ." In contrast, in the in-situ approach, the reinforcement particles are produced within the matrix owing to chemical reactions. In other words, the production of reinforcement is owing to the chemical reactions in the matrix during AM of MMCs. MMCs can be fabricated by different AM processes such as direct energy deposition (DED), powder bed fusion (PBF), binder-jet 3-D printing (BJ3-DP), and hybrid 3-D printing. Fig. 6.1 is illustrated schematic of AM processes for MMCs.

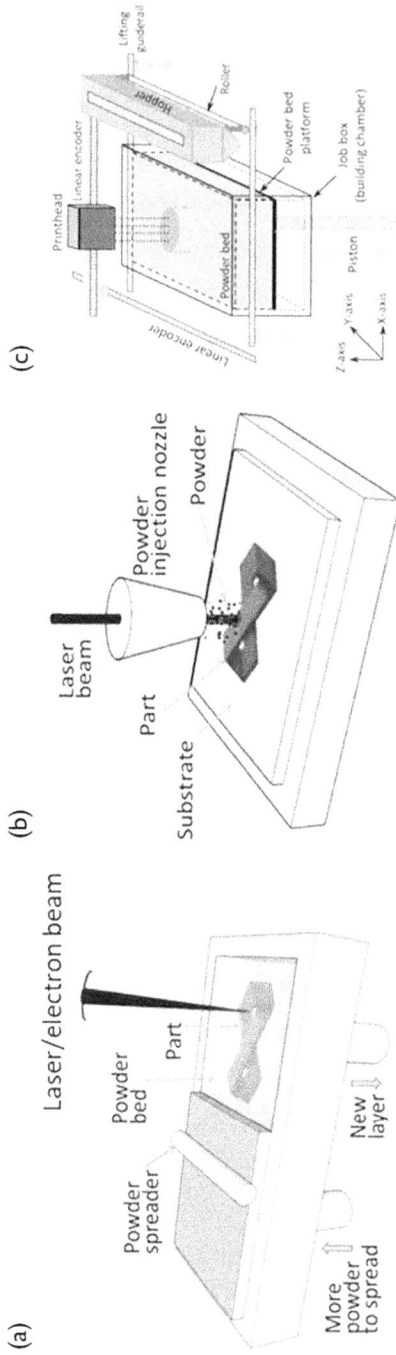

Figure 6.1 Diagram of (a) PBF, (b) DED, and (c) BJ3-DP [12].

6.2.1.1 AM Used for MMC Fabrication by Fusion-Based Methods

Numerous powder-based fusion techniques are used to fabricating MMCs using AM processes that are direct energy deposition (DED), electron beam melting (EBM), direct metal laser sintering (DMLS), and selective laser melting/sintering (SLM/SLS) can be used. In the electron beam PBF (E-PBF) or laser-based PBF (L-PBF) processes, at the area of interest, a focused beam of electron or laser is directed to join powders. Pre-mixed powder particles are sintered (partially melted) as a result ex-situ reinforced MMCs are fabricated in the DMLS and SLS methods. On other hand, in EBM and SLM, powdered materials are melted fully. Thus, the in-situ reinforced MMCs can be fabricated using these methods. An inert chamber is needed to operate L-PBF and the laser beam having raster motion is used to process the powder, as a result, melting and quick solidification. L-PBF and E-PBF have few differences. E-PBF works under vacuum conditions. Further, there is a two-step sequence of the E-PBF process. First, the powder bed is pre-heated, and each layer of powder is used to repulsion of the powder particles and prevent electrostatic charging. Second, the electron beam is used to fuse powder particles. The beam of electrons is allowed a fast-scanning speed owing to pre-sintering whereas L-PBF is not needed preheating. In powder-based DED, a nozzle is used to fed powder particles into the desired spot and a laser or electron beam is used to create the melt path and molten pool. Various DED processes are laser metal deposition (LMD), direct laser deposition (DLD), laser consolidation (LC), laser engineered net shaping (LENS), and laser cladding. A CAD model is used to begins the deposition process and production happens layer-by-layer on a build piece. In the molten pool, powder flow takes place better way in presence of a shielding gas such as argon and protects the molten materials from oxidation. DED processes are used to fabricated ex-situ as well as in-situ reinforced MMCs. The PBF process has a better surface finish as compared to the DED process. The DED process fabricated samples needed a further surface finish to obtained the desired surface quality.

The main advantage of AM processes using powder bed fusion (PBF) can incorporate reinforcements into the metal matrix phase either in-situ or ex-situ. The premixed powder conditions are used to identify the shape and size of reinforcements in ex-situ reinforced MMCs. In in-situ reinforced MMCs, the desired reinforcement is synthesized from the chemical reactions. MMCs parts are fabricated by PBF AM have generally used different reinforced particles such as Al_2O_3 ([13]), SiC([14,15]), TiC ([16,17]), TiB_2 ([18,19]), and carbon ([20,21]). Besides, ex-situ reinforced MMCs have two other drawbacks such as the weak interfacial bonding between the metal matrix and reinforcement, and the poor wettability at the interface between the matrix and reinforcement.

6.2.1.2 AM for MMC Fabrication by Non-Beam-Based Methods

As compared to a non-beam-based AM process such as binder jetting, fusion-based AM processes such as DED, and PBF methods where powder particles are melted by the heating source. Based on a CAD model, printhead sprays and a roller binder are used as a layer of powder to spread on the outward in this method (powder bed technique) to fabricated parts. The printing step is finished by this process repeating multiple times. In the binder jetting method, powder layer thickness, drying time, print speed, and binder saturation are the principal print process constraints. In this method, solid-state processing, densifications (sintering and infiltration), and curing (to increase green part strength) are post-processing after 3-D-printed pre-mixed powder. As compared to the 3-D printing matrix, it uses a lower melting temperature alloy. It achieves a full density and eliminates residual porosity with insignificant dimensional changes. Powder particles are nearby melted in the sintering of densification method. The final resulting microstructure and properties depend on holding time, temperature, and atmosphere during sintering (Mostafaei, 2018). Binder composition is the main factor in curing generally used at 180 C–200 C for up to 8 hours.

6.2.1.3 3-D Printing

MMCs fabricated by 3-D printing are not common as compared to pure metals and alloys. There are two ways of processing materials in binder jet 3-D printing. First, a powder mix has printed the composite and a sintering post-processing step is used for finishing. Second, the reinforcement material is printed and a melt infiltration step is used to finish the metal matrix material.

6.2.2 MMCs by AM Processes

Some common MMCs by AM are aluminum-based MMCs, nickel-based MMCs, stainless steel-based MMCs, and titanium-based MMCs. Other matrices used are Ag, Be, Co, Cu, Fe, and Mg. The common reinforcements used are B_4C, Al_2O_3, graphite, C, BeO, NBC, Mo, SiC, TiB, TaC, TiC, $TiBl_2$, WC, and W. The most commonly used reinforcement is SiC and followed by TiC and Al_2O_3.

6.2.3 Challenges of AM-Based MMC Fabrication

A few challenges of MMC fabrication are as follows ([5,6,22,23]):

1. Optimum process parameters cannot be found by AM of MMC fabrication.
2. During the fusion-based AM process in the inter-layer areas, the residual stresses and thermal gradient might be induced that affect mechanical properties.

3. In the 3-D-printed parts, different defects may form depending on the processing parameters and printing method.
4. Between pre-mixed powders maybe occur unwanted chemical reactions which change the microstructure and designed composition of MMCs. Ex-situ reinforced MMCs are more challenging owing to chemical reactions as compared to in-situ reinforced MMCs.
5. Owing to hard reinforcements materials, the reinforcement/matrix interface may have defects like micro-cracking which affect the performance and functionality of MMCs.
6. Loss of specific materials, dissolution of alloying element(s), and micro-segregation of enforcing materials maybe occur during AM MMC fabrication.

6.3 ADDITIVE MANUFACTURING (AM) FOR POLYMER MATRIX COMPOSITES (PMCS)

The better materials requirement in terms of density, strength, lower cost, and stiffness with improved sustainability is rising. Polymers are one of them that has taken the attention of fabrications owing to their inimitable features such as lightweight, availability, low cost, ease of production, long life, and ductility [24]. Pure polymers have limitations such as poor mechanical properties, load-bearing applications, and fully functional parts [25]. To overcome these challenges, PMCs have developed as one of the most wanted solutions [24]. PMCs fabricated by AM have better properties than either matrix or reinforcement that eliminate the problems associated with pure polymers [25]. In PMCs, reinforcement may be polymeric inclusions, ceramic, metallic in the form of particles, fibers, platelets, whiskers. Various types of PMCs available in the literature are fiber-reinforced PMCs, particle-reinforced PMCs, and nanocomposites [25].

6.3.1 PMCs with Fiber-Reinforced

AM has altered fiber-reinforced composites into the manufacturing unit of a robust model. PMCs have excellent properties such as damping property, durability, high strength to weight ratio, flexural strength, stiffness, wear resistance to corrosion, and impact [26]. Synthetic fibers used are commonly Kevlar fibers, graphene fibers, glass fibers, carbon fibers, and basalt fibers. Synthetic fibers are not as economical as natural fibers. Natural fibers have properties like low cost, low weight, not toxicity, ease of availability, and recyclability [27]. Acrylonitrile butadiene styrene (ABS) composites have used short fibers reinforced for 3-D printing. As an increase of fiber content, the modulus and tensile strength increase [28]. Fatigue life and static strength of basalt fiber-reinforced polymer (BFRP) composites are rise with a decrease in temperature at a certain maximum stress [26].

6.3.2 PMCs with Particle Reinforced

The functionalities and properties of the PMCs are better in the case of particle reinforcements owing to their low cost and isotropic properties as compare to the polymer matrix. Stereolithography (SLA) is used to mixed reinforcement particles in the liquid polymer matrix. Selective laser sintering (SLS) is used to mixed particles reinforcement in powder form polymer matrix whereas fused deposition modeling (FDM) is used to mixed reinforcement particles in the form of the printable filaments' polymer matrix. PMCs have enhanced mechanical properties, better wear resistance, and improved dielectric permittivity [29]. The size of particle reinforcement may be micro or nano. In nanoparticle reinforcement, PMCs have improved the material properties owing to the full dense structure. A thermally conducting PMC material was developed using the SLA technique having 30% micro-diamond particles in acrylate resins [30]. The cooling coils and heat sink are fabricated by these materials as rapidly 3-D print prototypes.

6.3.3 Nano-Based PMCs

Nano-based PMCs have added considerable amount attention of to engineers and researchers [31]. In nano-based PMCs, nanofillers are distributed in the polymer matrix. Nanofillers used are in the form of reinforcements and functional additives. Nanofillers improve properties marginally such as chemical, mechanical, thermal, and other [32]. Both nano-based PMCs and pure polymers are fabricated by AM in the same method and machine [32]. In nano-based PMCs, incorporated nanofillers into a 3-D print product can be obtained in two ways. The first way, manually or automatically adding nanofillers in the matrix phase of 3-D printing by intermittent stopping the process. The second way, before the 3-D printing, nanofillers are added into the matrix phase [31]. Nano PMCs were prepared using graphene oxide (GO)/polyurethane/polylactic acid nanoparticles [33]. Solvent-based mixing process is used for the FDM process and filament is obtained and extruded. The thermal stability and mechanical property of nano-based PMCs are improved largely by the addition of GO.

6.3.4 Applications of PMCs Using AM

Applications of PMCs are growing faster ways among all the materials for several applications. Fig. 6.2 shows applications of polymers and their composite parts fabricated using AM. AM-fabricated polymer composites are used in the field of automobile industries, aerospace, electronics, and biomedical applications. Recent applications of AM in biomedical applications are focused on tissue engineering, regenerative medicine, orthopaedic applications, drug delivery, medical instrument fabrication, prosthesis, 3-D bioprinting, dental applications, etc. [34].

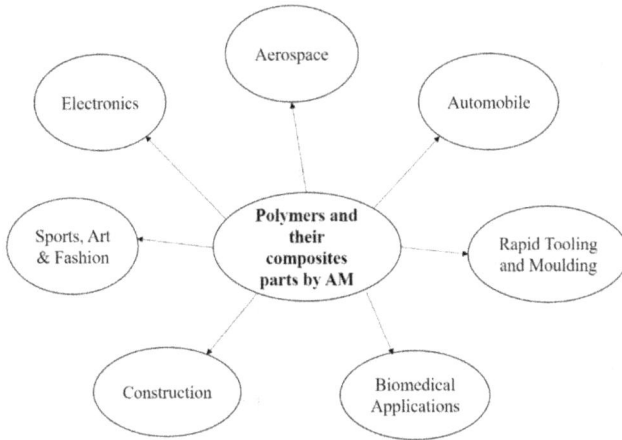

Figure 6.2 Polymers and their composite applications fabricated by AM 3-D printed.

6.3.5 Sustainability PMCs by AM and Their Recycling

Polymers and PMCs fabricated by AM make them smart and sustainable manufacturing. Extended product life, reconfigured value chains, and improved resource efficiency are some of the sustainability benefits. The materials used play the key factors in the sustainability of AM PMCs. Polymers and PMCs are rapidly consumed owing to high demand and their waste is an enormous impact on sustainability. AM process, used, disposal, and reused of the polymers have made PMCs sustainable and have minimal impact on the environment [35].

6.4 CERAMIC MATRIX COMPOSITES (CMCS) BY ADDITIVE MANUFACTURING

In CMCs, reinforcements used are ceramic in the form of whiskers, continuous fibers, and nanoparticles owing to significantly better thermomechanical properties such as fracture toughness and thermal shock resistance than the ceramic matrix. In demanding service conditions, thermo-structural components for high-performance applications have been used. Table 6.1 shows high-temperature turbomachinery components [36]. Owing to highly specific properties, CMCs are highly desirable for many industrial applications as compared to single-phase ceramics [37].

Several conventional or advanced methods have been used for CMC fabrication. Common processes are sol-gel, solid-state, hydrothermal synthesis, and co-precipitation routes [38]. However, currently, CMCs have the most commonly used reinforcements be continuous fibers that are being used commercially mostly use as reinforcement and the

Table 6.1 For turbomachinery components, specific strength of materials vs working
       temperature [36]

Composite	Specific Strength (MPa)	Temperature (°C)
Metal matrix composite	200–400	50–500
Polymer matrix composite	400–800	50–400
Ceramic matrix composite	50–100	1000–1550

composites are carbon fibers/silicon carbide matrix (C_f/SiC), SiC$_f$/SiC, and C_f/C [39,40].

Advanced ceramic components, CMCs, are fabricated by AM which are still under developing stage. As compared to polymer and metal industries, execution of AM process in CMCs have been slower owing to the poor mechanical properties, resolution, surface quality, and scalability, and now considerable interest in its development. CMCs by AM have reduced production costs, expensive tooling, and lead times. They are provided increasing design freedom. CMC fabrication can be used in a large variety of AM technologies, but different results have been found so far. Based on applications and requirements of CMCs, select the correct AM process.

AM for CMCs provides the economical manufacturing of low-volume productions, even individual customized parts, prototypes, and without the usage of high-cost mold tooling [41]. For small production volumes, AM for CMCs is a particularly economical and attractive technology to replace injection molding. The fabrication cost of CMCs by AM processes is not related to design complexity which is only associated with labor cost, and equipment power consumption and material used.

6.4.1 Additive Manufacturing for Fabrication of Ceramic Matrix Composites (CMCs)

Ceramic AM processes can be classified as "multi-step" and "single-step" processes. In AM of multi-step processes, the final CMCs parts are obtained by successive debinding and sintering. Most AM technologies using a multi-step process have all been applied up until now: freeze form extrusion fabrication (FEF) [42], fused deposition of ceramics (FDC) based on extrusion [43], sheet lamination processes [44,45] and robocasting (RC) [46], direct inkjet printing (DIP) [47], vat photo polymerization-based technologies [48,49], binder jetting [50], and indirect laser sintering (LS) [51]. In a single step, currently used AM processes are directed energy deposition (DED) [52] and selective laser melting (SLM)/direct laser sintering (dLS) [53]. The commonly used AM processes for advanced ceramics parts are given in Table 6.2.

Alternatively, negative ceramic AM is another ceramic AM process that generally used polymer molds that are similar to gel casting or investment

Table 6.2 AM processes used for ceramic materials [54]

AM of Advanced Ceramics								
Single-step processes		Multi-step processes						
Bed	Bedless	Bed			Bedless			
Powder Bed	Direct Energy Deposition	Sheet Lamination	Binder Jetting	Fusion	Material Jetting	Vat Photopolymerization	Material Extrusion	
							Water-based	Wax-based
Powder-dLS	LENS	LOM	Powder-BJ	Powder-iLS	Solvent-DIP	SL	RC/DIW	FDC
Slurry-dLS	–	CAM-LEM	Slurry-BJ	Slurry-iLS	Wax-DIP	DLP/LCM	FEF	MJS
						SPPW	CODE	T3-DP
						2PP	3-DGP	PHASE

BJ: Binder Jetting
LOM: Laminated Object Manufacturing
LENS: Lens Engineered Net Shaping
dLS/iLS: Direct/Indirect Laser Sintering
CAM-LEM: Computer Aided Manufacturing of Laminated Engineering Materials
SPPW: Sell-Propagating Photopolymer Waveguide
2PP: Two-Photon Photopolymerization
SL: Stereolithography Extrusion
LCM: Lithography-based Ceramic Manufacturing

RC: Robocasting
DIW: Direct Ink Writing
FDC: Fused Deposition Ceramics
MJS: Multiphase Jet Solidification
CODE: Ceramic On-Demand Extrusion
3-DGP: 3-D Gel Printing
DIP: DirectInkjet Printing
T3-DP: Thermoplastic 3-D Printing
FEF: Freeze Form Extrusion Fabrication
PHASE: Photopolymerization Assisted

casting and a layer of polymer removed from casting parts by thermal burn-out or dissolution [55]. Polymer AM processes such as material jetting [56,57], FDM [58], SL [59,60], and LS [61] are used to prepared polymer molds for negative ceramic AM.

6.4.2 Applications and Benefits of AM in the Fabrication of CMCs

AM used for CMCs are providing low cost, high performance, and relia-bility. AM could benefit from that no need for machining for ceramic composite parts. Machining costs can be as high as 70% of total fabrication costs [62]. Moreover, AM can lead to much shorter production lead times, significant cost reductions, low-volume production, prototyping, and low mold/die tooling cost. They permit the rapid fabrication of custom and fully personalized ceramic composite components. Commercialization of new products can be faster and provides improved design freedom without any extra cost. Applications of AM for ceramics and CMCs are listed in Table 6.3.

Table 6.3 Applications where ceramics and their composites by AM [54]

Application	Ceramic materials	Potential benefits of AM
Scaffolds for tissue engineering	HA, TCP, Bioglass	P, V, M, C, F, L
Dentistry, dental crowns	ZrO_2, Al_2O_3, $Li_2Si_2O_5$	P, V, L, C
Blood valves	Al_2O_3	C, L, V
Orthopaedic implants, spinal components, shoulder buttons	HA, TCP, Al_2O_3, ZrO_2, Bioglass	P, L, V, C
Blades, bearings, gear wheels	SiC, Si_3N_4, Al_2O_3	V, A, G, L
Cutting tools	Si_3N_4, Al_2O_3, B_4C	V, G, A, L
Vehicle panels, ballistic armor, and personal protection	Al_2O_3, B_4C, SiC, TiB_2	P, V, F, M, L, A, C
High performance valve components for corrosive and abrasive fluid flows	Si_3N_4, ZrO2	L, M, V
Nozzle for slurry pumping, water jet cutting and grit blasting	B_4C	V, G, L, M
Waveguides for microwave applications	Al_2O_3	C, L, V
Filters	Al_2O_3, ZrO_2	V, G, L, F
Electrical components, spark plug insulators, connectors	Al_2O_3, SiC	V, G, L, C
Control rods, shielding, and shut down pellets in nuclear power plants	B_4C	L, V
Laser reflectors, chambers, waveguides, tubes, and spacers	AlN, Al_2O_3	L, C, V

V: low-volume production; **G**: geometrical complexity; **M**: material grading; **P**: personalized design; **F**: functional grading; **C**: lower cost; **A**: less assembly; **L**: reduced lead time

Although CMC by AM have made fabulous progress in the last three decays, nevertheless, owing to several limitations such as scalability, limited material selection, materials-process-microstructure-properties relationship, surface defects, and the occurrence of bulk, etc. have limited AM for CMC fabrication so far.

6.5 CONCLUSION

The composite materials fabricated by the AM process have been studied and focus on past, present, and future scopes. Metal matrix composites (MMCs), polymer matrix composites (PMCs), and ceramic matrix composites (CMCs) are three types of composite materials that have generally been fabricated by AM. As compared to a conventional method, the composites fabricated by AM have fabulous advantages like design complexity freedom, low cost, short lead time to develop new product to market, etc. Some challenges have been discussed in this smart manufacturing process.

REFERENCES

[1] M.P. Behera, T. Dougherty, and S. Singamneni, "Conventional and additive manufacturing with metal matrix composites: A perspective," *Procedia Manuf.*, vol. 30, pp. 159–166, 2019, doi: 10.1016/j.promfg.2019.02.023.

[2] K.V. Wong and A. Hernandez, "A Review of Additive Manufacturing," *ISRN Mech. Eng.*, vol. 2012, pp. 1–10, Aug. 2012, doi: 10.5402/2012/2 08760.

[3] N. Guo and M.C. Leu, "Additive manufacturing: Technology, applications and research needs," *Front. Mech. Eng.*, vol. 8, no. 3, pp. 215–243, Sep. 2013, doi: 10.1007/s11465-013-0248-8.

[4] I. Gibson, D. Rosen, and B. Stucker, *Additive Manufacturing Technologies.* New York, NY: Springer New York, 2015.

[5] H.A. Hegab, "Design for additive manufacturing of composite materials and potential alloys: a review," *Manuf. Rev.*, vol. 3, p. 11, Jul. 2016, doi: 10.1051/mfreview/2016010.

[6] Z. Quan *et al.*, "Additive manufacturing of multi-directional preforms for composites: opportunities and challenges," *Mater. Today*, vol. 18, no. 9, pp. 503–512, Nov. 2015, doi: 10.1016/j.mattod.2015.05.001.

[7] S. Kumar and J.-P. Kruth, "Composites by rapid prototyping technology," *Mater. Des.*, vol. 31, no. 2, pp. 850–856, Feb. 2010, doi: 10.1016/j.matdes. 2009.07.045.

[8] S.P. Rawal, "Metal-matrix composites for space applications," *JOM*, vol. 53, no. 4, pp. 14–17, Apr. 2001, doi: 10.1007/s11837-001-0139-z.

[9] A.M. Russell and K.L. Lee, *Structure-Property Relations in Nonferrous Metals.* Hoboken, NJ, USA: John Wiley & Sons, Inc., 2005.

[10] A. Mostafaei, E.L. Stevens, J.J. Ference, D.E. Schmidt, and M. Chmielus, "Binder jetting of a complex-shaped metal partial denture framework,"

Addit. Manuf., vol. 21, pp. 63–68, May 2018, doi: 10.1016/j.addma. 2018.02.014.

[11] J. Giannatsis and V. Dedoussis, "Additive fabrication technologies applied to medicine and health care: a review," *Int. J. Adv. Manuf. Technol.*, vol. 40, no. 1–2, pp. 116–127, Jan. 2009, doi: 10.1007/s00170-007-1308-1.

[12] T. DebRoy *et al.*, "Additive manufacturing of metallic components – Process, structure and properties," *Prog. Mater. Sci.*, vol. 92, pp. 112–224, Mar. 2018, doi: 10.1016/j.pmatsci.2017.10.001.

[13] Q. Han, R. Setchi, and S.L. Evans, "Synthesis and characterisation of advanced ball-milled $Al-Al_2O_3$ nanocomposites for selective laser melting," *Powder Technol.*, vol. 297, pp. 183–192, Sep. 2016, doi: 10.1016/j.powtec. 2016.04.015.

[14] L.C. Astfalck, G.K. Kelly, X. Li, and T.B. Sercombe, "On the breakdown of SiC during the selective laser melting of aluminum matrix composites," *Adv. Eng. Mater.*, vol. 19, no. 8, p. 1600835, Aug. 2017, doi: 10.1002/adem.201600835.

[15] B. Song *et al.*, "Microstructure and tensile behavior of hybrid nano-micro SiC reinforced iron matrix composites produced by selective laser melting," *J. Alloys Compd.*, vol. 579, pp. 415–421, Dec. 2013, doi: 10.1016/j.jallcom. 2013.06.087.

[16] D. Gu, H. Wang, D. Dai, P. Yuan, W. Meiners, and R. Poprawe, "Rapid fabrication of Al-based bulk-form nanocomposites with novel reinforcement and enhanced performance by selective laser melting," *Scr. Mater.*, vol. 96, pp. 25–28, Feb. 2015, doi: 10.1016/j.scriptamat.2014.10.011.

[17] B. AlMangour, D. Grzesiak, and J-M. Yang, "Selective laser melting of TiC reinforced 316L stainless steel matrix nanocomposites: Influence of starting TiC particle size and volume content," *Mater. Des.*, vol. 104, pp. 141–151, Aug. 2016, doi: 10.1016/j.matdes.2016.05.018.

[18] B. Zhang, G. Bi, S. Nai, C. Sun, and J. Wei, "Microhardness and microstructure evolution of TiB_2 reinforced Inconel 625/TiB_2 composite produced by selective laser melting," *Opt. Laser Technol.*, vol. 80, pp. 186–195, Jun. 2016, doi: 10.1016/j.optlastec.2016.01.010.

[19] I. Shishkovsky, N. Kakovkina, and V. Sherbakov, "Graded layered titanium composite structures with TiB_2 inclusions fabricated by selective laser melting," *Compos. Struct.*, vol. 169, pp. 90–96, Jun. 2017, doi: 10.1016/ j.compstruct.2016.11.013.

[20] X. Zhao, B. Song, W. Fan, Y. Zhang, and Y. Shi, "Selective laser melting of carbon/AlSi10Mg composites: Microstructure, mechanical and electronical properties," *J. Alloys Compd.*, vol. 665, pp. 271–281, Apr. 2016, doi: 10.1016/j.jallcom.2015.12.126.

[21] P. Wang *et al.*, "Microstructural characteristics and mechanical properties of carbon nanotube reinforced Inconel 625 parts fabricated by selective laser melting," *Mater. Des.*, vol. 112, pp. 290–299, Dec. 2016, doi: 10.1016/ j.matdes.2016.09.080.

[22] M. Yakout, A. Cadamuro, M.A. Elbestawi, and S.C. Veldhuis, "The selection of process parameters in additive manufacturing for aerospace alloys," *Int. J. Adv. Manuf. Technol.*, vol. 92, no. 5–8, pp. 2081–2098, Sep. 2017, doi: 10.1007/s00170-017-0280-7.

[23] S.K. Ghosh and P. Saha, "Crack and wear behavior of SiC particulate reinforced aluminium based metal matrix composite fabricated by direct metal

laser sintering process," *Mater. Des.*, vol. 32, no. 1, pp. 139–145, Jan. 2011, doi: 10.1016/j.matdes.2010.06.020.

[24] S. Singh, S. Ramakrishna, and F. Berto, "3-D Printing of polymer composites: A short review," *Mater. Des. Process. Commun.*, vol. 2, no. 2, Apr. 2020, doi: 10.1002/mdp2.97.

[25] X. Wang, M. Jiang, Z. Zhou, J. Gou, and D. Hui, "3-D printing of polymer matrix composites: A review and prospective," *Compos. Part B Eng.*, vol. 110, pp. 442–458, Feb. 2017, doi: 10.1016/j.compositesb.2016.11.034.

[26] D. Rajak, D. Pagar, P. Menezes, and E. Linul, "Fiber-reinforced polymer composites: Manufacturing, properties, and applications," *Polymers (Basel).*, vol. 11, no. 10, p. 1667, Oct. 2019, doi: 10.3390/polym11101667.

[27] A.D. Valino, J.R.C. Dizon, A.H. Espera, Q. Chen, J. Messman, and R.C. Advincula, "Advances in 3-D printing of thermoplastic polymer composites and nanocomposites," *Prog. Polym. Sci.*, vol. 98, p. 101162, Nov. 2019, doi: 10.1016/j.progpolymsci.2019.101162.

[28] H.L. Tekinalp *et al.*, "Highly oriented carbon fiber–polymer composites via additive manufacturing," *Compos. Sci. Technol.*, vol. 105, pp. 144–150, Dec. 2014, doi: 10.1016/j.compscitech.2014.10.009.

[29] I. Blanco, "The use of composite materials in 3-D printing," *J. Compos. Sci.*, vol. 4, no. 2, p. 42, Apr. 2020, doi: 10.3390/jcs4020042.

[30] U. Kalsoom, A. Peristyy, P.N. Nesterenko, and B. Paull, "A 3-D printable diamond polymer composite: A novel material for fabrication of low cost thermally conducting devices," *RSC Adv.*, vol. 6, no. 44, pp. 38140–38147, 2016, doi: 10.1039/C6RA05261D.

[31] T.A. Campbell and O.S. Ivanova, "3-D printing of multifunctional nanocomposites," *Nano Today*, vol. 8, no. 2, pp. 119–120, Apr. 2013, doi: 10.1016/j.nantod.2012.12.002.

[32] H. Wu *et al.*, "Recent developments in polymers/polymer nanocomposites for additive manufacturing," *Prog. Mater. Sci.*, vol. 111, p. 100638, Jun. 2020, doi: 10.1016/j.pmatsci.2020.100638.

[33] Q. Chen, J.D. Mangadlao, J. Wallat, A. De Leon, J.K. Pokorski, and R.C. Advincula, "3-D Printing biocompatible polyurethane/poly(lactic acid)/graphene oxide nanocomposites: Anisotropic properties," *ACS Appl. Mater. Interfaces*, vol. 9, no. 4, pp. 4015–4023, Feb. 2017, doi: 10.1021/acsami. 6b11793.

[34] C. Mota, D. Puppi, F. Chiellini, and E. Chiellini, "Additive manufacturing techniques for the production of tissue engineering constructs," *J. Tissue Eng. Regen. Med.*, vol. 9, no. 3, pp. 174–190, Mar. 2015, doi: 10.1002/ term.1635.

[35] P. Ramesh and S. Vinodh, "State of art review on Life Cycle Assessment of polymers," *Int. J. Sustain. Eng.*, vol. 13, no. 6, pp. 411–422, Nov. 2020, doi: 10.1080/19397038.2020.1802623.

[36] F. Klocke *et al.*, "Turbomachinery component manufacture by application of electrochemical, electro-physical and photonic processes," *CIRP Ann.*, vol. 63, no. 2, pp. 703–726, 2014, doi: 10.1016/j.cirp.2014.05.004.

[37] MarketsandMarkets, "Ceramic Matrix Composites Market by Matrix Type (Oxide/Oxide, C/SiC, C/C, SiC/SiC), End-Use Industry (Aerospace & Defense, Automotive, Energy & Power, Industrial), Region (North America, Europe, APAC, Middle East & Africa,) – Global Forecast to 2029," 2019.

[38] N. Singh, R. Mazumder, P. Gupta, and D. Kumar, "Ceramic matrix composites: Processing techniques and recent advancements," *J. Mater. Environ. Sci.*, vol. 8, no. 5, pp. 1654–1660, 2017.

[39] R.R. Naslain, Buschow K.H. Jürgen, Flemings Merton C., Veyssière Patrick, Kramer Edward, Mahajan Subhash, Cahn Robert, Ilschner Bernhard, editors, "Ceramic matrix composites: Matrices and processing," in *Encyclopedia of Materials: Science and Technology*, Elsevier, Pergamon, 2001, pp. 1060–1066.

[40] D. Kopeliovich, Low I.M., editor, "Advances in the manufacture of ceramic matrix composites using infiltration techniques," in *Advances in Ceramic Matrix Composites*, Wood Head Publishing Elsevier, UK, 2014, pp. 79–108.

[41] I. Gibson, D.W. Rosen, and B. Stucker, *Additive Manufacturing Technologies*. Boston, MA: Springer US, 2010.

[42] T. Huang, M.S. Mason, G.E. Hilmas, and M.C. Leu, "Freeze-form extrusion fabrication of ceramic parts," *Virtual Phys. Prototyp.*, vol. 1, no. 2, pp. 93–100, Jun. 2006, doi: 10.1080/17452750600649609.

[43] S. Onagoruwa, S. Bose, and A. Bandyopadhyay, "Fused Deposition of Ceramics (FDC) and Composites," *Proc. Solid Free. Fabr. Symp.*, pp. 224–231, 2001.

[44] H. Windsheimer, N. Travitzky, A. Hofenauer, and P. Greil, "Laminated Object manufacturing of preceramic-paper-derived Si-SiC composites," *Adv. Mater.*, vol. 19, no. 24, pp. 4515–4519, Dec. 2007, doi: 10.1002/adma. 200700789.

[45] J.D. Cawley, "Computer-aided manufacturing of laminated engineering materials (CAM-LEM) and its application to the fabrication of ceramic components without tooling," Jun. 1997, Vol 78712, p. V004T13A019.

[46] J. Cesarano, B.H. King, H.B. Denham, J. Cesarano III, and H.B. Denham, "Recent developments in robocasting of ceramics and multimaterial deposition," *Proc. Solid Free. Fabr. Symp.*, pp. 697–703, 1998.

[47] A.M. Wätjen, P. Gingter, M. Kramer, and R. Telle, "Novel prospects and possibilities in additive manufacturing of ceramics by means of direct inkjet printing," *Adv. Mech. Eng.*, vol. 6, p. 141346, Jan. 2014, doi: 10.1155/2 014/141346.

[48] T. Chartier, C. Chaput, F. Doreau, and M. Loiseau, "Stereolithography of structural complex ceramic parts," *J. Mater. Sci.*, vol. 37, no. 15, pp. 3141–3147, 2002.

[49] Z.C. Eckel, C. Zhou, J.H. Martin, A.J. Jacobsen, W.B. Carter, and T.A. Schaedler, "Additive manufacturing of polymer-derived ceramics," *Science (80).*, vol. 351, no. 6268, pp. 58–62, Jan. 2016, doi: 10.1126/science.aad2688.

[50] B. Leukers *et al.*, "Biocompatibility of ceramic scaffolds for bone replacement made by 3-D printing," *Materwiss. Werksttech.*, vol. 36, no. 12, pp. 781–787, Dec. 2005, doi: 10.1002/mawe.200500968.

[51] J.P. Deckers, K. Shahzad, L. Cardon, M. Rombouts, J. Vleugels, and J.-P. Kruth, "Shaping ceramics through indirect selective laser sintering," *Rapid Prototyp. J.*, vol. 22, no. 3, pp. 544–558, Apr. 2016, doi: 10.1108/RPJ-10-2014-0143.

[52] V.K. Balla, S. Bose, and A. Bandyopadhyay, "Processing of bulk alumina ceramics using laser engineered net shaping," *Int. J. Appl. Ceram. Technol.*, vol. 5, no. 3, pp. 234–242, May 2008, doi: 10.1111/j.1744-7402.2008. 02202.x.

[53] E. Juste, F. Petit, V. Lardot, and F. Cambier, "Shaping of ceramic parts by selective laser melting of powder bed," *J. Mater. Res.*, vol. 29, no. 17, pp. 2086–2094, Sep. 2014, doi: 10.1557/jmr.2014.127.

[54] Y. Lakhdar, C. Tuck, J. Binner, A. Terry, and R. Goodridge, "Additive manufacturing of advanced ceramic materials," *Prog. Mater. Sci.*, vol. 116, p. 100736, 2021, doi: 10.1016/j.pmatsci.2020.100736.

[55] R. Lu, S. Chandrasekaran, W.L. Du Frane, R.L. Landingham, M.A. Worsley, and J.D. Kuntz, "Complex shaped boron carbides from negative additive manufacturing," *Mater. Des.*, vol. 148, pp. 8–16, Jun. 2018, doi: 10.1016/j.matdes.2018.03.026.

[56] R. Detsch, F. Uhl, U. Deisinger, and G. Ziegler, "3-D-Cultivation of bone marrow stromal cells on hydroxyapatite scaffolds fabricated by dispense-plotting and negative mould technique," *J. Mater. Sci. Mater. Med.*, vol. 19, no. 4, pp. 1491–1496, Apr. 2008, doi: 10.1007/s10856-007-3297-x.

[57] A. Ortona, C. D'Angelo, S. Gianella, and D. Gaia, "Cellular ceramics produced by rapid prototyping and replication," *Mater. Lett.*, vol. 80, pp. 95–98, Aug. 2012, doi: 10.1016/j.matlet.2012.04.050.

[58] S. Bose, J. Darsell, M. Kintner, H. Hosick, and A. Bandyopadhyay, "Pore size and pore volume effects on alumina and TCP ceramic scaffolds," *Mater. Sci. Eng. C*, vol. 23, no. 4, pp. 479–486, Jun. 2003, doi: 10.1016/S0928-4 931(02)00129-7.

[59] H. Yin, S. Kirihara, and Y. Miyamoto, "Fabrication of ceramic photonic crystals with diamond structure for microwave applications," *J. Am. Ceram. Soc.*, vol. 87, no. 4, pp. 598–601, Apr. 2004, doi: 10.1111/j.1551-2916. 2004.00598.x.

[60] A. Woesz *et al.*, "Towards bone replacement materials from calcium phosphates via rapid prototyping and ceramic gelcasting," *Mater. Sci. Eng. C*, vol. 25, no. 2, pp. 181–186, Apr. 2005, doi: 10.1016/j.msec.2005.01.014.

[61] D. Guo, L. Li, K. Cai, Z. Gui, and C. Nan, "Rapid prototyping of piezoelectric ceramics via selective laser sintering and gelcasting," *J. Am. Ceram. Soc.*, vol. 87, no. 1, pp. 17–22, Jan. 2004, doi: 10.1111/j.1151-2916.2004. tb19938.x.

[62] Ioan D. Marinescu, editor, *Handbook of Advanced Ceramics Machining*, 1st edition. CRC Press, Taylor & Francis Group, FL, USA, 2007.

Chapter 7

Fabrication of Shape Memory Polymers

S.V. Satya Prasad

Department of Production and Industrial Engineering, National Institute of Technology Jamshedpur, Jharkhand, India

P. Prasanna Kumari

Department of Mechanical Engineering, Vignan's Institute of Engineering for Women, Visakhapatnam, Andhra Pradesh, India

CONTENTS

7.1 INTRODUCTION

The materials responding to stimuli are considered smart since they are capable of sensing their surroundings and respond in a straight forward manner [1]. Research on such materials came into existence after deriving inspiration from organisms' bionic behavior. The SMPs since their discovery were considered to be polymeric stimuli-responsive materials. The discovery of SMPs was as early as 1940s as there was a mention of "shape memory" of dental materials by Vernon et al. in their U.S. patent. The material involved was methacrylic ester resin [2]. Subsequently, SMPs' evolution took place in 1960 when polyethylene (PE), heat shrinkable, was developed for tubing films application. The reports suggested the exhibition of memory effect γ-ray lightened polyethylene (PE) under low-high temperature cycles [3]. As time passed, in countries like the USA and Japan, much research was carried out on SMPs, particularly in the last

20 years [4]. It is the industrial-oriented research that has focused on the principles of design and important mechanisms of the SMPs.

The SMPs have the ability to remember an everlasting shape that can be worked upon such that a particular shape, which is momentary, can be fixed in certain conditions. Moreover, an external stimulus such as a light or heat could be used to transform that momentary shape into the permanent shape which was earlier remembered [5]. The permanent shape is retained at ambient temperature and deformation occurs at higher temperatures which, upon further cooling, reverts to its initial permanent shape. This phenomenon is termed "shape memory effect." For example, if SMPs responding to temperatures are considered, the basic programing involves an initial process where the shape needs to be fixed and deformed along with the removal of external stress. Prior to the programming, deformation of the polymers is necessary through various processes like pressing, extrusion etc. This needs to be done above a certain temperature, termed "switch temperature" (T_{sw}). The reason for this is elevated temperatures enhance the polymer's entropy, lower the energy barrier, and facilitate easier mobility of the molecular chain. Therefore, the SMP can be comfortably manipulated. Once the temporary deformation occurs within the polymer, it is secured with a temperature decrease (below T_{sw}) and upheld due to an external mechanical stress. There is a possibility of both chemical as well as physical modifications, which have the ability to restrict the motion of molecular chains within the polymer. This way the momentary shape is achieved and held for a long duration post-cooling. There will be no change in it as the external force is removed. As the entropy within the polymer increases, there will be storage of energy and stress. Thus, momentary shape is unstable in comparison to the permanent shape. When the SMPs are subjected to elevated temperatures (above T_{sw}) there will be liberation of the stored energy and internal stress which will allow the movement of molecules within the polymer chain. The SMPs in such case will revert to their permanent, original shape post-stimulation [6,7]. This cyclic process of programming to a temporary shapes and restoration to permanent shapes can be repeated several times without any issues. Hence, SMPs can have various momentary shapes but will have just a single, everlasting shape. Fig. 7.1a shows a typical shape memory cycle and Fig. 7.1b describes the principle behind shape memory effects within a polymer.

This shape memory effect is seen in diverse polymer types with significant chemical compositions such as semi-crystalline/amorphous polymers as well as liquid crystalline elastomers. SMPs comprise of molecular switches as well as netpoints. The interactions among molecules as well as covalent bonds aid in netpoints formation which are physical or chemical by nature. SMPs with chemical cross-linking are thermosets. SMPs with physical cross-linking and two separated domains are thermoplastics. Hence the SMPs network chains are crystalline or amorphous. So, the transition

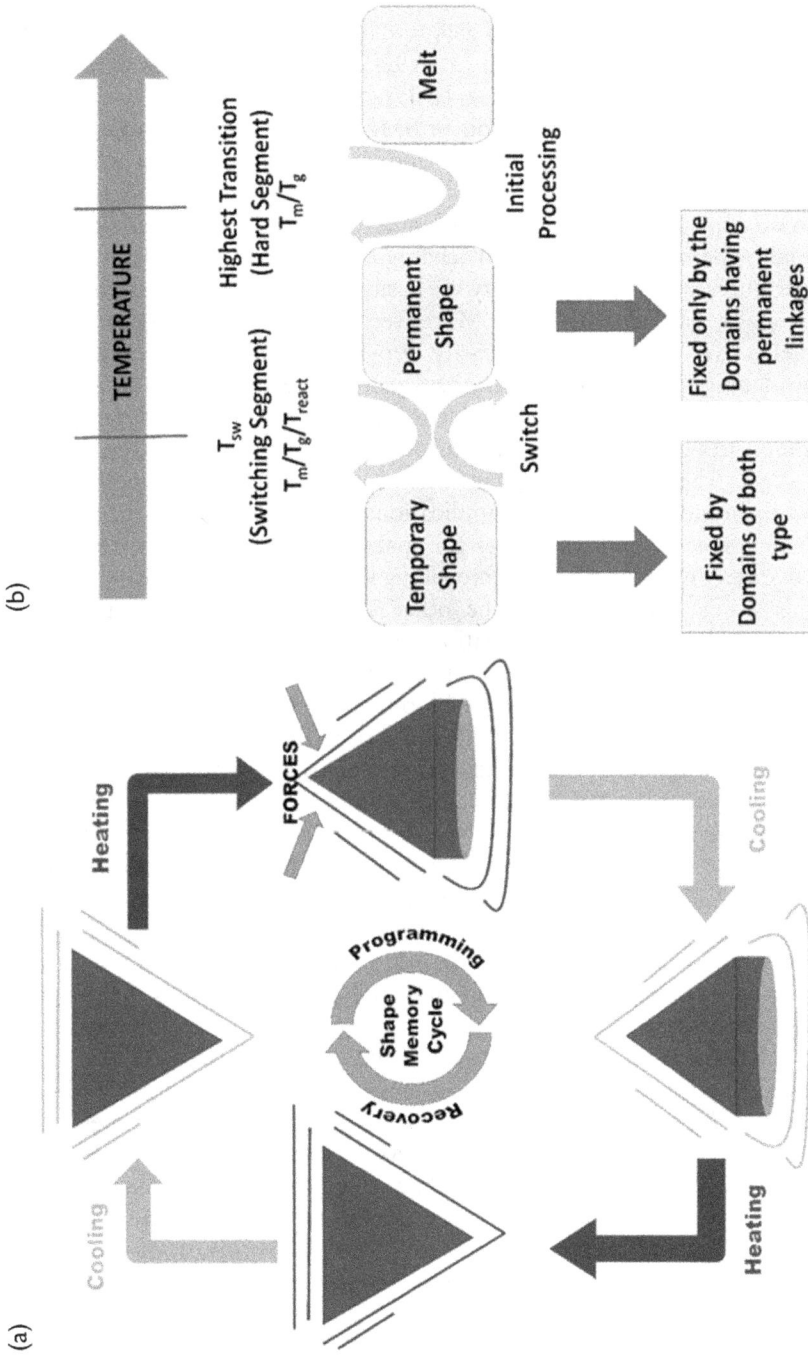

Figure 7.1 (a) Shape memory cycle; (b) schematics of the shape-memory effect principle in polymers.

temperature (T_t) is glass (T_g) or melting temperature (T_m) [8]. At glass temperature, within network chains, the micro-Brownian motion freezes at a temperature lower than T_g and gets actuated when reheated at $T_t \geq T_g$. At the melting point, crystallization of switching segments occurs at $T_t < T_m$ and shape recovery occurs at $T_t \geq T_m$ [9].

Apart from heat, the activation of SMPs can be done through various other stimuli like light, electricity, and moisture, as well as chemicals (pH change) [10]. The SMPs possess multiple advantages in terms of technical aspects such as low strength and low density, and their shape can be highly deformable and their characteristics like stiffness, transition temperature, and biodegradability are easily tailored. The SMPs can also be functionally graded with ease. Moreover, the SMPs are biocompatible, cost effective, biodegradable, easily programmable, and their recovery controllability is exceptional as they generate recoverable strains about greater than 200% [5], [11]. It is due to these advantages that development of composites of SMPs as well as various SMPs which are multi-functional was done. The development has further integrated nanotechnology and SMPs to broaden its range of applications. The applications of SMPs include aerospace where structures for space like trusses, hinges, reflectors, mirrors, etc. are prepared. Also, morphing skins for variable/folding aircraft camber wings are prepared. The other areas of SMPs applications are biomedical devices and bioinspirational instruments such as vascular stents with smart drug delivery systems, surgical sutures that are smart, micro-actuators activated with lasers for blood clot removal in vessels, artificial muscles, etc. [12]. Some other applications of SMPs are bionics engineering, civil engineering, electronics engineering, energy, wearable devices, automobile actuators, textiles, self-healing systems, and various other smart household products [8], [13].

In spite of so many advantages, the applications of SMPs have been restricted by the present, conventional, fabrication techniques. The conventional manufacturing methods restrict the active structures of SMPs and do not allow complicated shapes. The traditional manufacturing methods are not at all suitable for smart polymer materials as they cannot exploit the complete potential of such complex, advanced, tailored materials. This scenario is presently changing with the advent of the latest and highly advanced manufacturing methods like 3-D printing and 4-D printing, which are being integrated to SMPs. In recent times, 3-D printing or additive manufacturing (AM) is known to us as one of the most advanced techniques of manufacturing which can produce largely customizable components of intricate designs with exceptional quality. It is a method where fabrication is carried out by deposition of layers one over the other until the entire product is printed. Initially, a solid 3-D CAD model is generated using a modelling software, tomographic imaging data, or through math equations that is read by the 3-D printing equipment in the form of a CAD file. As per the given input data, sequential layers of polymer material in the form of powder, liquid resin, sheets or filament are deposited to fabricate the final 3-D object. The highly accurate, autonomous nature of 3-D printing, where a huge range of materials can be manufactured with utmost, infinite, complex shapes very easily, signifies the superiority of this method in comparison to the conventional, constructive/subtractive fabrication methods [14]. Based on the raw material's physical state and the material fusing techniques (thermal radiation/UV, electron/laser beams), 3-D printing comprises 18 kinds of processes categorized as powder-, solid-, and liquid-based techniques. Among these, Fused Deposition Modeling (FDM), Stereolithography (SLA), Digital Light Processing (DLP), Selective Laser Melting (SLM), Selective Laser Sintering (SLS), and Direct Ink Writing (DIW) are the most commonly used 3-D/4-D printing techniques. It is the advancements in this technique that have made direct printing of SMPs feasible [15,16]. The existence of SMPs is in the foam, film or bulk material forms. The 3-D printing of SMPs objects in a flawless manner is influenced by parameters like rate of printing, filling of the resin, temperatures of platform and nozzle, resin's T_g, compactness as well as thickness of printing, print resolution, improper adhesion, curing beam's potency, product deformation due to laser beam's extreme thermal energy, orientation due to dissimilarity in dimensions, large fabrication time, and the inability to interact with the ruling environment [17,18]. These parameters complicate and impede the process of 3-D printing. The SMPs are a unique set of materials used in engineering which have created an identity for themselves because of their shape memory effect so, to overcome the complications in AM, it is essential to look beyond the 3-D printing. As in integrating these unique materials with an advanced version of 3-D printing termed 4-D printing, having an extra dimension in the form of time to enhance the 3-D printed structure's functionality to meet the various demands of engineering applications [19]. In the 4-D printing technique, the 3-D printed objects undergo active transformation of their configurations with respect to time when subjected to an external stimulus. As in, an SMP component which is printed flat will be given a 3-D shape, for example, by the process of heat treatment [20]. With increasing research on the 4-D printing technology, it has been defined in a more complete manner. The 4-D printing has been described as a directed evolution of a 3-D structure printed in the aspects of its functionality, characteristics, and geometry. In 2013, there was a revolution in the field of research when Tibbits presented the world with 4-D printing technology, which is efficient in terms of cost and can print resolutions of about 50 μm. 4-D printing is different from 3-D printing as it is predictable, dependable on time, and autonomous to the printer. The basic difference in the techniques between 3-D and 4-D printing is represented in Fig. 7.2 [20].

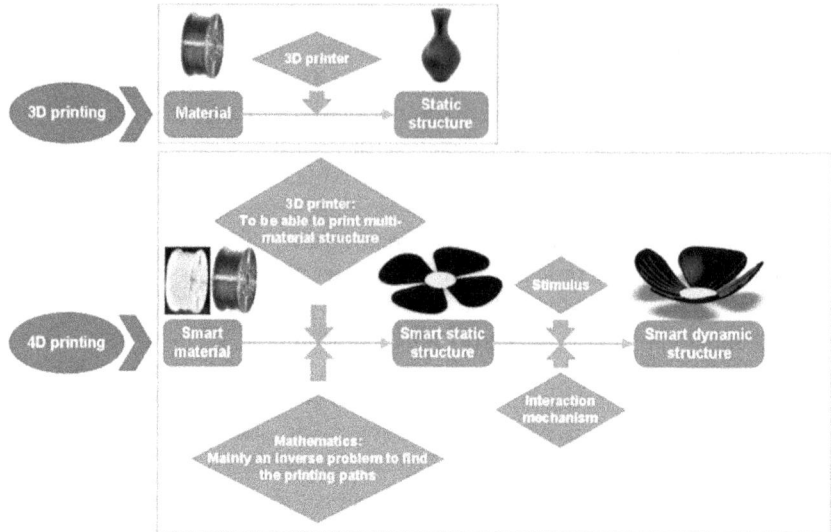

Figure 7.2 Difference between techniques of 3-D and 4-D printing [20].

The technique of 4-D printing is ideal for fabricating complicated components with SMPs, which have the ability to exhibit multifunctionality, which have the capability of self-healing/self-repairing and adapt well to diverse environments. The printed components can be used in applications pertaining to flexible electronics as well as biomedical sensing [21]. Moreover, properties like quicker rate of response, elevated stiffness, easy fabrication, and an influential response strain (approx. 800%) SMPs are extremely preferable for the technique of 4-D printing in comparison to other smart materials [22]. These, along with cheaper material as well as manufacturing costs, make SMPs extremely advantageous for 4-D printing. The major advantage of 4-D printing is that it endorses printing of simple geometries through structured, serial folding, which can later be converted into intricate geometries through programming that is synonymous to the 3-D printing technique [23]. So the applications of 4-D printing can be expanded to bioprinting, soft robotics, and microscale, thereby making it an important technique for advanced engineering applications.

7.2 DIFFERENT 3-D/4-D PROCESSING TECHNIQUES

There is a growing popularity of 3-D printing techniques using materials such as metals, polymers, as well ceramics due to an added demands for finer resolutions and complicated structures with intricate geometries across various engineering fields. The applications include fabrication of devices, smart robotics, and tissue engineering where 3-D printing plays a key role influencing the design, modelling and expenses on the products thereby influencing the flow of business related to them [22,24]. With development in research, there has been an innovation related to 3-D printing at the nanoscale, termed "FP-TPL (femtosecond projection two photon lithography)." This is a novel time-related method that controls the light emitted from a rapid laser and produces miniscule structures at a much quicker rate in comparison to the typical TPL (two photon lithography) with no compromises in the print resolution. The process is capable of printing 90° structures and much greater resolutions of 175 nm, higher than the conventional ones, making it ideal for advanced materials for various applications such as micro-optics, biological scaffolds, flexible electronics, electrochemical interfaces, and various functional structures at micro or nano levels [25]. In this 3-D/4-D printing method, a flawless output product of the CAD-modelled designs is produced. The design, a 2-D model of the final object, is converted into a STL file and this is scanned by the printer. When material is fed, the printing equipment scans the data and the final product is printed as per the input designs without any deviations or errors. An overview of various 3-D/4-D printing processes is represented in Table 7.1 [26].

7.3 3-D PRINTING OF SMPs

7.3.1 SMPs Printed through FDM

According to the report given by Wholer, the FDM holds a 41.5% market share in the USA and is highly employed technique in 3-D printing of thermoplastic filaments, since its inception in 1990s.

Bottom-up extrusion is done in the FDM in which the material is fed via roller is melted within a liquefier (temperature > T_m) and extruded through a nozzle in the form of strands, over a platform that is pre-heated. This happens with the help of a static filament which can be lowered once material deposition is complete. After which, the next layer will get deposited as desired via mobile extruder sustained at low temperatures, whose head is capable of moving along the horizontal plane. The platform is capable of moving along the perpendicular axis of the horizontal plane. The single filament acts like a piston to push out the material in molten form to be deposited as extremely thin successive layers in a set location. FDM is advantageous compared to SLA since layer solidification doesn't occur due to the energy from a photon (UV radiation) [27]. For the polymer materials like polylactic acid (PLA) and polyethylene (PE), PTFE (polytetrafluoroethylene) as well as Cu are utilized in calibrating nozzle and for sealing heat. But in case of polymers with high temperatures like Acrylonitrile Butadiene Styrene (ABS), Al is employed in a calibrating nozzle. Usually the FDM technique is ideal for thermoplastics and its

Table 7.1 An overview of 3-D/4-D printing processes [26]

Process	Modelling	Solidification Route	Materials	Methods	Overall Accuracy	Post Processing	Applications
FDM	Continuous Extrusion Deposition	Cooling	Thermoplastics, composites	FDM	0.5% Dimensional accuracy (Reliable)	Best surface Finish. Post-processing for support structure removal.	Stents, scaffolds, aerospace, automotive.
DLP	Liquid Layer Deposition	UV radiation	Metamaterials, elastomers	Photopolymerization	0.3% Dimensional accuracy (highly accurate)	Smooth surface with high resolution. Post-processing is needed.	Aerospace, automotive, medicines, soft robotics, and micro fluids.
SLA	Liquid Layer Deposition	Photopolymerization	ABS, ceramics, composites, semi-flexible materials	Laser Illumination	0.5% Dimensional accuracy (The best)	Exceptional surface Finish. Post-processing for support structure removal.	Same as DLP.
SLS	Powder Deposition	Laser Sintering	Filled polymers, PA, metals, and ceramics	Laser based	0.3% Dimensional accuracy (not very accurate)	Smooth surface. Only post-processing treatments.	Cartilage repairs, scaffolds, bone tissue repairs.
INKJET	Drop on demand Deposition	Cooling	Polymers	Milling layers	0.1% Dimensional accuracy	Better surface. Only post-processing treatments.	Injection moulded prototypes, medical models.

composites like PLA, ABS, polyurethane (PU), PC (polycarbonate), and polymers filled with glass. Among these, PLA is mainly 3-D printed using the FDM process due to its lower temperatures of printing, easy usage, lower warping w.r.t other polymers, and its lower levels of toxicity in comparison to ABS.

7.3.1.1 Pure PLA

PLA exhibits SME and its characteristics are because of crystallization and physical cross-links. But without having any breakages, it's not possible to extend the PLA in pure form over 10%, approximately [28]. Hence, 3-D printing becomes indispensable to print intricate shapes of PLA-based SMPs and overcome the above limitations. In this context, to get rid of the limitations, origami structures were employed by Langford et al. to facilitate folding/unfolding of the structure from small to large and vice versa [29]. This characteristic is beneficial for biomedical applications like induction of scaffolds within the human body. When compared, it was observed that a waterbomb had an unreliable deformation whereas herringbone tessellation, in the form of a tube could be intensely compressed with major restoration afterwards (Fig. 7.3(a) and (b)) [29].

The recovery of shape was constant in PLA filament (approx. 61%), irrespective of the deformation whereas herringbone tessellated tube, with some cracks, exhibited an increase in the rate of recovery (approx. 96%) post-compression (Fig. 7.3(c)). There was no change in the rate of recovery even after the addition of an infill structure that is porous and bone-mimicking [29].

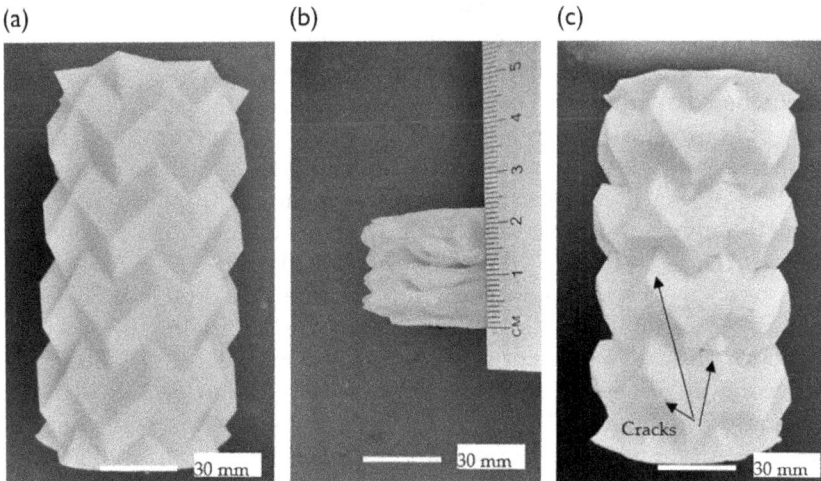

Figure 7.3 Herringbone tessellated tube: **(a)** post-3-D printing; **(b)** post-compression; and **(c)** post-recovery [29].

Investigations were carried out on tubular structured, vascular stents, self-expanding in nature, and made of PLA using FDM. It had hexagon nested structure, 4 mm diameter, 40 mm long with the thickness of its wall being 0.4 mm as its dimensions [30]. Compression as well as recovery inside a water bath were done at 70°C but subsequently quenched. The momentary shape post-quenching was held for a week at ambient temperature whereas initial shape recovery was seen in a duration of 5 seconds at the temperature of 70°C. The rate of shapes recovered were around 87% and 96% after the first and second cycles, respectively, because of superior orientation of molecules after the initial test. Apart from biomedical-related applications, 3-D printed, pure PLA-based SMPs can also be utilized for smart textiles as well as antennas that can be deployed. To enhance the SME in PLA, fibers or nanoparticles can be added or PLA could be blended with other polymers.

7.3.1.2 PLA Composites

PLA is combined with various materials of diverse geometries to enhance its SME. In this context, Zhao et al. fabricated PLA/Fe_3O_4 SMP, which is capable of recovering its shape in the presence of an externally place magnetic field [31]. Such stimuli can substitute heat and ideal for applications like *in vivo* where us of higher temperatures is not possible. The porous structures of this study were imitations of bone trabecular (gyroid structure) [32] and lotus rhizomes having around 50%–60% of porosities; 80% PLA and 20% Fe_3O_4 were combined together to make solid filaments for the printer of FDM. This later was shredded after which was extruded within twin-screw extruder [31]. Recovery in the shape was over 95% and obtained in a duration of 14 o 24 seconds when an alternating magnetic field was applied. Apart from this the structures exhibited proliferation and adhesion of cells along with excellent biocompatibility. The differences were only because of distribution of various sizes of pores. Also, when hydroxyapatite (HA) is combined with PLA, its biocompatibility enhances and these composites of PLA/HA can be great substitutes for implants in applications where the defects in the bone are smaller [33]. This has, therefore, encouraged the researchers to focus of the composites of PLA/HA. As a part of one such research, various quantities of powdered HA (4–8 wt.%) was combined to granules of PLA by Singh et al. The entire mixture was ball milled to fabricate a filament using twin-screw extruder [34]. The rate of recovery was observed to be in the range of 72% to 96% within the temperature range of 60°C to 70°C when a comparison was done between various infill percentages, outer perimeter numbers as well as post-heating steps for enhancing mechanical strength.

7.3.1.3 PLA Blends

Blending of PLA and various other materials to enhance the SME or to add certain advanced characteristics is possible. PLA was blended with PCL (poly (ε -caprolactone)) by Liu et al. to observe that PCL's phase is reversible in nature, whereas PLA has a fixed phase [35]. Based on the quantity of PCL (10% to 60%), ratios of shape recovery were obtained in the range of 59% to 84%. But the fixity of shape was over 95%. The characteristic of shape memory was greatly influenced by raster angle, thickness of layers and various parameters of printing but the SME was almost constant w.r.t. the infill density. In another study, the filament of PLA was melt-blended with 10 wt.%/30 wt.% of PEG (poly (ethylene glycol)), which facilitated the tuning of actuation temperatures along diverse parts within a sample. Pure PLA exhibited recovery above 60°C whereas blended PEG exhibited recovery at temperatures less than or equal to 55°C [36]. Within a thermomechanic actuator, such objects can be utilized in which temperature of recovery is influenced by the position. Apart from the above PLA illustrations, there are plenty other SMPs fabricated through FDM method.

7.3.1.4 Other SMPs

Barring PLA, PU is commonly used for FDM printing. PU SMP was commercially printed by Cersoli et al. through a pellet extruder that was placed on FDM printer that enabled printing directly using pellets [37]. With the usage of hot water/heat gun, more than 96% recovery of shape was observed. This SMP was endorsed for electronics and authentication related applications by creating a QR code which could not be read when deformed but was readable post subjection to a thermal stimulus. Apart from this, hybrid actuator replicating a thermal switch was fabricated by the combination of this PU-based SMP with a SMA. Unlike PU SMPs, the literature on nylon is extremely limited. A study on polypropylene (PP)/nylon 6 (PA6) blend comprising of the quantities of PA6 in the range 0 to 30 wt.%, excellent shape recovery was observed for 175°C temperature (temporary shape setting) with a subsequent reheating to 175°C after a 3 minute duration [38].

The PVA, poly (vinyl alcohol), utilized as a supporting material in 3-D printing and soluble in water, also exhibits shape memory characteristics. Its ability to slowly dissolve in water makes it suitable for delivery of drugs within the body. One such study was carried out by Melocchi et al. who blended glycerol and PVA and plasticized it prior to addition of drug-containing materials. Its actual, spiral geometry, was compressed to be easily consumed orally and was observed to be retain its actual shape in 0.1 M HCL solution after a duration of 3 hours at a normal body temperature of 37°C. The process of releasing a drug from 3-D printed samples was decelerated through the coating process [39].

Various materials like ABS, PEEK, or PETG (polyethylene terephthalate glycol) can be printed through FDM but they don't exhibit SME. Apart from the above SMPs, some others can be fabricated through other methods of 3-D printing like SLA.

7.3.2 SMPs Printed through SLA

Stereolithography, one of the older technologies, is a 3-D printing photo-chemical process that fabricates products through in an ultra-thin, successive layer, where a stabilized 3-D structure is obtained by chemical monomers or oligomers cross-links due to the impingement of ultraviolet (UV) radiation. SLA discovered by Hideo Kodama in 1981, quickly reconstructs smaller sized 3-D models polymers of satisfactory resolutions with good surface quality by the help of a holographic technique [26]. SLA/vat polymerization process facilitates solidification of liquid precursor's layers with the vat are made susceptible to UV radiations. The molecules of photo-initiator (PI) within the resin respond to the incoming light that activates the reaction of chemical polymerization when irradiated and cures the exposed areas in an increasing manner for every layer. The easily available acrylate resins which have great building speeds along with excellent reactivity are ideal for SLA. Their resistance to temperatures and mechanical characteristics can be enhanced through tuning and altering the quantity of reactive groups [40]. Also, epoxies undergoing the reaction of cationic polymerization are stable in the presence of oxygen, but their reaction times are longer in comparison to acrylates and moisture is a constraint. The applications include aerospace, medicines, microfluids, and automotive sectors.

Radchenko et al. proposed and fabricated cycloaliphatic epoxy-functionalized ionic liquids using imidazolium units (one/two) for curing through SLA printing [41]. The diaryliodonium hexafluoroantimonate (thermoacid generator) was utilized for enhancing the rate of crosslinking resulting in a rubbery nature of poly-ionic liquid. In this study, all fabricated networks exhibited shape memory characteristics with a decent temporary retention of shape and an exceptional recovery.

Kumagai et al. initially developed a highly transparent shape memory gels of exceptional strength which can be printed through SLA whose refractive index was influenced by the concentration of monomer but autonomous to photoinitiator as well as crosslinker [42]. It was the rate of progression of gelation that was influenced by both photoinitiator and crosslinker. The shape memory gels of diverse refractive indices were fabricated with a method which had their application in the gel replicas of intraocular lenses.

Liquid resin comprising of monomer units of acrylic was utilized in a study for SLA printing. Hollow tubes with an octet lattice arrangement were fabricated, evacuated, and later filled the tubes with gallium [43]. The material exhibited recovery of its shape at 90°C because of its core and not due to the polymer shell with a potential damage.

Miao et al. conducted a study by combining SLA and photolithography for producing hierarchical micropatterns through acrylate inks epoxidized by smart soybean oil. The applications of it pertaining to the growth of mesenchyme, stem cell in bone marrow as well as its alignment [44]. Triggering of the thin scaffolds to self-assemble as rolling structures was possible by immersing them into ethanol attributing this to a gradient of cross-link density because of the photolithographic process.

Coming to fabrication of SMPs by other 3-D printing processes, very little work has been cited in regards to DLP, whereas zero work has been cited on SMPs fabricated by SLS/SLM and inkjet printing [45]. It is clearly evident that 3-D printing of SMPs is extremely new to be for it to be carried out across various 3-D printing methodologies. It can be confidently said that this leaves a wide scope for exploration in this particular aspects of 3-D printing.

7.4 4-D PRINTING OF SMPS

The immobile quality of functionally developed 3-D components through the methods of SLM and SLS leading to formation of residual stresses excluding movable segments like encapsulated bearings, live hinges, and ball-and-socket joints brought about certain complications like exceeding dimensional values of the formed cracks and thereby hampering parameters like fracture toughness as well as fatigue life. This ruled them out for structural applications. This highlighted a new aspect of 3-D printing by adding an additional dimension in the form of time which evaded the complications of 3-D printing. This 4-D printing endorsed evolution of 3-D printed components w.r.t time in aspects like properties, geometries as well as functionality [46]. Increasing demand for novel engineering application of 3-D printing as well as research advancements resulted in the development of 4-D printing in which better features like exceptional quality and higher performance along with better efficiency are seen in comparison to the typical fabrication methods. One of the major aspects of 4-D printing is the ability of materials to exhibit SME. The final components in 4-D printing are capable of self-transmuting their behavior and characteristics according to the variations of the subjected external stimuli of temperature, light, pH, pressure etc. 4-D printing is a sustainable manufacturing technique as there is extremely less amounts of material used at the time of printing and more of usage post the object disposition. This enhances the projection of market in future. Optimized capabilities are witnessed in 4-D printing because of novel material such as grains of wood, custom composites of textiles and fibers of carbon which are capable of exhibiting sensing, programmable actuation and self-transmutation. 4-D printing applications may be categorized as self-assembly/self-repair and multi-functionality and therefore adopted in medical, military, aerospace, and automotive sectors. The present research is concentrated on utilizing the shape shifting capabilities of 4-D printed objects and their behavioral programming which can attain controlled geometrical changing through anisotropy of the final printed material [47]. The SME is dependent on a material's crystallinity, structure as well as surface properties.

4-D printed SMPs can be utilized for structural changes based on light induction. It is possible to reuse thermal stimulus actuated SMPs through thermomechanical programming to fabricate multicolored and complicated shapes having predefined responses [48]. It is possible to redesign a microscale crystalline or amorphous based material properties into macro-scaled 3-D-printed layers and achieve optimum 4-D performance. An example for this would be the functionally graded layers despite the inability to characterize diverse functional layers. In this context, studies related to fabrication of multi layered SMP thermosets with dissimilar T_g of each layer and fabrication of carbon nanofiber's as well as boron nitride's functionally graded layers over an SMP for achieving electroactive ability can be noted [49]. As compared to SMPs, fibers or wood as well as botanical cellulose are ideal for 4-D printing. Hence, it is essential to know the mechanical characteristics of ideal 4-D printing suitable materials to achieve smart efficiency. As per SMPs behavioral model or the theory on classical lamina, thermomechanical characteristics have been experimentally defined. A 2-D model can be utilized to project the actuation mechanism of the 4-D printed object but a 3-D archetype is needed for optimum design as well as self-collision. The SMPs distortion model for its behavior was utilized for studying active motion produced due to SME [26]. The idea of initially designing and gauging of process parameters through simulation prior to printing strongly influences the imagination of the final stimulus response as well as the behavioral and characteristics of life cycle, microstructure and mechanics of the object to be printed (Hu et al.). FDM-printed structures of ribbon and cone as well as double curvature underwent 2-D/3-D transformation from their initial configurations of 1-D ribbons as well as 2-D (double curves, cones).

The 2-D and 3-D curvatures were majorly impacted by the pre-strain formed during printing post-impetus action, because of the nozzle's speed variation. Therefore, the differential structural and proportional speed dependency on curvatures were attributed to the strain variations along the structure's dimensions [50].

Distinct petal patterns were printed from hydrogel ink (acrylic matrix comprising of grafted fibers of cellulose) in a study to determine diverse curvatures of a biomimetic flower in the presence of a hydro stimulus. Apart from printing speed, structure orientation's impact over the printed structure's curvature was reported. The orientation's anisotropy as well as contiguous layers (0°/90°, –45°/45° or pattern) are related to the aspects of rolling, spiral, differential helix, bending, or twisting of the structures [51]. In another study, FDM 4-D printing was done to depict the SME within PLA preforms of circular braided tubes along with a corresponding composites of silicone elastomer matrix. Dynamic mechanical analysis was employed for quantifying the shape recovery forces in both the materials. The influence of braided microstructure and shape recovery temperature over preform's shape memory behavior was determined. There was enhancement in shape recovery force and ratio along with radial compressive failure load within the composite upon addition of silicone elastomer in the form of a matrix [52].

The SMPs can easily control parameters like the alteration strain forces, weight factor, available feedstock, surface framework and transformational boundaries of the temperatures which makes them the prime research target in terms of SME under the influence of a stimulus. So considering the responsiveness of a SMP in the presence of body temperature, a polymer-nanocomposite of 4-D printed PLA and carbon nanotube (CNT) can be fabricated due to its constant distribution of temperature with an impeccable recoverability [53]. In similar terms, polymer nanocomposite of PLA/Fe_2O_3 was shown to have extraordinary applications and be an ideal replacement of metals in remote controlled occlusion devices. This is due to PLA/Fe_2O_3 exhibiting characteristics of histocompatibility and cytocompatibility w.r.t. adhesion in cells and also because of its responsiveness to magnetic stimulus [54].

Compounds of cellulose are generally fabricated through 4-D printing. In this context, study was conducted to 4-D print and analyze the capability of a composite's computational ultra-sonication microstructure (the ability to change w.r.t time as per pre-modelled design in response to moisture stimulus). Analysis was done on extracted cross-sectional views, waviness profiles, as well as roughness of the specimen's surface by comparing with the Gaussian filter to understand particle decomposition with a higher precision. Along with the process of mixed sedimentation, characterization of thermoplastic nanoparticles was done to produce the structure of a polymer composite. The pulp tissues were segregated by hydrogel which optimally disperses the fibers, thereby giving satisfactory roughness. Moreover, the particles' leakage potential in the component, after the processes of hydration as well as dehydration influences the shape memory [55].

7.5 CONCLUSIONS

Shape memory polymers (SMPs) are a distinctive set of stimuli-responsive polymers that are able to recall their original geometry from an existing provisional geometry and vice versa when subjected to external stimuli like heat, light, electricity, moisture, magnetism, etc. This exciting phenomenon is called the shape memory effect. The discovery of this has become an advancement in the field of materials. With growing technology and increasing demands in various fields, it is essential for the manufacturing sector to come up with new techniques in fabricating these advanced materials.

So with the discovery and existence of 3-D printing since its inception in the 1980s, this process has become extremely popular for fabrication of high-quality materials at much lower costs, especially while fabricating through advanced materials. In recent years, 3-D printing of SMPs has garnered much interest. There are different methods of 3-D printing like FDM, SLA, SLS/SLM, DLP and Inkjet printing. Among these, the fabrication of a majority of SMPs is through processes like fused deposition modelling (FDM) and stereolithography (SLA). Among all SMPs, PLA is the best suited for FDM process in which different studies endorse the use of PLA in pure form, PLA composites or blends of PLA with other polymers to fabricate using the FDM, 3-D printing process. Apart from PLA, polyurethane is suited for the FDM process. Not only through FDM but there are many SMPs that can be fabricated through other 3-D printing methods like SLA. Very little research has been done on SMP printing through DLP but absolutely no work has been done on 3-D printing of SMPs via SLS/SLM and inkjet printing. This clarifies that 3-D printing of SMPs is extremely new for exploring various 3-D printing methodologies. This strongly provides a scope for exploration 3-D printing of SMPs in these aspects.

The advancements in 3-D printing resulted in the advent of 4-D printing technology for fabrication of smart materials. The main idea behind 4-D printing is connected with SMPs that exhibit SME when subjected to an external stimuli. So it can be said that 4-D printing is 3-D printing with an addition dimension which is the time factor. It speaks of smart materials regaining their permanent shape w.r.t. time once they are 3-D printed. This has broadened the 4-D printing application into areas of robotics, electronics, smart textiles, and biomedical applications along with automotive and aerospace sectors. Therefore, polymers capable of self-healing have interested researchers because the defects can be avoided through programming and enhance the life of 3-D-printed components. This is beneficial, especially in the area of space research.

REFERENCES

[1] Zhang X, Chen L, Lim KH, Gonuguntla S, Lim KW, Pranantyo D, Yong WP, Yam WJ, Low Z, Teo WJ, Nien HP. The pathway to intelligence: Using stimuli-responsive materials as building blocks for constructing smart and functional systems. *Advanced Materials*. 2019 Mar; 31(11): 1804540.

[2] Lester B, Vernon B, Vernon HM. Process of manufacturing articles of thermoplastic synthetic resins. US 2234993, 1941.

[3] Charlesby A. *Atomic radiation and polymers: International series of monographs on radiation effects in materials*. Elsevier; 2016 Jun 6.

[4] Hager MD, Bode S, Weber C, Schubert US. Shape memory polymers: Past, present and future developments. *Progress in Polymer Science*. 2015 Oct 1; 49: 3–33.

[5] Liu C, Qin H, Mather PT. Review of progress in shape-memory polymers. *Journal of Materials Chemistry*. 2007; 17(16): 1543–1558.

[6] Xia Y, He Y, Zhang F, Liu Y, Leng J. A review of shape memory polymers and composites: Mechanisms, materials, and applications. *Advanced Materials*. 2020 Sep 23; 33(6): 2000713.

[7] Lendlein A, Gould OE. Reprogrammable recovery and actuation behavior of shape-memory polymers. *Nature Reviews Materials*. 2019 Feb; 4(2): 116–133.

[8] Miaudet P, Derre A, Maugey M, Zakri C, Piccione PM, Inoubli R, Poulin P. Shape and temperature memory of nanocomposites with broadened glass transition. *Science*. 2007 Nov 23; 318(5854): 1294–1296.

[9] Hu J. *Shape memory polymers and textiles*. Elsevier; 2007 Apr 30.

[10] Leng J, Lan X, Liu Y, Du S. Shape-memory polymers and their composites: Stimulus methods and applications. *Progress in Materials Science*. 2011 Sep 1; 56(7): 1077–1135.

[11] Sabahi N, Chen W, Wang CH, Kruzic JJ, Li X. A review on additive manufacturing of shape-memory materials for biomedical applications. *JOM*. 2020 Mar; 72(3): 1229–1253.

[12] Meng H, Li G. A review of stimuli-responsive shape memory polymer composites. *Polymer*. 2013 Apr 19; 54(9): 2199–2221.

[13] Ge Q, Sakhaei AH, Lee H, Dunn CK, Fang NX, Dunn ML. Multimaterial 4-D printing with tailorable shape memory polymers. *Scientific Reports*. 2016 Aug 8; 6(1): 1–11.

[14] Izdebska-Podsiadły J, Thomas S, editors. *Printing on polymers: fundamentals and applications*. William Andrew; 2015 Sep 24.

[15] Hopkinson N, Hague RJ, Dickens PM. *Rapid manufacturing: An industrial revolution for the digital age*. Chichister, England: John Wiley and Sons, Ltd; 2006.

[16] Rastogi P, Kandasubramanian B. Breakthrough in the printing tactics for stimuli-responsive materials: 4-D printing. *Chemical Engineering Journal*. 2019 Jun 15; 366: 264–304.

[17] Lee JY, An J, Chua CK. Fundamentals and applications of 3-D printing for novel materials. *Applied Materials Today*. 2017 Jun 1; 7: 120–133.

[18] Nazan MA, Ramli FR, Alkahari MR, Abdullah MA, Sudin MN. An exploration of polymer adhesion on 3-D printer bed. InIOP Conference Series: Materials Science and Engineering 2017 Jun 1 (Vol. 210, No. 1, p. 012062). IOP Publishing.

[19] Li Z, Loh XJ. Four-dimensional (4-D) printing: Applying soft adaptive materials to additive manufacturing. *Journal of Molecular and Engineering Materials*. 2017 Jun 15; 5(02): 1740003.

[20] Momeni F, Liu X, Ni J. A review of 4-D printing. *Materials & Design*. 2017 May 15; 122: 42–79.

[21] Liu T, Liu L, Zeng C, Liu Y, Leng J. 4-D printed anisotropic structures with tailored mechanical behaviors and shape memory effects. *Composites Science and Technology*. 2020 Jan 20; 186: 107935.

[22] Pearce JM. Building research equipment with free, open-source hardware. *Science*. 2012 Sep 14; 337(6100): 1303–1304.

[23] Yu K, Ritchie A, Mao Y, Dunn ML, Qi HJ. Controlled sequential shape changing components by 3-D printing of shape memory polymer multi-materials. *Procedia Iutam*. 2015 Jan 1; 12: 193–203.

[24] Zarek M, Layani M, Cooperstein I, Sachyani E, Cohn D, Magdassi S. 3-D printing: 3-D printing of shape memory polymers for flexible electronic devices (Adv. Mater. 22/2016). *Advanced Materials*. 2016 Jun; 28(22): 4166.

[25] Lee M, Kim HY. Toward nanoscale three-dimensional printing: Nanowalls built of electrospun nanofibers. *Langmuir*. 2014 Feb 11; 30(5): 1210–1214.

[26] Subash A, Kandasubramanian B. 4-D printing of shape memory polymers. *European Polymer Journal*. 2020 May 18; 133: 109771.

[27] Jasveer S, Jianbin X. Comparison of different types of 3-D printing technologies. *International Journal of Scientific and Research Publications (IJSRP)*. 2018 Apr; 8(4): 1–9.

[28] Yahia L. Introduction to shape-memory polymers for biomedical applications. In *Shape memory polymers for biomedical applications*. Lendlein, Andreas (pp. 3–8). Woodhead Publishing; 2015 Jan 1.

[29] Langford T, Mohammed A, Essa K, Elshaer A, Hassanin H. 4-D printing of origami structures for minimally invasive surgeries using functional scaffold. *Applied Sciences*. 2021 Jan; 11(1): 332.

[30] Jia H, Gu SY, Chang K. 3-D printed self-expandable vascular stents from biodegradable shape memory polymer. *Advances in Polymer Technology*. 2018 Dec; 37(8): 3222–3228.

[31] Zhao W, Huang Z, Liu L, Wang W, Leng J, Liu Y. Porous bone tissue scaffold concept based on shape memory PLA/Fe$_3$O$_4$. *Composites Science and Technology*. 2021 Feb 8; 203: 108563.

[32] Ehrmann G, Ehrmann A. Shape-memory properties of 3-D printed PLA structures. In Multidisciplinary Digital Publishing Institute Proceedings 2020 (Vol. 69, No. 1, p. 6).

[33] Senatov FS, Zadorozhnyy MY, Niaza KV, Medvedev VV, Kaloshkin SD, Anisimova NY, Kiselevskiy MV, Yang KC. Shape memory effect in 3-D-printed scaffolds for self-fitting implants. *European Polymer Journal*. 2017 Aug 1; 93: 222–231.

[34] Singh G, Singh S, Prakash C, Kumar R, Kumar R, Ramakrishna S. Characterization of three-dimensional printed thermal-stimulus polylactic acid-hydroxyapatite-based shape memory scaffolds. *Polymer Composites*. 2020 Sep; 41(9): 3871–3891.

[35] Liu H, He H, Huang B. Favorable thermo responsive shape memory effects of 3-D printed poly (lactic acid)/poly (ε-caprolactone) blends fabricated by fused deposition modeling. *Macromolecular Materials and Engineering*. 2020 Nov; 305(11): 2000295.

[36] Sun YC, Wan Y, Nam R, Chu M, Naguib HE. 4-D-printed hybrids with localized shape memory behaviour: Implementation in a functionally graded structure. *Scientific Reports*. 2019 Dec 10; 9(1): 1–3.

[37] Cersoli T, Cresanto A, Herberger C, MacDonald E, Cortes P. 3-D printed shape memory polymers produced via direct pellet extrusion. *Micromachines*. 2021 Jan; 12(1): 87.

[38] Peng X, He H, Jia Y, Liu H, Geng Y, Huang B, Luo C. Shape memory effect of three-dimensional printed products based on polypropylene/nylon 6 alloy. *Journal of Materials Science*. 2019 Jun; 54(12): 9235–9246.

[39] Melocchi A, Uboldi M, Inverardi N, Briatico-Vangosa F, Baldi F, Pandini S, Scalet G, Auricchio F, Cerea M, Foppoli A, Maroni A. Expandable drug delivery system for gastric retention based on shape memory polymers: Development via 4-D printing and extrusion. *International Journal of Pharmaceutics*. 2019 Nov 25; 571: 118700.

[40] Stampfl J, Baudis S, Heller C, Liska R, Neumeister A, Kling R, Ostendorf A, Spitzbart M. Photopolymers with tunable mechanical properties processed by laser-based high-resolution stereolithography. *Journal of Micromechanics and Microengineering*. 2008 Nov 6; 18(12): 125014.

[41] Radchenko AV, Duchet-Rumeau J, Gérard JF, Baudoux J, Livi S. Cycloaliphatic epoxidized ionic liquids as new versatile monomers for the development of shape memory PIL networks by 3-D printing. *Polymer Chemistry*. 2020; 11(34): 5475–5483.

[42] Kumagai H, Arai M, Gong J, Sakai K, Kawakami M, Furukawa H. Modeling the transparent shape memory gels by 3-D printer Acculas. In Nanosensors, Biosensors, and Info-Tech Sensors and Systems 2016 2016 Apr 16 (Vol. 9802, p. 98020K). International Society for Optics and Photonics.

[43] Zhang W, Chen J, Li X, Lu Y. Liquid metal-polymer microlattice metamaterials with high fracture toughness and damage recoverability. *Small*. 2020 Nov; 16(46): 2004190.

[44] Miao S, Cui H, Nowicki M, Lee SJ, Almeida J, Zhou X, Zhu W, Yao X, Masood F, Plesniak MW, Mohiuddin M. Photolithographic-stereolithographic-tandem fabrication of 4-D smart scaffolds for improved stem cell cardiomyogenic differentiation. *Biofabrication*. 2018 May 2; 10(3): 035007.

[45] Ehrmann G, Ehrmann A. 3-D printing of shape memory polymers. *Journal of Applied Polymer Science*. 2021 Sep 10; 138(34): 50847.

[46] Momeni F, Hassani NSMM, Liu X., Ni J. A review of 4-D printing. *Mater.* 2017 Dec; 122: 42.

[47] Zhang Z, Demir KG, Gu GX. Developments in 4-D-printing: A review on current smart materials, technologies, and applications. *International Journal of Smart and Nano Materials*. 2019 Jul 3; 10(3): 205–224.

[48] Jeong HY, Woo BH, Kim N, Jun YC. Multicolor 4-D printing of shape-memory polymers for light-induced selective heating and remote actuation. *Scientific Reports*. 2020 Apr 10; 10(1): 1–11.

[49] Sun YC, Wan Y, Nam R, Chu M, Naguib HE. 4-D-printed hybrids with localized shape memory behaviour: Implementation in a functionally graded structure. *Scientific Reports*. 2019 Dec 10; 9(1): 1–3.

[50] Hu GF, Damanpack AR, Bodaghi M, Liao WH. Increasing dimension of structures by 4-D printing shape memory polymers via fused deposition modeling. *Smart Materials and Structures*. 2017 Nov 10; 26(12): 125023.

[51] Gladman AS, Matsumoto EA, Nuzzo RG, Mahadevan L, Lewis JA. Biomimetic 4-D printing. *Nature Materials*. 2016 Apr; 15(4): 413–418.

[52] Alam MA, Al Riyami K. Shear strengthening of reinforced concrete beam using natural fiber reinforced polymer laminates. *Construction and Building Materials*. 2018 Feb 20; 162: 683–696.

[53] Liu Y, Zhang F, Leng J, Fu K, Lu XL, Wang L, Cotton C, Sun B, Gu B, Chou TW. Remotely and sequentially controlled actuation of electro-activated carbon nanotube/shape memory polymer composites. *Advanced Materials Technologies*. 2019 Dec; 4(12): 1900600.

[54] Lin C, Lv J, Li Y, Zhang F, Li J, Liu Y, Liu L, Leng J. 4-D-printed biodegradable and remotely controllable shape memory occlusion devices. *Advanced Functional Materials*. 2019 Dec; 29(51): 1906569.

[55] Oladapo BI, Adebiyi AV, Elemure EI. Microstructural 4-D printing investigation of ultra-sonication biocomposite polymer. *Journal of King Saud University-Engineering Sciences*. 2019 Dec 10; 33(1): 54–60.

Chapter 8

Electrochemical Discharge Machining for Hybrid Polymer Matrix Composites

Girija Nandan Arka and Shashi Bhushan Prasad

Department of Production & Industrial Engineering, National Institute of Technology Jamshedpur, Jharkhand, India

Subhash Singh

Department of Mechanical and Automation Engineering, Indira Gandhi Delhi Technical University for Women, New Delhi, India

CONTENTS

8.1 INTRODUCTION

Polymer matrix composites (PMCs) have gained significant attention to accomplish the need for prosperous industrial civilization which outclass among numerous novel advanced materials owing to its better strength to weight ratio signify favorable mechanical and tribological property. The effective quantity of reinforcements with qualitative incorporation involves abrasives such as glass fiber, SiC, and carbon fibers into amiable polymer matrix potentially responsible to concoct outshine mechanical characteristics. PC has heartened its extensive and versatile applications in marine,

DOI: 10.1201/9781003327370-8

aviation, and automobile industries and these involve precise machining for magnanimous operative performance. The PC machining conduct usually differs from monolithic material as it involves types of polymer matrix, types of reinforcement, along with volume fraction with its orientation, physical bonding strength, etc. Chronological development of PC extended its physical properties to superior level by employing hybrid composites for efficient enduring applications. These potentials could be attributed by specific characteristics called bonding agent as interphase region structured with multi-phased composite structure consisting of reinforcement phase enclosed with matrix phase [1]. Preferably the hybrid polymer matrix composite consisting of an organic polymer matrix infused with one or more molecular or nano-dimension-level inorganic reinforcements integrate to pronounce superior mechanical, electrical, thermal and optical outreach. However, the presence of abrasives in a polymer matrix composite confronted with deteriorated drilling behavior that depreciated the quality characteristics by increasing tool wear, cutting force, and primary surface texture that hindered its adaptability widespread application [2]. Moreover, PC machined products resulted in damaged structure in terms of delamination, splintering, fiber pull-out, and spalling. These collective associated problems with the conventional approach can be effectively altered by nontraditional machining processes. However, nontraditional machining classified as laser beam machining encompasses severe thermal impairment, electro discharge machining and electrochemical machining comprise conductive nature workpieces that are potentially inappropriate for polymer composite machining, ultrasonic machining and abrasive water jet machining became ostracized due to the occurrence of delamination [3], inherent moisture absorption property [4] and burning of fibers [5]. Electrochemical discharge machining, abbreviated ECDM, could be an efficient and economical choice to machine PCs. Therefore, to comprehend the ECDM potential, the author was encouraged to explore ECDM from the basic level.

ECDM is a hybrid nontraditional machining process exponentially eminent due to its intrinsic machining potential towards application in micro- and meso-dimension fabrication of non-conductive materials [6]. The ECDM unique among nontraditional machining, deliberately combines electro discharge machine (EDM) and electrochemical machine (ECM) technology. EDM yields discharge phenomenon liberated thermal energy and ECM produces chemical etching phenomenon contribute for removal of materials [7]. Owing to these combinational exertions of chemical and thermal energy make the machining flexible for machining hardened, conductive, nonconductive, and reflective materials [8]. The potential of ECDM in terms of material removal rate narrated 5 to 50 times higher relative to ECM and negligible tool wear rate relative to EDM [9]. The foremost advantage of ECDM over EDM is involvement of most stable machining process due to conductive nature of electrolyte rather than

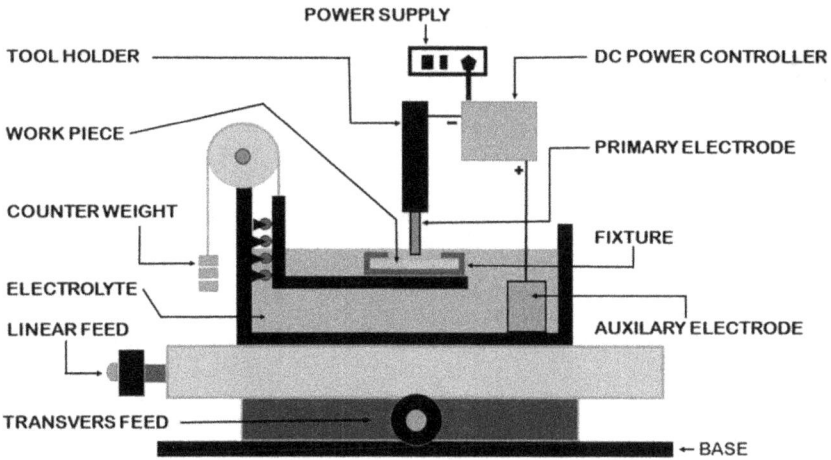

Figure 8.1 Overview of ECDM architecture for the potential application.

dielectric fluid working medium. Literature available to machine glass reflected a lot of attention invested for machine nonconductive material but hardly any work depicted to machine polymer-based composites. Therefore, the author narrated the ECDM potential and to comprehend the machining behavior associated with productive machining.

The basic architect of ECDM is replicated in Fig. 8.1. Ideally, the machine construction includes primary electrode, auxiliary electrode, electrolyte, counterweight to facilitate feed, linear, and transverse feed, DC power supply, respectively.

8.1.1 Working Principle of ECDM

Electrolyte functioning as a conductive medium which flexible close the electrical circuit to form an electrochemical cell (ECC). Ideally, electrolyte composed of an aquatic solution could be either concentrated blending of alkaline or acidic or neutral incorporated to motivate the machining process. Primary electrode is a tool connected to DC negative power (cathode) whose shape will be replicate over workpiece and auxiliary electrode connected to DC positive power (anode) to initiate the electrolysis process.

Generally, size of auxiliary electrode approximately 100 times larger than primary electrode to ensure good amount of gas generated at primary electrode side [10]. For effective machining the tool must immersed 2 to 3 mm into electrolyte relative to electrolyte level [11]. Since higher electrolyte level relative to immerse tool depth expose higher surface contact area which further need more gas bubbles to coat thin gas film which consequence unstable gas film and deteriorate machining process [12].

When a voltage potential difference is applied between the primary electrode and auxiliary electrode the electrolysis process occurs in an electrolytic solution. Concurrently, the flow of electron starts The result of the electro-lysis process produces hydrogen gas bubbles and oxygen gas bubbles at the exposed surface area of the primary electrode and auxiliary electrode, respectively, due to the following reactions:

At cathode:

$$2H_2O + 2e^- \rightarrow H_2 + 2(OH)^- \tag{8.1}$$

At anode:

$$2(OH)^- \rightarrow 2H_2O + O_2 + 4e^- \tag{8.2}$$

Along with dissolution, takes place from anode due to combine effect of hydroxide ion $(OH)^-$ from electrolyte and metallic ion A^{k+} from anodic metal are insoluble to electrolytic solution presented in equations (3) and (4).

$$A \rightarrow A^{k+} + ke^- \tag{8.3}$$

$$A^{k+} + k(OH)^- \rightarrow A(OH)_k \tag{8.4}$$

where k represents the number of ions and $A(OH)_k$ represents anodic metal hydroxide. Thus, the production of small gas bubbles is accumu-lated and coat over the exposed cathode surface area act as a dielectric medium. The above overall process primarily depends upon applied voltage and electrolyte concentration broadly discussed with graphical articulation, illustrated in Fig. 8.2. The whole mechanism could be ex-pressed as region 2: initiation of electrolysis, region 3: formation of re-latively smaller hydrogen gas bubbles, region 4: coalescence of relatively larger gas bubbles and gas film formation, and region 5: spark formation. As the voltage increases, current also increases, resulting formation of small to relatively larger hydrogen gas bubbles through the electrolysis process around the tool. Further increase in voltage reaches a critical level in which the gas bubbles convert to a thin stable gas film around the tool. The thin stable gas film act as insulator further intensified as the flow of current gets obstructed. Further, an increase in voltage produced an in-tensified spark initiated from the tool edge is responsible to liberate high thermal energy. Since a higher electric field intensity is generated over a tool edge; therefore, the spark initiated from the edge, however minor discharge may generate from flat surface of tool. If the workpiece is kept closer to the tool, the liberated thermal energy potentially responsible to remove material as well as the machining zone temperature can drive chemical etching phenomenon to remove material parallelly.

The discharge mechanism further investigated by many novel researchers for conceptualizing the backend mechanics to contribute towards ECDM

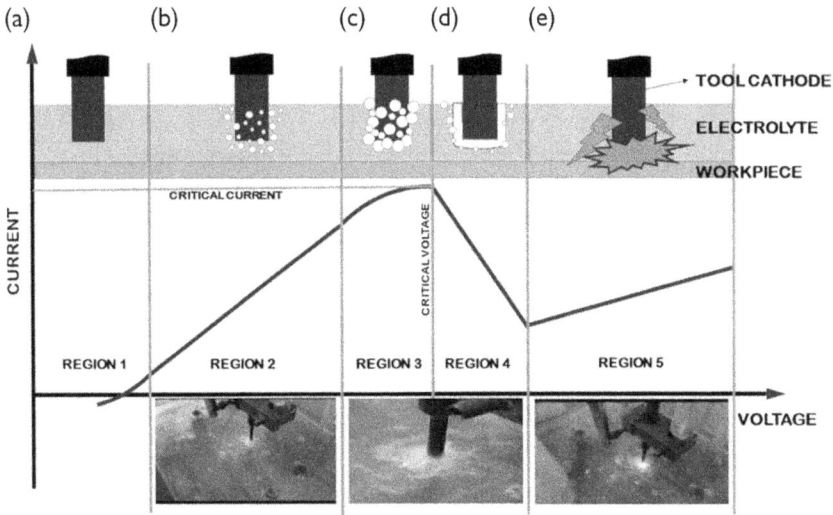

Figure 8.2 ECDM behavior while on operation (a) starting of machining at low voltage and current, (b) formation of relatively small size hydrogen gas bubble with increase in voltage and current, (c) formation of relatively higher gas bubbles due to relatively higher voltage and current, (d) formation of thin gas film around the tool after critical voltage in which it block the flow of current and intensify the machining environment, (e) Initiation of discharge after breaking the insulation strength of the gas film.

development. Basak and Ghosh detected few narrow conducting bridges across the tool electrode at the machining zone and at high current density it triggers instant boiling causes discharge [13]. Jain et al. narrated individual gas bubbles as a valve and produce identical discharge by breakdown of each valve under an intensified electric field environment [14]. Since the electrolyte gets heated while discharged, consequence electrochemical etching contributes further towards removal of material. However, no literature expressed to differentiate the material removal contribution quantitatively. Hence, the collective material removal actions are responsible for producing remarkable microscale features. Further, these combined actions magnanimously depend on machine input parameters, as illustrated in Fig. 8.3.

Fig. 8.3 portrays the root causes of the ECDM process to govern high-performance potential. Therefore, it is necessary to establish a control and optimize combination for having quality machining for the productive work.

8.2 ECDM POTENTIAL

The gas film formation was contributed by combined effect of gas bubbles coalescence and exerted buoyance force on bubble. Since uniform stable gas

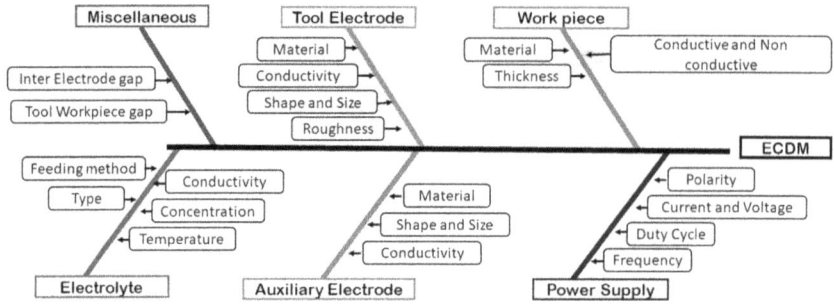

Figure 8.3 Cause-and-effect diagram for ECDM performance.

film highly encourage to create low frequency high intensity discharge which favor efficient material removal rate, work overcut, surface finish etc. There are many working conditions on which quality of the hydrogen gas layer (film) depends like percentage of electrolyte, quality of electrolyte, electrolyte temperature and shape of the cathode tool electrode, etc. and is crucial for the surface quality [15]. Visual observation of the gas film formation is quite difficult but with the help of high-speed cameras visual observation of size and thickness of the gas film can be obtained. In an electrochemical discharge machining process, there are several parameters which influence the gas film quality such as tool immersion depth, electrolyte resistivity, thermo-capillary flow between electrolyte and cathode tool electrode because of temperature gradients, current density, cathode tool radius, and force of electrostatic attraction between cathode tool electrode and hydrogen gas bubbles. Han et al. observed that by increasing the diameter of the tool (cathode) electrode from 50 to 300 mm and immersed depth of tool tip inside electrolyte from 8 to 303 μm, as a consequence the gas film thickness altered from 3 to 45 μm [16]. Guelcher observed that formation of a hydrogen gas layer (film) originates because of the movement of hydrogen bubbles towards each other, which is caused by thermos capillary flow [17]. Hence, for efficient machining, control of the gas film is highly essential. Further research narrated that tool wettability and surface tension between electrolyte and bubble affect the gas film thickness [18]. Research reported that the gas bubble experienced inertia force, surface tension force and buoyancy force and remain attached to tool surface until threshold bubble size where forces remain balanced [19]. Since the generated gas bubble experiences surface tension force to attach with tool surface, whereas buoyancy force enthralls to eject bubbles towards upward direction [19] and therefore to produce stable discharge, optimum surface tension force should be predominant relative to buoyancy force. Hence, many authors incorporated numerous novel technological alterations to establish thin stable hydrogen gas film for having stable discharge. Moreover, as the interelectrode gap increases, the electrochemical resistance increases which as a consequence reduction in electrochemical activity is responsible for the reduction gas volume and buoyancy force conveyed that stability of gas film encouraged by interelectrode gap as reported [20].

8.2.1 By Altering the Tool

Tool electrode shape and material are ideally acknowledged for having effective utilization of discharge energy for efficient machining. Research reported that discharge should be initiated from the bottom portion of the tool electrode for significant availability of thermal energy for machining. Thus, a side insulated electrode was introduced to produce bottom discharge by avoiding side discharge had registered relatively smaller hole taper angle 3.3° for 600 µm machining depth [21]. Similarly, Singh and Dvivedi engineered textured tool made of up chromium and nickel coat surgical stainless steel created micro gaps among tool and workpiece triggered thin and stable gas film for 20% vol concentration of NaOH electrolyte produced low-intensity high-frequency discharge which persuade for the high efficacy [6]. This phenomenon creates significant thermal energy and deteriorate tool shape by accumulating heat which technically named as tool wear rate. An experiment was conducted by altering tool material by tungsten, brass, and steel and revealed tungsten as an excellent choice relative to others owing to a higher melting point temperature around 2870°C [22]. Further study reported that at deeper machining of the debris and sludge remain over the machined surface due to improper flushing encountered, which limits the ingress of fresh electrolyte circulation and hinder the electrochemical discharge [23]. Therefore, rotation of the tool electrode was attempted and revealed the formation of stable thin gas film exerted due to the action of the centrifugal force created an ample opportunity for improving the ECDM potential [9]. Centrifugal force not only established a thin stable gas film but also assisted in removal of struggled sludges present inside the machining zone. Although many tool electrode shapes and materials attempted but still struggled to find the best result. For example, coaxial-jet nozzle shaped tungsten carbide tool electrode incorporated to reach good result and resulted reduction of axial wear by 39.29% and radial wear by 84.09% [24].

Spring-feed mechanism against the tool electrode at the pressure window from 1.5 to 2.5 N/mm^2 was attempted by Singh et al. to establish thin and stable gas film. Their result registered significant machining performance at high pressure owing to breakdown of the gas film constituted consistence high-frequency low-intensity discharge [25].

8.2.2 By Altering the Electrolyte

Electrolyte type and concentration portray a major role for the ECDM as the electrical conductivity of the electrolyte rises with the increase in electrolyte concentration favoured to reduce critical voltage [26]. Since the gas bubble formation crucially depends upon critical voltage, which further distinguished through electrolyte concentration, therefore optimum selection of electrolyte is essential for locating stable discharge. Moreover, research conveyed that increase in electrolyte temperature caused to form thin and more uniform hydrogen gas film accredited for resourceful discharge [27]. Most prominently pragmatic electrolytes incorporated in ECDM popularized are NaOH, KOH, and NaCl consisting of electrostatic force bonded with ionic crystal lattices. Moreover, the addition of solvent instigates the ions to move apart and increase the mobility of ions attribute to have greater conductance which improves machining efficacy. However, at very high concentrations, the mobility of ions gets confined, owing to attraction of the oppositely charged ions closer to each other [28]. Meanwhile, both KOH and NaOH electrolytes possess the same breakdown voltage but found imparity on discharge current contributed by KOH is superior to NaOH owing to higher mobility of $K+$ ion in aquas solution relative to $Na+$ [29]. Further study revealed KOH is superior to NaOH due to relatively higher specific conductance of KOH encouraged a faster electrolysis rate during dissolution of K into $K+$ and rapid formation of hydrogen bubbles assistances to enhance discharge density [30]. Moreover, the amalgamate of NaOH and KOH at equal proportion reported superior conductivity at 20% weight percentages concentration attributed reduction in surface tension due to excessive OH^- and high mobility of $K+$ ion [31]. Furthermore, anionic and cationic surfactants such as sodium dodecyl sulphate and cetyl-trimethyl-ammonium bromide also been incorporated to increase the wettability responsible for lowering down of surface tension due to weak van der waals bonding force between molecules which encouraged for the generation of thin gas film [32].

Moreover, low electrolyte level always favors for construction of stable hydrogen gas film and increase productivity but is difficult to establish due to continuous vaporization of electrolyte caused by raised temperature in the machining zone during machining [33]. The specific reason was identified as increased electrolyte level caused accumulation gas film at the top of the surface level encouraged side discharge in the horizontal direction rather than vertical discharge should have originated from the tool electrode bottom surface [34]. Therefore, many novel provisions were acknowledged by researchers to eliminate gas bubble accumulation at the top and accomplish electrolyte compensation that eventually evaporated. Hence, Singh and Dvivedi incorporated titanate electrolyte flow to initiate drag force towards tool tip and achieved a 45 µm/s penetration rate [35]. Tool rotation with coaxial electrolyte jet flow accessorized ECDM-enabled machining depth reached 193 µm [24].

Hence, to achieve outclass ECDM machining, the generation of thinner gas film is most desirable as it indeed required low critical voltage and current. The technicality to produce thin film is either by reducing surface tension or by increasing electrolyte density [19].

8.3 QUANTIFYING ECDM POTENTIAL

The various quantifying parameters such as material removal rate, over dimension cut, hole wall surface texture, and delamination on drill hole quantify for quality appraisal for machining hybrid PCs, discussed below.

8.3.1 Delamination on Drilled Hole

ECDM-machined hybrid PCs articulate the utmost irregular features integrated with number of hitches as overcut, fibrous residual, heat affected zone, tearing of matrix, fiber pull-outs, fiber tearing, etc. These hitches convicted delamination at the entrance and exit of a feature that hampers quality of machining and structural integrity of PC. Davim et al. narrated the prime damage delamination into linear delamination quantify as delamination factor expressed in equation (5) [36].

$$Delamination\ Factor = \frac{Maximum\ Delamination\ diameter}{Inner\ diameter} \qquad (8.5)$$

Further, this delamination factor is refined by considering area of irregular delamination and quantified as equation (6) [37].

$$Delamination\ Factor = a\frac{Maximum\ Delamination\ diameter}{Inner\ diameter}$$
$$+ b\frac{Maximum\ delamination\ area}{Inner\ area} \qquad (8.6)$$

where a and b customized assigned weight.

8.3.2 Over Dimension Cut (ODC)

Over dimension cut is an inherent machining characteristic used to analysis quantitative appraisal by calculating diametric difference through measuring instrument. In micro- to meso-machining process ODC depicted a significant role to define a quality product through eminence and precision. Therefore, it is necessary to comprehend technical glitches committed towards overcut dimension for polymer matrix. Parameters that affect ODC are electrolyte concentration, tool configuration, interelectrode gap, voltage, etc. Since low applied voltage instigate electrochemical action, whereas high applied voltage is fruitful for thermal action [38] and increase in voltage increases high density hydrogen bubble encouraged stray or side sparking, which is further responsible for the ODC and microcracks [39]; hence, optimum voltage is crucial for the dominant effect. Since low feed rate derived long contact with discharges depicted enhanced hole wall surface and high feed rate reflected short contact with discharges in the machining zone depicted diminished ODC [40]; thus, optimum feed rate is essential for the feature parity.

8.3.3 Material Removal Rate

The feasibility and productivity of a machining process is always justified through rate of material removal parameter. However, ECDM deviated from other machining process owing to nonuniform rate of material removal mechanism. Research reported the volume of material removal discovered decreased with increasing machining depth [39]. This phenomenon consequence due to the ECDM process depicted dual nature of material removal methods and confined with two disparate regimes along machining depth direction ascribed as discharge regime and hydrodynamic regime. From which discharge regime is a region where fresh electrolyte easily available in machining zone for electrolysis and establish stable gas film consequence drilling speed as a function of voltage, whereas hydrodynamic regime is a region where instability of gas film depicted due to inadequate electrolyte circulation at higher machining depth consequence differ in machining behavior [23]. It is quite obvious that at higher voltage the availability of thermal energy is adequate to remove more material but due to excessive heat generation few micro cracks may appear on machined surface [41].

Thus, to unfold the ECDM potential, Table 8.1 depicted numerous alterations to achieve superior result for founding its candidature against competitive machining domain.

Above briefing describe the ECDM potential to machine numerous materials irrespective of material and mechanical properties. Hence, ECDM has great potential to machine hybrid PC with economical prospective.

8.4 MACHINING POTENTIAL TO MACHINE HYBRID PC

Hybrid PC is recognized for achieving superior mechanical properties by incorporating e-glass fiber or silicon carbide or any other reinforcement material into polymer matrix. Moreover, the constituents present in composite are physically differ from each other. Although composites have reached the intended property but mostly confronted with machinability. Many researchers potentially applied ECDM for machining polymer composites are addressed for the future development.

Manna and Narang conducted experiment to machine glass fiber epoxy composite by ECDM and found the concentration of electrolyte as the utmost significant parameter followed by voltage towards material removal rate and radial overcut. This result attributed due to the conductive property of electrolyte concentration encourage to produce quick electrochemical process

Table 8.1 ECDM potential

Tool configuration	Work material	Experimental condition	Depth and overcut in μm	References
Rotated 700 μm stainless steel (SS-304) solid cylindrical tool	Al 6063/SiC 10% metal matrix composite	Applied voltage: 65 V; Electrolyte concentration 17% wt./vol; Pulse on: pulse off: 3:1 ms; Tool rotation: 901 rpm.	1307 and 138	[9]
Tungsten tool having 300 μm diameter	Borosilicate glass	Concentration: 0.5 M; Inter electrode gap: 4 cm; Level of electrolyte: 1 cm; Time of machining: 0.5 min.	1100 and 494	[20]
4 μm-thick diamond coat side insulated tool over 250 μm dia electrode	Quartz	Applied voltage: 40 V; NaOH Electrolyte concentration: 6 M; Pulse on time (Ton): 2 ms; Pulse off time (Toff): 1 ms; Rotational speed: 1000 rpm.	600 and <350	[21]
Coaxial-jet nozzle electrode	Sapphire	Applied voltage: 53 V; KOH electrolyte concentration: 5 M; Spindle speed: 300 rpm; Nozzle pressure: 0.2 kpa.	193 and –	[24]
SS304 tubular electrode having 500 μm outside diameter and 260 μm inside diameter	Borosilicate glass	Applied voltage: 50 V; Electrolyte concentration: 20% wt./vol; Ton/Toff: 5:1; Electrolyte flow rate: 4 ml/hr; Machining time 150 s.	1052 and 724	[33]
Taper angle 11.88° SS-304 tool	Borosilicate glass	Applied voltage: 60 V; NaOH electrolyte concentration: 20% wt./vol; pulse on time of 3 ms; electrolyte flow rate: 3.30 ml/min.	1350 and 516	[35]
Drill bit made up of Tungsten carbide having 300 μm diameter	Silicon wafer	Applied voltage: 60 V; NaOH Electrolyte concentration: 10% wt./Vol; Tool feed rate: 250 μm/min.	– and 180	[40]
Stainless steel (SS 304) solid cylindrical tool having 600 μm diameter	Borosilicate glass	Applied voltage: 70 V; NaOH Electrolyte concentration: 25% wt./Vol; Pulse duty cycle: 4:1 ms; Machining time: 120 s.	920 and –	[42]
Spring-fed needle tapered tool made up of Tungsten carbide having 404 μm diameter with tip size 77 μm	Quartz glass Thickness 10 mm	Applied voltage: 50 V; Peak current: 1 A, Electrolytic solution KOH, Working-flow pressure: 30 kPa.	1000 and 441	[43]
Stainless steel tool having 300 μm diameter	Sintered ceramic sample of Silicon Carbide	Applied voltage: 25 V; Electrolyte concentration: 20 wt. %; Inter-electrode gap: 40 mm.	– and 80.5	[44]

for quick discharge. However, the surface created during micro drilling was very poor due to the reinforced fibres were not entirely cut during machining, rather they were burnt and combined. These causes rough fabric surface which appeared in the form of debris. It was also observed during machining that the generated hole along its depth was conical in nature [45]. Improper flushing in hydrodynamic region may cause to produce such architectural hole. Additionally, the cracking of matrix propagated from the hole entrance was revealed by Antil et al. [46] while investigating the machining behavior for SiC reinforced polymer matrix composite. These cracks came into picture may be due to applied voltage that may cause uneven discharge resulted from stray sparking encourage thermally unbalanced heat-affected zone. Moreover, SiC particles detached from the matrix and leave space cavity on the machined surface, which makes the surface pore.

Moreover, delamination is one of the prime responses evolved on entrance and exit of hole after machining owing to fiber tearing, fiber pull-outs, cracking of matrix, etc. Antil et al. [47] studied the effect of machining parameters on hybrid PC and distinguished delamination zone by help of scanning electron microscopy. They revealed fibrous residual, heat-affected zone at the edge of the hole and tearing of matrix at hole entrance remain present. Their analysis reflected applied voltage as a most significant parameter for improving MRR, whereas the inter-electrode gap served as the most significant parameter to contribute radial overcut. Since increased voltage consequence high-energy discharge due to high density of supply energy beneficial for improving MRR and increased inter electrode gap consequence decreased heat affected zone due to increased gap resistance. Moreover, increased interelectrode gap resistance resulted increase in critical voltage and affect the conductivity of the channel [48]. This phenomenon thrills to produce many fins scattered spark causing irregularity on surface by melting the upper layer of PC and leads to undesirable radial overcut. Moreover, non uniform cutting with fractured fibers and removal of matrix from fibers caused from thermal spalling fascinated from high temperature of the plasma. Therefore, these fractures of fibers and non-uniform machining led to producing an irregular hole surface. Moreover, chemical etching phenomenon contributed by optimal interelectrode gap as well. Apart from applied voltage, electrolyte concentration, interelectrode gap, etc. there are other parameters such as pulse on-time and pulse off-time also contributes significantly for the quality discharge. Pulse on-time regulates the machining time for the quality hydrogen bubble formation, which was encouraged to produce quality discharge, whereas pulse off-time regulates the replenishment time of electrolytes for removal of sludge.

Finding an optimal setting of machining parameters is a crucial task to flourish superior results. Since each machining parameter contributes towards improvement of the quality characteristics, thus finding the best combinational setting urged research to achieve synergistic effect for the same. Therefore, many researches applied many novel optimization techniques to get the best optimized process parameters to contribute for the remarkable result. Antil et al. [46] implemented the Taguchi gray relational analysis to predict optimum result and found efficient technique relative to Taguchi by clinching 1.118 mg/min MRR. Thus, it is necessary to adjust the machining parameters for the revival of individual contribution along with their interaction contribution. Since ECDM is a complicated process and required a nonlinear optimization technique, therefore Antil et al. [49] incorporated population-based technology as differential evolution; genetic algorithm; simulated annealing algorithm, etc. and found differential evolution optimization superior relative to other by reaching 1.792 mg/min at Electrolyte concentration 109.054 g/L, interelectrode gap 139.298 mm, duty factor 0.791, applied voltage 59.091 V, respectively. These parametric combinations reflected a smooth hole surface that may be caused by the uniform discharge contributed by synergistic efforts of individual parameters. Table 8.2 shows the potential of ECDM to machine different hybrid composites reported elsewhere.

Table 8.2 Machining potential to machine hybrid PCs

Reinforcement	Alteration	MRR (mg/min)	Overcut	References
e-glass–fibers	Applied voltage: 70 V; Electrolyte concentration: 80 g/l; Inter electrode gap: 180 mm.	2.97525	0.1200	[45]
Silicon carbide particles	Applied voltage: 60 V; Electrolytic concentration: 110 g/l; Inter-electrode gap: 120 mm; Duty factor: 0.66.	1.095	0.167 mm	[46]
Glass fibers and SiC	Applied voltage: 70 V; Electrolytic concentration: 100 g/l; linterelectrode gap: 100 mm	1.041	0.136	[47]
Carbon fiber	Applied voltage: 40 V; NaOH electrolyte concentration: 20% (wt/V); Inter-electrode Gap: 30 mm; Duty cycle 80% s; Tool feed rate: 100 mm/min; Tool rotation: 100 rpm	1.712	122 μm	[50]
SiC/glass fiber	Applied voltage: 65 V; Electrolyte concentration: 90 g/l; Inter electrode gap: 70 mm; Duty factor: 0.875.	1.789		[51]
Carbon fibers	Applied voltage: 70 V; NaOH electrolyte concentration: 50% mass; Interelectrode gap 70 mm.	1.310		[52]

Hence, this informative discussion acknowledges ECDM as a proficient method to machine hybrid polymer composites. It also directs damaged hole surface after machining and is a crucial challenge to attend excellent surface. For making the surface excellent, it is necessary to control the machining parameters for uniform quality discharge. Furthermore, removal of sludge from the machining zone is critical and instigate the advancement of ECDM. Thus, to create quality discharge and sludge removal from the machining zone, researchers incorporated several technological alterations by making the ECDM hybrid.

8.4.1 Hybrid ECDM

From the above discussion, ECDM still an underdog machining process confronted with damaged hole wall surface, non-uniform machining of re-inforcements, damaged surroundings, etc. Therefore, many novel thinking brains technologically advance a ECDM by making it hybrid by incorporating two or more principles to get synergistic productive activity to fabricate sound product. Technically, the idea to make it hybrid is to create thin stable gas film that encourages stable uniform discharge and favors a sound product.

8.4.1.1 LASER-Assisted ECDM

It is well known that electrolyte temperature serves the activation energy required for electrochemical reactions. Therefore, a low-power LASER beam could harvest on electrolyte near to the tool surface can raise electrolyte tool interface temperature and cause booming of an electrochemical reaction that may further combine with high temperature initiate thin and stable gas film. Furthermore, laser assists more beneficial for mechanical precision of etching operation acknowledge sharper profiles [53]. Singh et al. integrated LASER to attend precative catalyst for electrochemical reactions and experimented to enhance process capability of ECDM for carbon fiber–reinforced PC registered a superior machining characteristic with unform cutting of fibers by improving spark discharge attributed by stable and thinner gas film [54]. They revealed a smooth cutting of PC resulted due to uniform machining of fibers encouraged by assisted laser to ECDM. Moreover, they potentially found 720 μm smooth hole at an optimum combination of 450 μm tool diameter, 0.9 m/min feed, 300 rpm tool rotation and 60% duty cycle, respectively. These combinations provided stable spark discharge and prolonged chemical etching after machining mutually contributes to excel the hole quality.

8.5 CONCLUSION

A systematic and inclusive study was carried out for ECDM to explore its potential for machine nonconductive materials. Furthermore, an informative discussion was scripted for ECDM to machine hybrid polymer composites. From the discussion, it was revealed that stable thin film generation is prominently responsible for the quality discharge for which many researchers attempted to alter tool configuration, electrolyte concentration, and machining parameters, respectively. While machining of hybrid composite, residual reinforced material, unburned fibrous constituents, and matrix cracking make the surface damaged and stray discharge encourage to form wider overcut hole entrance. Significant efforts have been incorporated to establish synergistic effect of individual machining parameters towards quality characteristics by various optimization techniques. However, very few investigations were undertaken for hybrid polymer composite by altering engineered electrolyte and tool. Therefore, it urges research to produce quality discharge for the quality hole surface of hybrid polymer composites. Furthermore, delamination and cracking of polymer are crucial challenge to enrich the quality of machining. Thus, thermal balancing of heat among electrolytes, tools, and workpiece analysis could assist the problem of cracking and could make the solution accordingly. Thus, it could be a new direction of brainstorming for the development of ECDM to machine hybrid polymer composites.

ACKNOWLEDGEMENT

The author(s) disclosed receipt of the following financial support for the research, authorship, and/or publication of this article: This research work had the financial support of NIT Jamshedpur and Ministry of Human Resource and Development (MHRD), Government of India.

REFERENCES

[1] Rahman M, Ramakrishna S, Prakash JRS, Tan DCG. Machinability study of carbon fiber reinforced composite. *J Mater Process Technol*. 1999; 89–90: 292–297. 10.1016/S0924-0136(99)00040-0.

[2] Hatt O, Crawforth P, Jackson M. On the mechanism of tool crater wear during titanium alloy machining. *Wear*. 2017; 374–375: 15–20. 10.1016/j.wear.2016.12.036.

[3] Liu J, Zhang D, Qin L, Yan L. Feasibility study of the rotary ultrasonic elliptical machining of carbon fiber reinforced plastics (CFRP). *Int J Mach Tools Manuf*. 2012; 53: 141–150. 10.1016/j.ijmachtools.2011.10.007.

[4] Azmir MA, Ahsan AK. Investigation on glass/epoxy composite surfaces machined by abrasive water jet machining. *J Mater Process Technol*. 2008; 198: 122–128. 10.1016/j.jmatprotec.2007.07.014.

[5] Kumar R, Agrawal PK, Singh I. Fabrication of micro holes in CFRP laminates using EDM. *J Manuf Process*. 2018; 31: 859–866. 10.1016/j.jmapro.2018.01.011.

[6] Singh T, Dvivedi A. On performance evaluation of textured tools during micro-channeling with ECDM. *J Manuf Process*. 2018; 32: 699–713. 10.101 6/j.jmapro.2018.03.033.

[7] Mehrabi F, Farahnakian M, Elhami S, Razfar MR. Application of electrolyte injection to the electro-chemical discharge machining (ECDM) on the optical glass. *J Mater Process Technol*. 2018; 255: 665–672. 10.1016/j.jmatprotec. 2018.01.016.

[8] Singh T, Dvivedi A. Developments in electrochemical discharge machining: A review on electrochemical discharge machining, process variants and their hybrid methods. *Int J Mach Tools Manuf*. 2016; 105: 1–13. 10.1016/ j.ijmachtools.2016.03.004.

[9] Singh T, Arya RK, Dvivedi A. Experimental investigations into rotary mode electrochemical discharge drilling (RM-ECDD) of metal matrix composites. *Mach Sci Technol*. 2020; 24: 195–226. 10.1080/10910344.2019. 1636270.

[10] Bindu Madhavi J, Hiremath SS. Machining of micro-holes on borosilicate glass using micro-electro chemical discharge machining (μ-ECDM) and parametric optimisation. *Adv Mater Process Technol*. 2019; 5: 542–557. 10.1080/2374068X.2019.1636187.

[11] Huang SF, Liu Y, Li J, Hu HX, Sun LY. Electrochemical discharge machining micro-hole in stainless steel with tool electrode high-speed rotating. *Mater Manuf Process*. 2014; 29: 634–637. 10.1080/10426914.2014.901523.

[12] Kumar N, Mandal N, Das AK. Micro-machining through electrochemical discharge processes: A review. *Mater Manuf Process*. 2020; 35: 363–404. 10.1080/10426914.2020.1711922.

[13] Basak I, Ghosh A. Mechanism of spark generation during electrochemical discharge machining: A theoretical model and experimental verification. *J Mater Process Technol*. 1996; 62: 46–53. 10.1016/0924-0136(95)02202-3.

[14] Jain VK, Dixit PM, Pandey PM. On the analysis of the electrochemical spark machining process. *Int J Mach Tools Manuf*. 1999; 39: 165–186. 10.1016/ S0890-6955(98)00010-8.

[15] Abou Ziki JD, Fatanat Didar T, Wüthrich R. Micro-texturing channel surfaces on glass with spark assisted chemical engraving. *Int J Mach Tools Manuf*. 2012; 57: 66–72. 10.1016/j.ijmachtools.2012.01.012.

[16] Han MS, Min BK, Lee SJ. Modeling gas film formation in electrochemical discharge machining processes using a side-insulated electrode. *J Micromechanics Microengineering*. 2008; 18(4); 18. 10.1088/0960-1317/18/4/045019.

[17] Guelcher SA, Solomentsev YE, Sides PJ, Anderson JL. Thermocapillary phenomena and bubble coalescence during electrolytic gas evolution. *J Electrochem Soc*. 1998; 145: 1848–1855.

[18] Liu Y, Zhang C, Li S, Guo C, Wei Z. Experimental study of micro electro-chemical discharge machining of ultra-clear glass with a rotating helical tool. *Processes*. 2019; 7: 195–209. 10.3390/pr7040195.

[19] Jiang B, Lan S, Wilt K, Ni J. Modeling and experimental investigation of gas film in micro-electrochemical discharge machining process. *Int J Mach Tools Manuf*. 2015; 90: 8–15. 10.1016/j.ijmachtools.2014.11.006.

[20] Kolhekar KR, Sundaram M. Study of gas film characterization and its effect in electrochemical discharge machining. *Precis Eng*. 2018; 53: 203–211. 10.1016/j.precisioneng.2018.04.002.

[21] Tang W, Kang X, Zhao W. Enhancement of electrochemical discharge machining accuracy and surface integrity using side-insulated tool electrode with diamond coating. *J Micromechanics Microengineering.* 2017; 27: 065013. 10.1088/1361-6439/aa6e94.

[22] Behroozfar A, Razfar MR. Experimental study of the tool wear during the electrochemical discharge machining (ECDM). *Mater Manuf Process.* 2016; 31: 574–580. 10.1080/10426914.2015.1004685.

[23] Singh M, Singh S. Electrochemical discharge machining: A review on preceding and perspective research. *Proc Inst Mech Eng Part B J Eng Manuf.* 2019; 233: 1425–1449. 10.1177/0954405418798865.

[24] Ho CC., Chen JC. Micro-drilling of sapphire using electro chemical discharge machining. *Micromachines.* 2020; 11: 377–391. 10.3390/MI11040377.

[25] Singh T, Rathore RS, Dvivedi A. Experimental investigations, empirical modeling and multi objective optimization of performance characteristics for ECDD with pressurized feeding method. *Meas J Int Meas Confed.* 2020; 149: 107017. 10.1016/j.measurement.2019.107017.

[26] Sankar M, Gnanavelbabu A, Rajkumar K, Thushal NA. Electrolytic concentration effect on the abrasive assisted-electrochemical machining of aluminium-boron carbide composite. *Mater Manuf Process.* 2017; 32: 687–692. 10.1080/10426914.2016.1244840.

[27] Leyva-Bravo J, Chiñas-Sanchez P, Hernandez-Rodriguez A, Hernandez-Alba GG. Electrochemical discharge machining modeling through different soft computing approaches. *Int J Adv Manuf Technol.* 2020; 106: 3587–3596. 10.1007/s00170-019-04766-z.

[28] Ladeesh VG, Manu R. Grinding-aided electrochemical discharge drilling in the light of electrochemistry. *Proc Inst Mech Eng Part C J Mech Eng Sci.* 2019; 233: 1896–1909. 10.1177/0954406218780129.

[29] Mishra DK, Verma AK, Arab J, Marla D, Dixit P. Numerical and experimental investigations into microchannel formation in glass substrate using electrochemical discharge machining. *J Micromechanics Microengineering.* 2019; 29(7): 075004. 10.1088/1361-6439/ab1da7.

[30] Mallick B, Biswas S, Sarkar BR, Doloi B, Bhattacharyya B. On performance of electrochemical discharge micro-machining process using different electrolytes and tool shapes. *Int J Manuf Mater Mech Eng.* 2020; 10: 49–63. 10.4018/IJMMME.2020040103.

[31] Sabahi N, Razfar MR. Investigating the effect of mixed alkaline electrolyte (NaOH + KOH) on the improvement of machining efficiency in 2D electrochemical discharge machining (ECDM). *Int J Adv Manuf Technol.* 2018; 95: 643–657. 10.1007/s00170-017-1210-4.

[32] Sabahi N, Razfar MR, Hajian M. Experimental investigation of surfactant-mixed electrolyte into electrochemical discharge machining (ECDM) process. *J Mater Process Technol.* 2017; 250: 190–202. 10.1016/j.jmatprotec.2017.07.017.

[33] Arya RK, Dvivedi A. Investigations on quantification and replenishment of vaporized electrolyte during deep micro-holes drilling using pressurized flow-ECDM process. *J Mater Process Technol.* 2019; 266: 217–229. 10.1016/j.jmatprotec.2018.10.035.

[34] Gupta PK. Effect of electrolyte level during electro chemical discharge machining of glass. *J Electrochem Soc.* 2018; 165: E279–E281. 10.1149/ 2.1021807jes.

[35] Singh T, Dvivedi A. On prolongation of discharge regime during ECDM by titrated flow of electrolyte. *Int J Adv Manuf Technol.* 2020; 107: 1819–1834. 10.1007/s00170-020-05126-y.

[36] Davim JP, Rubio JC, Abrão AM. Delamination assessment after drilling medium-density fibreboard (MDF) by digital image analysis. *Holzforschung.* 2007; 61: 294–300. 10.1515/HF.2007.066.

[37] Davim JP, Rubio JC, Abrao AM. A novel approach based on digital image analysis to evaluate the delamination factor after drilling composite laminates. *Compos Sci Technol.* 2007; 67: 1939–1945. 10.1016/j.compscitech. 2006.10.009.

[38] Jha NK, Singh T, Dvivedi A, Rajesha S. Experimental investigations into triplex hybrid process of GA-RDECDM during subtractive processing of MMC's. *Mater Manuf Process.* 2019; 34: 243–255. 10.1080/10426914. 2018.1512126.

[39] Ranganayakulu J, Hiremath SS, Paul L. Parametric analysis and a soft computing approach on material removal rate in electrochemical discharge machining. *Int J Manuf Technol Manag.* 2011; 24: 23–39. 10.1504/ IJMTM.2011.046758.

[40] Singh M, Singh S, Kumar S. Experimental investigation for generation of micro-holes on silicon wafer using electrochemical discharge machining process. *Silicon.* 2020; 12: 1683–1689. 10.1007/s12633-019-00273-8.

[41] Bhattacharyya B, Doloi BN, Sorkhel SK. Experimental investigations into electrochemical discharge machining (ECDM) of non-conductive ceramic materials. *J Mater Process Technol.* 1999; 95: 145–154. 10.1016/S0924-013 6(99)00318-0.

[42] Rathore RS, Dvivedi A. Sonication of tool electrode for utilizing high discharge energy during ECDM. *Mater Manuf Process.* 2020; 35: 415–429. 10.1080/10426914.2020.1718699.

[43] Ho CC, Wu DS, Chen JC. Flow-jet-assisted electrochemical discharge machining for quartz glass based on machine vision. *Meas J Int Meas Confed.* 2018; 128: 71–83. 10.1016/j.measurement.2018.06.031.

[44] Sarkar BR, Doloi B, Bhattacharyya B. Investigation on electrochemical discharge micro-machining of silicon carbide. *Int J Mater Form Mach Process.* 2017; 4: 29–44. 10.4018/ijmfmp.2017070103.

[45] Manna A, Narang V. A study on micro machining of e-glass-fibre-epoxy composite by ECSM process. *Int J Adv Manuf Technol.* 2012; 61: 1191–1197. 10.1007/s00170-012-4094-3.

[46] Antil P, Singh S, Manna A. Electrochemical discharge drilling of SiC reinforced polymer matrix composite using Taguchi's grey relational analysis. *Arab J Sci Eng.* 2018; 43: 1257–1266. 10.1007/s13369-017-2822-6.

[47] Antil P, Singh S, Manna A. Experimental investigation during electrochemical discharge machining (ECDM) of hybrid polymer matrix composites. *Iran J Sci Technol – Trans Mech Eng.* 2020; 44: 813–824. 10.1007/s4 0997-019-00280-5.

[48] Goud M, Sharma AK, Jawalkar C. A review on material removal mechanism in electrochemical discharge machining (ECDM) and possibilities

to enhance the material removal rate. *Precis Eng.* 2016; 45: 1–17. 10.101 6/j.precisioneng.2016.01.007.

[49] Antil P, Singh S, Singh S, Prakash C, Pruncu CI. Metaheuristic approach in machinability evaluation of silicon carbide particle/glass fiber-reinforced polymer matrix composites during electrochemical discharge machining process. *Meas Control* (United Kingdom). 2019; 52: 1167–1176. 10.1177/ 0020294019858216.

[50] Priti SM, Singh S. Micro-Machining of CFRP composite using electrochemical discharge machining and process optimization by Entropy-VIKOR method. *Mater Today Proc.* Jan 1; 2021; 44: 260–265. 10.1016/j.matpr. 2020.09.463.

[51] Antil P. Modelling and multi-objective optimization during ECDM of silicon carbide reinforced epoxy composites. *Silicon.* 2020; 12: 275–288. 10.1007/ s12633-019-00122-8.

[52] Singh M, Singh S. Machining of carbon fibre reinforced polymer composite by electrochemical discharge machining process. *IOP Conf Ser Mater Sci Eng.* 2019; 521: 012007. 10.1088/1757-899X/521/1/012007.

[53] Lescuras V, André JC, Lapicque F, Zouari I. Jet electrochemical etching of nickel in a sodium chloride medium assisted by a pulsed laser beam. *J Appl Electrochem.* 1995; 25: 933–939. 10.1007/BF00241587.

[54] Singh M, Singh S, Kumar S. Investigating the impact of LASER assistance on the accuracy of micro-holes generated in carbon fiber reinforced polymer composite by electrochemical discharge machining. *J Manuf Process.* 2020; 60: 586–595. 10.1016/j.jmapro.2020.10.056.

Chapter 9

Sustainable Manufacturing of Advanced Composites for Orthopaedic Application

Md Manzar Iqbal and Amaresh Kumar

Department of Production and Industrial Engineering, National Institute of Technology Jamshedpur, Jharkhand, India

Subhash Singh

Department of Mechanical and Automation Engineering, Indira Gandhi Delhi Technical University for Women, New Delhi, India

CONTENTS

9.1 INTRODUCTION

Millions of people are influenced by orthopaedic problems owing to accidental trauma, osteoporosis, and increased age [1]. Several orthopaedic issues are resolved without surgery requirement; nevertheless, they mostly need surgical operations with implant fixation at the fracture sites. Consequently, medical treatment associated with the bone is booming exponentially. Nowadays, bioinert orthopaedic metallic implants are employed for supporting purposes that cannot be degraded within the human body. However, it needs a minimum of two surgery. One for fixing the implant and the second for extraction of implant whenever fractured surface gets healed. Secondary surgery causes pain and extra financial burden to the patients. Therefore, researchers are endeavoring to cultivate the metallic implant that could be degraded in the human body after healing the fractured surface without causing any adverse impact.

DOI: 10.1201/9781003327370-9

Table 9.1 Comparative properties of tissue and orthopaedic metallic implant [3,4]

Material or tissue	Density g/cm³	Elastic modulus GPa	Yield stress MPa	Ultimate strength MPa
Cancellous bone	1.0–1.4	0.01–1.57	–	1.5–38
Cortical bone	1.8–2.0	5–23	130–180	35–283
Stainless steel	7.9–8.1	189–205	170–310	450–650
Titanium alloy	4.4–4.5	110–117	758–1117	930–1140
Mg alloy	1.74–2.0	41–45	65–100	135–285

Magnesium-based biodegradable materials are prominent for orthopaedic implants owing to their biodegradability in the human body [2]. The degradation product could be emitted from the human body via metabolism. Moreover, the mechanical properties of natural bone and several metallic implants are listed in Table 9.1. Moreover, the elastic modulus of Mg-based materials (1.74–2.0 g/cm³) corresponding to human bone (1.0–1.4 g/cm³) eliminates the stress shielding effect facilitating the orthopaedic implant. However, rapid or uncontrolled degradation restricted their application in the orthopaedic. This could be addressed through alloying, surface amendment, purification, and biodegradable magnesium matrix composites (BMMCs).

Among these elucidations, the BMMC technique is encouraging solutions due to its remarkable properties. The purpose of advanced composites is to enhance biocompatibility, alleviate the degradation rate, and avail the tunable mechanical properties [5]. Various types of reinforcement such as zirconium oxide (ZrO_2), hydroxyapatite (HA), tri-calcium phosphate (TCP), and titanium oxide (TiO_2) could be employed for orthopaedic applications [3]. HA is the most favorable reinforcement and most commonly used due to its crystallographic and chemical composition corresponding to the human bone [6,7]. Furthermore, it reduces the degradation rate of the Mg alloy by forming the protective layer of apatite [8]. The advanced composites could be fabricated through the solid-state process, liquid state process, and additive manufacturing. Among these, additive manufacturing (AM) processes are an expeditiously emerging novel sustainable fabrication process due to lower cost (reducing the multiple steps), minimum wastage of materials, and no harmful effect on the environment [9]. A few research works have been performed regarding the fabrication of Mg composite through AM, and it is still in the infant stage. This present book chapter is intended to illustrate an overview of the sustainable manufacturing process for orthopaedic implants. The fabrication of biodegradable Mg through the additive manufacturing process is highlighted. Furthermore, the changes faced during the process are addressed.

9.2 SUSTAINABLE MANUFACTURING

Humans have been using resources at a startling momentum these days, and we are unable to regenerate the material at that speedy rate because the process adopted is certainly not sustainable. If we use the same way, then our future generations will not be able to utilize those resources. The adverse impact of our utilization and its end effects on humankind are well known. Hence, decreasing the environmental influence of the manufacturing system and reducing resource consumption has become progressively more imperative [10]. Sustainable manufacturing (SM) is also referred to as green manufacturing. As per the U.S. Department of Commerce, sustainable manufacturing is stated as [11,12]: "The creation of manufactured products using processes that minimize negative environmental impacts, conserve energy and natural resources, are safe for employees, communities, and consumers, and are economically sound." The sustainability concept corresponds to the system stability. Six main elements significantly impact the sustainable manufacturing (SM) process, as depicted in Fig. 9.1. Waste management, energy consumption, and manufacturing cost are perceived as deterministic elements among these elements. However, operational safety, personal health, and environmental impact may not be exclusively recognized through manufacturing system parameters. Although, material and energy are essential inputs of the manufacturing process as well as systems. The outputs are considered as emissions and waste which are from the inputs in natural systems and industry. The effect of these can be felt economically, environmentally, and socially [13].

9.3 ADDITIVE MANUFACTURING

Additive manufacturing is a sustainable manufacturing process employed to fabricate the material at lower cost, minimum or no material wastage,

Figure 9.1 An overview of the elements of the sustainable manufacturing process [14].

and environmentally friendly. Additive manufacturing is also called 3-D printing and is employed to fabricate 3-D interconnected porous structures with uniform pore geometry [15]. It consists of layer-by-layer production processes from computer-aided design (CAD) models [16]. The kinds of AM [17] is represented in Fig. 9.2. AM could be categorized into two categories mainly based on the physical state of the raw materials: solid-, liquid-, and powder-based, and the second category is based on the mode of matter fused on a molecular level, which includes electron or laser beam ultraviolet light, thermal. The dimensional accuracy or net shape could be produced by powder-based fusion (PBF) [18]. Among the various AM techniques, the ones that sustain the orthopaedic applications are selective laser melting (SLS), selective laser sintering (SLS), electron beam melting (EBM), and wire arc additive manufacturing (WAAM).

These techniques yield enormously dense materials without the need for any further post-processing and comparable properties with other conventional processes. The SLM process could be employed to produce any materials. In SLM, very high energy is utilized to heat and melt the powdered material depicted in Fig. 9.3. A protective environment is provided for controlling or reducing the degradation and oxidation of the material.

However, EBM could be employed for producing only metals. A very high energy density electron beam is utilized to produce a highly dense, voids-free product. Although, the mechanical properties attained through friction stir additive manufacturing (FSAM) are more than conventional processes. However, WAAM is one of the emerging AM techniques, consisting of a combination of an electric arc to produce a heat source and a wire acts as a feedstock. The product could be achieved through depositions of weld beads layer by layer through arc welding technology.

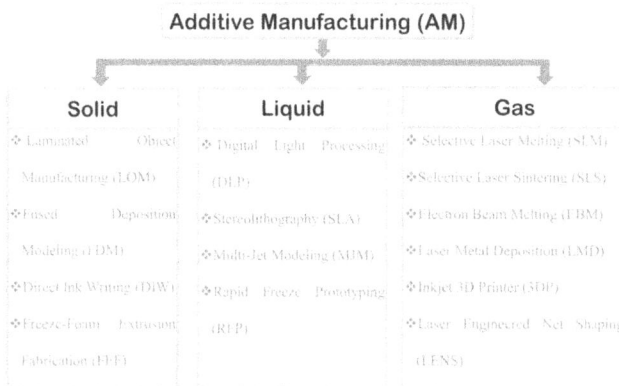

Figure 9.2 Categorization of additive manufacturing (AM) based on the raw material [17,19,20].

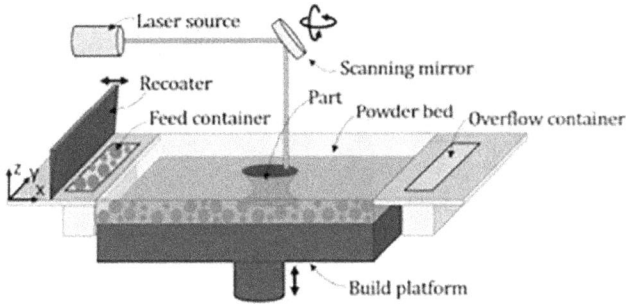

Figure 9.3 Schematic view of selective laser melting (SLS) machine setup [17].

9.4 NEED FOR ADDITIVE MANUFACTURING (AM) IN ORTHOPAEDIC APPLICATIONS

Bone is a complex, dense connective tissue that provides support and protects the organs. It is a natural composite comprised of inorganic and organic materials. The inorganic materials are primarily present in the form of hydroxyapatite (HA), with calcium and phosphorus along with some trace amount of Mg, potassium (K), chlorine (Cl), fluoride (F), and carbonate (CO_3^{2-}), and a few amounts of zinc (Zn), silicon (Si), copper (Cu), and iron (Fe) which provide strength to the bone [21]. The various constituents of bone are listed in Table 9.2. Moreover, natural bone mostly exhibited a multi-scale structure classified into the cancellous bone and cortical bone, as shown in Fig. 9.4. The main important part of the bone is cortical bone, situated around the surface of the bone. It comprises 99% of the calcium (Ca) and 90% of the phosphate in the human body. It is comparatively solid and dense, having 5% to 10% porosity [22]. However, cancellous bone is a spongy-like structure and circulated inside the bone. It has 50% to 90% porosity, and the surface area is about 20 times the cortical bone [23].

Table 9.2 The constituents of human bone [24,25]

Inorganic phase	Weight %	Organic phase	Weight %
Hydroxyapatite (HA)	60	Collagen	20
Carbonate	4	Water	9
Citrate	0.9	Non-collagenous proteins	3
Sodium (Na)	0.7	Other trace elements: lipids, polysaccharides, cytokines	
Magnesium (Mg)	0.5	Primary bone cells: osteoclast, osteocytes, osteoblast	
Other trace elements	Zn^{2+}, Fe^{2+}, Cl^-, F^-, K^+, Sr^{2+}, Pb^{2+}, Cu^{2+}		

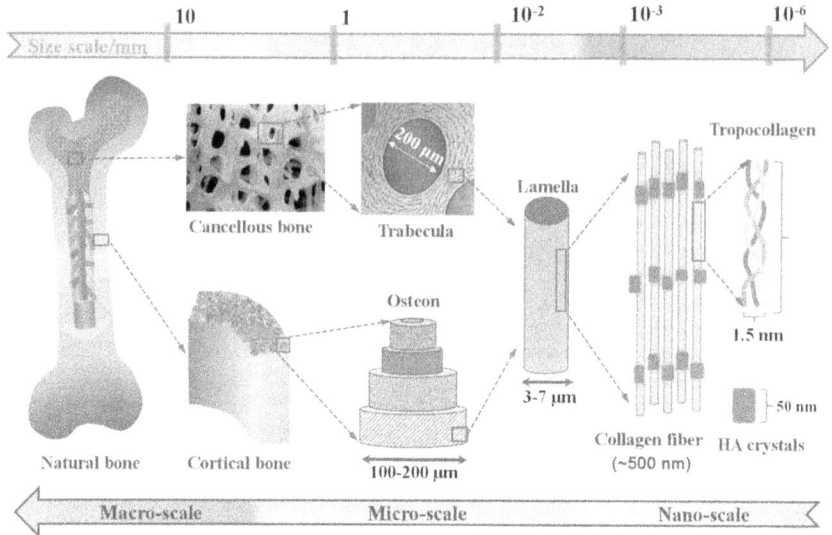

Figure 9.4 Overview of natural bone [21].

The structure and composition of bone varied according to the age, living standards of patients, genetic inheritance, and defect site addressed in various requirements for orthopaedic implants [26]. Hence, it is still a great confront to produce bone biomaterial that could fulfill all bone repair criteria. Bone biomaterial plays a vital role and links between native tissues and seeded cells [27]. The material and structure of biomaterials such as porosity play an essential role in regenerating tissue [28].

Additive manufacturing (AM), or three-dimensional printing (3DP), is a newer kind of emerging fabrication process of composite materials for orthopaedic application owing to produce complex geometry shapes that are impossible or very difficult to fabricate by other conventional processes. The intricate shape produced by AM process renders it promotes bone regeneration, cell proliferation, and growth [29–31]. Moreover, the fabricated implant corresponds to human anatomy geometry. The comparative study of AM and conventional manufacturing process are listed in Table 9.3. The conventional process takes multiple steps to produce the implant accountable for increasing the manufacturing time and cost, which is mostly eliminated in AM.

9.5 ADDITIVE MANUFACTURING FOR MG-BASED COMPOSITE

The biodegradable metal is prominently employed in orthopaedic applications. Although the fabrication of a 3-D porous structure with similar

Table 9.3 Comparative study of additive manufacturing and conventional manufacturing process [32]

S. No.	Parameters	Additive Manufacturing	Conventional manufacturing process
1	Time	AM takes significantly less time because it makes the product directly from the CAD model. Therefore, it minimized the time taken during the fabrication step, supply chain as well as inventory.	Manufacturing process times are more than AM due to dependency on dies, molds, inventory, etc.
2	Cost	The cost of the product is low when the product is produced in either small and medium batches.	The processes are very costly when the product is produced in batches.
3	Wastage of material	Wastage of material in AM process tends significantly less to none. It could be recycled.	Wastage of materials is very high in machining, finishing and is responsible for adverse environmental effects.
4	Post-production processing	There is less or no chance of post-production processing associated with the raw materials and fabrication process.	Post-production processing is mainly needed.
5	Product complexity	Complex or intricate shapes can be manufactured. It relies on the engineer's imagination.	It is challenging to make a complex shape. All the parts are fabricated individually and then assembled.

macro and micro geometry corresponds to natural bone, it is still very challenging through the conventional process such as casting, powder metallurgy, foaming, sintering. The conventional process can only produce uncontrollable porous structure means randomly oriented pore size, which could hamper the stiffness, mechanical strength, and permeability of the implant. Moreover, the non-uniformity of porous geometry could not demonstrate homogeneous mechanical and biological properties of the implant at micro-scale resulting in chances of implant failure due to twisting and bending [33]. However, AM possesses customized and uniformly oriented interconnected porous structures to enhance the remarkable biological and mechanical properties. Selective laser melting (SLM) is considered a laser-assisted additive manufacturing approach that could be engineered near-net shape products as per the CAD model without any moulds or tools in a comparatively short cycle [34,35]. In SLM, a small quantity of molten pools and heat-affected zones (HAZ) is encouraged using a fast-scanning laser beam under the monitoring of the galvanometer.

Consequently, high cooling rates $(10^3-10^4$ K/s) are generated inside the molten pools [36,37]. Therefore, grain growth could be restricted by an instant solidification at such a high cooling rate [38]. High cooling rates enable grain refinements and are accountable for enhancing vital properties required in medical implants. Therefore, SLM is a helpful process for fabricating the magnesium based materials. However, even at bulk quantity, Mg powders are highly reactive to oxygen [29]. The bulk powder form employed in SLM prominently upsurges the surface area, resulting in an explosion and exothermic reaction [39–41]. It is suggested to provide an oxidation layer on the surface of Mg to certify the safe handling. Therefore, very few researchers have been reported in this regard. The first analysis on AM was conveyed by Ng et al. [42]. They have successfully produced the Mg track by the SLM process. Moreover, the influence of the irradiation mode for the melting of pure Mg was identified. It was found that surface morphology, dimension, and oxygen pick up of the laser melted tracks firmly rely on process parameters and irradiation mode. SLM process is "net forming" of the product due to laser application to produce the product by placing layer by layer of powder [43]. The effect of process parameters on the Mg-HA composite was examined [44].

In this context, Mg (in micron size) and HA (in nano size) have been fabricated through SLM. As scan speed increased, severe particulate agglomeration was found in the micrograph. However, as laser intensity resulted in oxidation and evaporation of Mg during solidification leads to the formation of cauliflower-like grains. Therefore, an optimum quantity of two parameters was investigated for obtaining the condensed part. Shuai et al. [45] have examined the upshot of hydroxyapatite (HA) in Mg-Zn alloy using SLM. The samples were fabricated in multi-tracks having numerous layers by layer using the protective environment of argon gas. It was observed that grain size decreased and protective layers of bone-like apatite exhibited by incorporating HA, which is attributed to the slowing down the degradation rate of composites. Moreover, the incorporation of HA inhibited grain growth due to the increment of nucleation particles during solidification. Moreover, hardness increased due to the strengthening of grain as well as second-phase strengthening. HA particles are uniformly distributed in Mg alloy up to optimum quantity, and further increment causes agglomeration of HA particles in Mg alloy, as illustrated in Fig. 9.5. Furthermore, the homogeneity of HA has also been assessed by elemental mapping, as revealed in Fig. 9.6.

The morphology of the composite immersed in simulated body fluid (SBF) was studied, which is depicted in Fig. 9.7. It was visible that localized corrosion having deep corrosion happened in Mg-3Zn. The incorporation of HA resulted in spherical precipitates encouraged along with localized

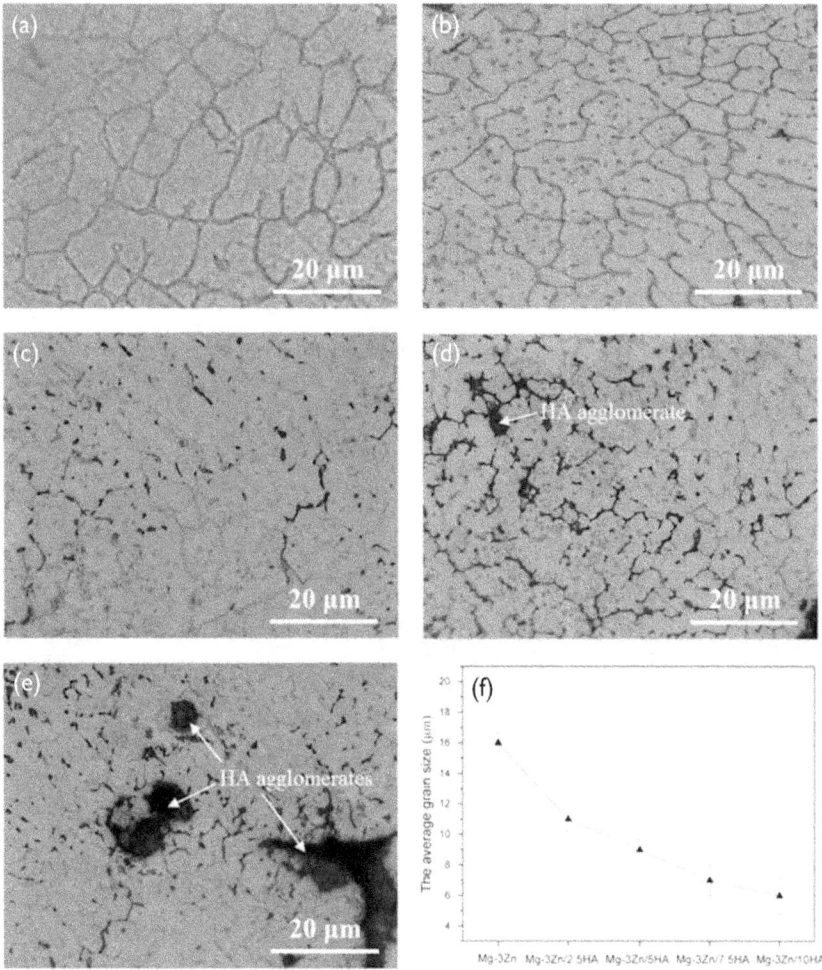

Figure 9.5 An overview of the crystal structure of the composites with varying the percentage of HA: (a) Mg-3Zn alloy, (b) Mg-3Zn/2.5HA composite, (c) Mg-3Zn/5HA composite, (d) Mg-3Zn/7.5HA composite, (e) Mg-3Zn/10HA composite, and (f) schematic representation of average grain size [45].

corrosion. However, excessive quantity of HA agglomerated on the surface of the composite resulted in the formation of some pores, which could be attributed to the decreasing the corrosion resistance.

The average hardness value was significantly increased as the content of HA increased, as shown in Fig. 9.8. This could be due to the refinement of grain that occurred as per the Hall-Petch equation.

Furthermore, HA acts as a barrier to dislocation movement. The degradation behaviour of the composite was displayed in Fig. 9.9. In the initial stage, the composite directly interacted with the SBF solution.

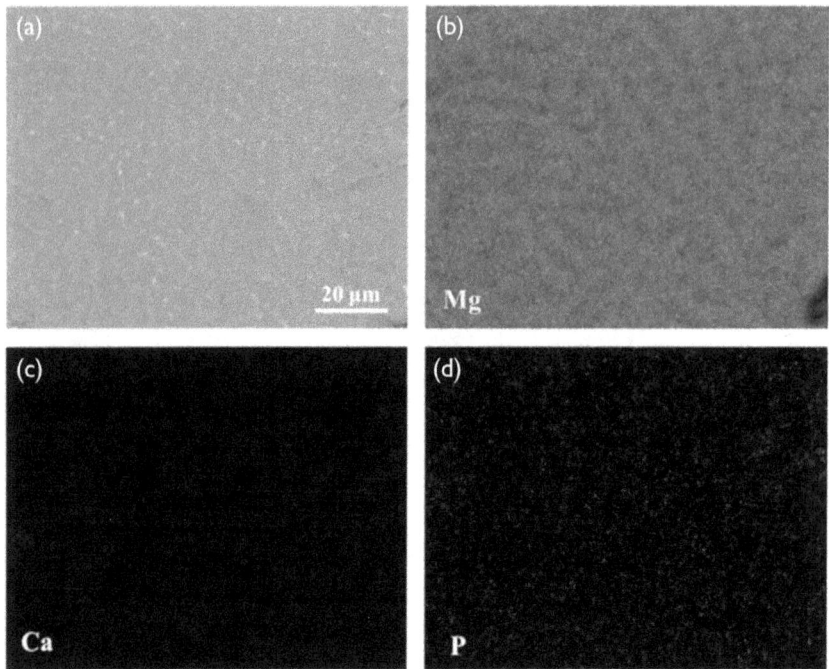

Figure 9.6 Elemental mapping of various constituents of composites: (a) Scanning electron microscopy image in back scattered electron mode, (b) magnesium, (c) calcium, and (d) Phosphorous [45].

After some interval of time, the reaction starts between the fluid and composite, resulted in the formation of Mg^{2+} and H_2. When the immersion time increased, Mg^{2+} combined with OH^-, the solution formed a protective film of $Mg(OH)_2$. However, the protective layer reacts with Cl^- and deformed into $MgCl_2$. The corrosion of composite has been slowly initiated. The Mg-Zn/HA composites could be promising candidates for the orthopaedic implant.

It is still a challenge to enhance the corrosion resistance of Mg alloy. For enhancing the corrosion resistance of Mg-6Zn-1Zr (ZK60), β-tricalcium phosphate (β-TCP) was incorporated by laser rapid solidification process [46]. Uniform distribution of β-TCP was found at the surface of the Mg alloy, which is responsible for the formation of the apatite layer at the surface of Mg. Furthermore, the apatite layer promoted the layer of compact corrosion product attributed to the hindering of the degradation rate. Moreover, the composite exhibited excellent cytocompatibility and enhanced mechanical properties. The composite can be employed as an implant in orthopaedic applications. In SLM, the chances of defects such as impurities, porosity, and segregation are minimal due to remelting the upper layer when placing the layer-by-layer

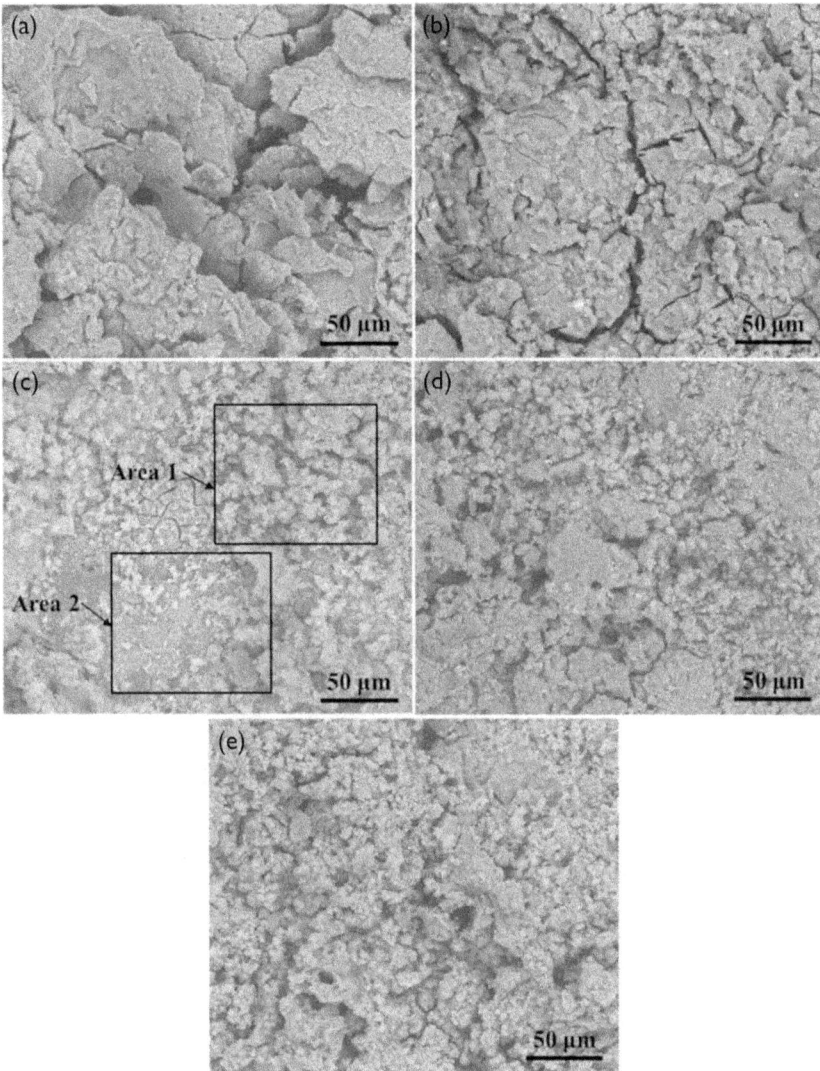

Figure 9.7 Effect of the morphology of the surface of Magnesium alloy immersed in SBF solution for 48 hours after addition of a variable percentage of HA: (a) Mg-3Zn alloy without HA, (b) Mg-3Zn/2.5HA composite, (c) Mg-3Zn/5HA composite, (d) Mg-3Zn/7.5HA composite, and (e) Mg-3Zn/10HA composite [45].

powder, which is attributed to the homogenization of the defects. Moreover, the development of very fine grains due to high melting speed resulted in a higher degree of supersaturation addressed to enhanced mechanical properties. Hence, the research of the SLM process for Mg-based material has significant attention outlooks [47].

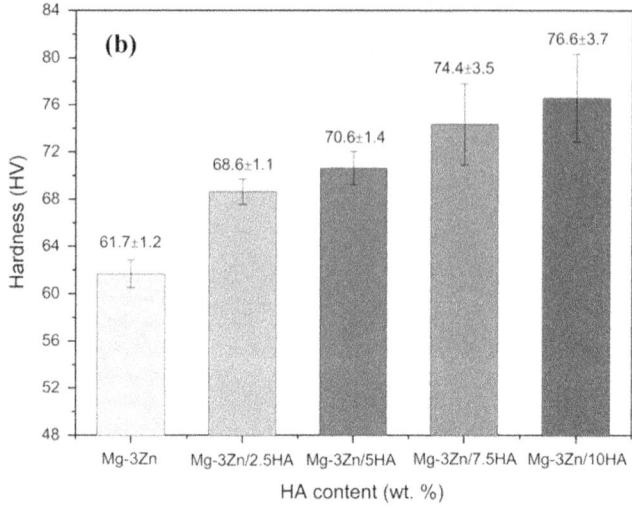

Figure 9.8 Influence of incorporation of HA at various percentages on Mg alloy in hardness value [45].

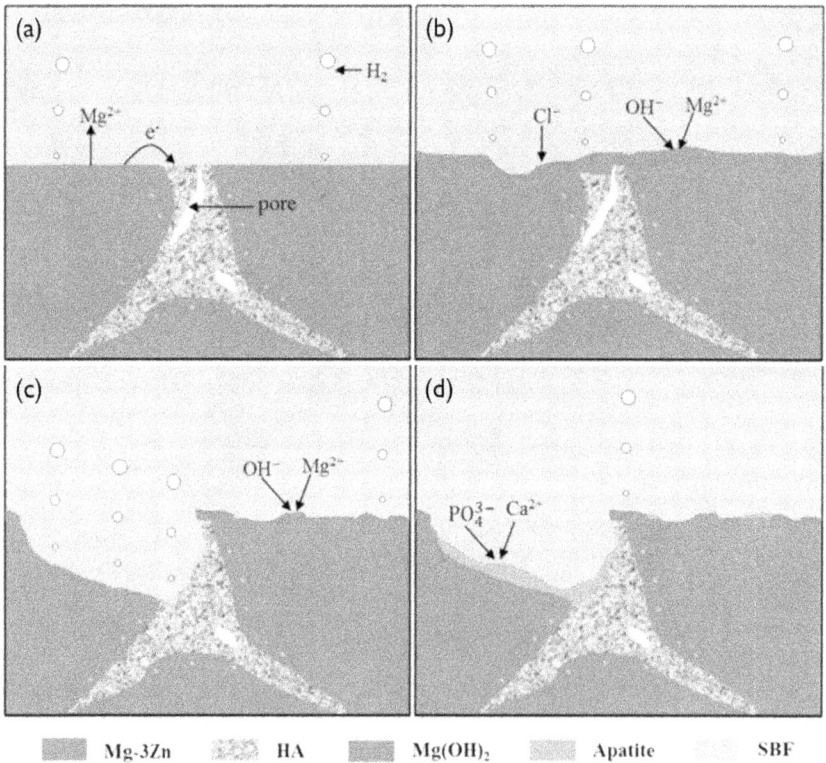

Figure 9.9 An overview of the illustration of the degradation behaviour of the Mg-based composites [45].

9.6 CONCLUSION AND FUTURE SCOPE

Sustainable manufacturing is the utmost significant aspect to be pondered by all production engineers, not because it is a trend but a need for gratitude to the sphere where we survive. Nowadays, product life cycle evaluation has to turn into a tool of choice to find the environmental influence of the products that we fabricate. Moreover, different parts within manufacturing can be helped significantly by implementing green manufacturing attempts. Nonetheless, the three fundamental principles to be perceived are the usage of environmentally friendly materials, reducing the resource utilization in the process, and reducing the waste material by recycling or reusing it as much as possible. Additive manufacturing is a kind of sustainable manufacturing process that is employed to fabricating the material. Magnesium-based composite can manufacture parts for the orthopaedic applications due to their biodegradability, superior bio-compatibility, non-toxicity, and similar elastic modulus to human bone. The additive manufacturing (AM) may facilitate removing the hurdles confronted when fabricating a Mg composite. The process of shaping Mg composite employed to date, such as casting, extrusion, forging, and powder metallurgy, has many restrictions, such as difficulties encountered to fabricate complex geometry, aging methods, and ignition of chips during machining deteriorates the mechanical and biological properties of the product. In noteworthy selective laser melting (SLM), the use of the AM process enables the Mg composite in a specific shape and complex porous structure in a single technology. Moreover, rapid solidification is attributed to the homogeneity of composite and inhibits grain growth, resulting in enhancement of mechanical properties and corrosion resistance. The main limitations in SLM are thermal fluctuation, oxidation, crack formation, balling, and loss of alloying elements. In order to overcome these inherent difficulties, researchers have used numerous approaches, which include melting the powders in helium or argon gas environment, optimum quantity of process parameters, and scanning strategy for minimizing the balling and preheating, and application of the overpressure process chamber that regulates the vaporization of powders. Since the structure and composition of human bone varies according to age, health conditions, and living standards, to speed up the clinical application of orthopaedic implants, much interdisciplinary research must be performed.

ACKNOWLEDGMENTS

The authors gratefully acknowledge the Ministry of Human Resource and Development (MHRD), Government of India, and National Institute of Technology, Jamshedpur, for providing financial support.

REFERENCES

[1] Yadav, D., Garg, R.K., Ahlawat, A., Chhabra, D. 3D printable biomaterials for orthopaedic implants: Solution for sustainable and circular economy. *Resour. Policy.* 68, 101767 (2020). 10.1016/j.resourpol.2020.101767

[2] Iqbal, M.M., Kumar, A., Singh, S. Biodegradable composite materials for orthopaedic implant: A review. In: AIP Conference Proceedings. p. 040031 (2021).

[3] Bommala, V.K., Krishna, M.G., Rao, C.T. Magnesium matrix composites for biomedical applications: A review. *J. Magnes. Alloy.* 7, 72–79 (2019). 10.1016/j.jma.2018.11.001

[4] Yusop, A.H., Bakir, A.A., Shaharom, N.A., Abdul Kadir, M.R., Hermawan, H. Porous biodegradable metals for hard tissue scaffolds: A review. *Int. J. Biomater.* 2012, 1–10 (2012). 10.1155/2012/641430

[5] Prasad, S.V.S., Prasad, S.B., Verma, K., Mishra, R.K., Kumar, V., Singh, S. The role and significance of magnesium in modern day research – A review. *J. Magnes. Alloy.* 10(1), 1–61 (2021). 10.1016/j.jma.2021.05.012

[6] Duer, M.J. The contribution of solid-state NMR spectroscopy to understanding biomineralization: Atomic and molecular structure of bone. *J. Magn. Reson.* 253, 98–110 (2015). 10.1016/j.jmr.2014.12.011

[7] Prasad, S.V.S., Singh, S., Prasad, S.B. A review on the corrosion process in magnesium. In: AIP Conference Proceedings. p. 040008 (2021).

[8] Campo, R. del, Savoini, B., Muñoz, A., Monge, M.A., Garcés, G. Mechanical properties and corrosion behavior of Mg–HAP composites. *J. Mech. Behav. Biomed. Mater.* 39, 238–246 (2014). 10.1016/j.jmbbm.2014.07.014

[9] Saleh Alghamdi, S., John, S., Roy Choudhury, N., Dutta, N.K. Additive manufacturing of polymer materials: Progress, promise and challenges. *Polymers* (Basel) 13, 753 (2021). 10.3390/polym13050753

[10] Posinasetti, N. Sustainable manufacturing: Principles, applications and directions. https://www.industr.com/en/sustainable-manufacturing-principles-applications-and-directions-2333598.

[11] Gupta, K., Laubscher, R.F. Sustainable machining of titanium alloys: A critical review. *Proc. Inst. Mech. Eng. Part B J. Eng. Manuf.* 231, 2543–2560 (2017). 10.1177/0954405416634278

[12] Jayal, A.D., Badurdeen, F., Dillon, O.W., Jawahir, I.S. Sustainable manufacturing: Modeling and optimization challenges at the product, process and system levels. *CIRP J. Manuf. Sci. Technol.* 2, 144–152 (2010). 10.1016/j.cirpj.2010.03.006

[13] Haapala, K.R., Zhao, F., Camelio, J., Sutherland, J.W., Skerlos, S.J., Dornfeld, D.A., Jawahir, I.S., Clarens, A.F., Rickli, J.L. A review of engineering research in sustainable manufacturing. *J. Manuf. Sci. Eng.* 135, 1–16 (2013). 10.1115/1.4024040

[14] Haapala, K.R., Zhao, F., Camelio, J., Sutherland, J.W. A review of engineering research in sustainable manufacturing. In: Proceedings of the ASME 2011 International Manufacturing Science and Engineering Conference MSEC2011 MSEC2011. Proceedings of the ASME 2011 International Manufacturing Science and, Corvallis, Oregon, USA (2014).

[15] Sing, S.L., An, J., Yeong, W.Y., Wiria, F.E. Laser and electron-beam powder-bed additive manufacturing of metallic implants: A review on processes, materials and designs. *J. Orthop. Res.* 34, 369–385 (2016). 10.1002/jor.23075

[16] Wang, Z., Wu, W., Qian, G., Sun, L., Li, X., Correia, J.A.F.O. In-situ SEM investigation on fatigue behaviors of additive manufactured Al-Si10-Mg alloy at elevated temperature. *Eng. Fract. Mech.* 214, 149–163 (2019). 10.1016/j.engfracmech.2019.03.040

[17] Abdulhameed, O., Al-Ahmari, A., Ameen, W., Mian, S.H. Additive manufacturing: Challenges, trends, and applications. *Adv. Mech. Eng.* 11, 168781401882288 (2019). 10.1177/1687814018822880

[18] Wei, K., Gao, M., Wang, Z., Zeng, X. Effect of energy input on formability, microstructure and mechanical properties of selective laser melted AZ91D magnesium alloy. *Mater. Sci. Eng. A.* 611, 212–222 (2014). 10.1016/j.msea.2014.05.092

[19] Bikas, H., Stavropoulos, P., Chryssolouris, G. Additive manufacturing methods and modelling approaches: A critical review. *Int. J. Adv. Manuf. Technol.* 83, 389–405 (2016). 10.1007/s00170-015-7576-2

[20] Huang, S.H., Liu, P., Mokasdar, A., Hou, L. Additive manufacturing and its societal impact: A literature review. *Int. J. Adv. Manuf. Technol.* 67, 1191–1203 (2013). 10.1007/s00170-012-4558-5

[21] Gao, C., Peng, S., Feng, P., Shuai, C. Bone biomaterials and interactions with stem cells. *Bone Res.* 5, 17059 (2017). 10.1038/boneres.2017.59

[22] Nguyen, L.H., Annabi, N., Nikkhah, M., Bae, H., Binan, L., Park, S., Kang, Y., Yang, Y., Khademhosseini, A. Vascularized bone tissue engineering: Approaches for potential improvement. *Tissue Eng. Part B Rev.* 18, 363–382 (2012). 10.1089/ten.teb.2012.0012

[23] McKittrick, J., Chen, P.-Y., Tombolato, L., Novitskaya, E.E., Trim, M.W., Hirata, G.A., Olevsky, E.A., Horstemeyer, M.F., Meyers, M.A. Energy absorbent natural materials and bioinspired design strategies: A review. *Mater. Sci. Eng. C.* 30, 331–342 (2010). 10.1016/j.msec.2010.01.011

[24] Murugan, R., Ramakrishna, S. Development of nanocomposites for bone grafting. *Compos. Sci. Technol.* 65, 2385–2406 (2005). 10.1016/j.compscitech.2005.07.022

[25] Shirdar, M.R., Farajpour, N., Shahbazian-Yassar, R., Shokuhfar, T. Nanocomposite materials in orthopaedic applications. *Front. Chem. Sci. Eng.* 13, 1–13 (2019). 10.1007/s11705-018-1764-1

[26] Hanson, M.A., Gluckman, P.D. Early developmental conditioning of later health and disease: Physiology or pathophysiology? *Physiol. Rev.* 94, 1027–1076 (2014). 10.1152/physrev.00029.2013

[27] Webber, M.J., Khan, O.F., Sydlik, S.A., Tang, B.C., Langer, R. A perspective on the clinical translation of scaffolds for tissue engineering. *Ann. Biomed. Eng.* 43, 641–656 (2015). 10.1007/s10439-014-1104-7

[28] Chaudhari, A., Vig, K., Baganizi, D., Sahu, R., Dixit, S., Dennis, V., Singh, S., Pillai, S. Future prospects for scaffolding methods and biomaterials in skin tissue engineering: A review. *Int. J. Mol. Sci.* 17, 1974 (2016). 10.3390/ijms17121974

[29] Sezer, N., Evis, Z., Koç, M. Additive manufacturing of biodegradable magnesium implants and scaffolds: Review of the recent advances and

research trends. *J. Magnes. Alloy.* 9, 392–415 (2021). 10.1016/j.jma.202 0.09.014

[30] Do, A.-V., Khorsand, B., Geary, S.M., Salem, A.K. 3D printing of scaffolds for tissue regeneration applications. *Adv. Healthc. Mater.* 4, 1742–1762 (2015). 10.1002/adhm.201500168

[31] Hutmacher, D.W. Scaffold design and fabrication technologies for engineering tissues—state of the art and future perspectives. *J. Biomater. Sci. Polym. Ed.* 12, 107–124 (2001). 10.1163/156856201744489

[32] Joshi, S.C., Sheikh, A.A. 3D printing in aerospace and its long-term sustainability. *Virtual Phys. Prototyp.* 10, 175–185 (2015). 10.1080/1745275 9.2015.1111519

[33] Noor, M.F., Hasan, F., Bhardwaj, S., Hasan, S. Reverse engineering in customization of products: Review and case studyIn Ergonomics for Improved Productivity. *Design Science and Innovation.* Springer, Singapore, (2021), pp. 587–592.

[34] Yadroitsev, I., Krakhmalev, P., Yadroitsava, I., Johansson, S., Smurov, I. Energy input effect on morphology and microstructure of selective laser melting single track from metallic powder. *J. Mater. Process. Technol.* 213, 606–613 (2013). 10.1016/j.jmatprotec.2012.11.014

[35] Yadroitsev, I., Gusarov, A., Yadroitsava, I., Smurov, I. Single track formation in selective laser melting of metal powders. *J. Mater. Process. Technol.* 210, 1624–1631 (2010). 10.1016/j.jmatprotec.2010.05.010

[36] Wang, L., Felicelli, S., Gooroochurn, Y., Wang, P.T., Horstemeyer, M.F. Optimization of the LENS® process for steady molten pool size. *Mater. Sci. Eng. A.* 474, 148–156 (2008). 10.1016/j.msea.2007.04.119

[37] Vilaro, T., Colin, C., Bartout, J.D. As-Fabricated and heat-treated microstructures of the Ti-6Al-4V alloy processed by selective laser melting. *Metall. Mater. Trans. A.* 42, 3190–3199 (2011). 10.1007/s11661-011-0731-y

[38] Xu, R., Zhao, M.-C., Zhao, Y.-C., Liu, L., Liu, C., Gao, C., Shuai, C., Atrens, A. Improved biodegradation resistance by grain refinement of novel antibacterial ZK30-Cu alloys produced via selective laser melting. *Mater. Lett.* 237, 253–257 (2019). 10.1016/j.matlet.2018.11.071

[39] Tan, X., Tan, Y.J. 3D Printing of metallic cellular scaffolds for bone implants. In: *3D and 4D printing in biomedical applications*, pp. 297–316. Wiley-VCH Verlag GmbH & Co. KGaA, Weinheim, Germany (2018).

[40] Zumdick, N.A., Jauer, L., Kersting, L.C., Kutz, T.N., Schleifenbaum, J.H., Zander, D. Additive manufactured WE43 magnesium: A comparative study of the microstructure and mechanical properties with those of powder extruded and as-cast WE43. *Mater. Charact.* 147, 384–397 (2019). 10.1016/ j.matchar.2018.11.011

[41] Qin, Y., Wen, P., Guo, H., Xia, D., Zheng, Y., Jauer, L., Poprawe, R., Voshage, M., Schleifenbaum, J.H. Additive manufacturing of biodegradable metals: Current research status and future perspectives. *Acta Biomater.* 98, 3–22 (2019). 10.1016/j.actbio.2019.04.046

[42] Chung Ng, C., Savalani, M., Chung Man, H. Fabrication of magnesium using selective laser melting technique. *Rapid Prototyp. J.* 17, 479–490 (2011). 10.1108/13552541111184206

[43] Olakanmi, E.O. Selective laser sintering/melting (SLS/SLM) of pure Al, Al–Mg, and Al–Si powders: Effect of processing conditions and powder

properties. *J. Mater. Process. Technol.* 213, 1387–1405 (2013). 10.1016/j. jmatprotec.2013.03.009

[44] Zhang, C., Lin, T., Xiaochun, L. Parametric study on selective laser melting of magnesium micropowder with hydroxyapatite (HA) nanoparticles. Univ. Calif. Los Angeles, CA, USA (2015).

[45] Shuai, C., Zhou, Y., Yang, Y., Feng, P., Liu, L., He, C., Zhao, M., Yang, S., Gao, C., Wu, P. Biodegradation resistance and bioactivity of hydroxyapatite enhanced Mg-Zn composites via selective laser melting. *Materials* (Basel). 10, 307 (2017). 10.3390/ma10030307

[46] Deng, Y., Yang, Y., Gao, C., Feng, P., Guo, W., He, C., Chen, J., Shuai, C. Mechanism for corrosion protection of β-TCP reinforced ZK60 via laser rapid solidification. *Int. J. Bioprinting.* 4, 1–11 (2017). 10.18063/ijb. v4i1.124

[47] Zhang, W., Wang, L., Feng, Z., Chen, Y. Research progress on selective laser melting (SLM) of magnesium alloys: A review. *Optik* (Stuttg). 207, 163842 (2020). 10.1016/j.ijleo.2019.163842

Chapter 10

Performance Evaluation and Characterization of EDM While Machining of AISI202 Using Cu-Cr-Zr Electrode

Subhash Singh

Department of Mechanical and Automation Engineering, Indira Gandhi Delhi
Technical University for Women, Delhi, India

Rajneesh Raghav

Department of Mechanical and Industrial Engineering, IIT Roorkee,
Uttarakhand, India

Girija N. Arka

Department of Production and Industrial Engineering, NIT Jamshedpur,
Jharkhand, India

Rahul S. Mulik and Kaushik Pal

Department of Mechanical and Industrial Engineering, IIT Roorkee,
Uttarakhand, India

CONTENTS

DOI: 10.1201/9781003327370-10

10.1 INTRODUCTION

Generally manufacturing industries face difficulty in machining hard material by conventional machining due to their conventional approach of performing machining. An advanced machining process is preferred in order to overcome certain limitations of conventional machining related to cutting tool, machining parameters, and complex nature of job. Electro discharge machining, known as EDM, was found suitable to do machining of such striving materials due to melting and vaporizing technology of removing material and hence it has a great scope of performing machining [1]. The EDM basic principle of working architecture is illustrated in Fig. 10.3. The quality characteristics of EDM, pulse current I, pulse on-time T_{on}, and pulse off-time T_{off} are responsible for producing quality discharge at an inter electrode gap in the presence of a dielectric fluid environment, ensuring industrial productive requirements. It adopts a simple approach of forming a pulsed spark at the inter-electrode gap by breaking the dielectric strength of surrounded dielectric fluid using pulse generator whose function is to convert continuous AC to pulse DC [2].

Since different materials require different machining conditions of electrical energy variables to get thermal erosion, because of that electrical process parameters are required to be optimized [3]. During machining, there is always an optimum gap maintained between the electrode and workpiece and is monitored by a servo controller, thus eliminating possibilities of mechanical stress or any type of vibration [4]. Owing to its unconventional nature of doing machining, this can be used in many highly demanded application materials like ceramics, heat-treated materials, case-hardened materials, and composites [1,5,6]. This also can be applied to machine extrusion and forging dies along with other operations like drilling, milling, and grinding.

If focused on EDM, the electrode reflected as a vital element from which a spark energized and is designed by considering good machinability, electrically conductive, thermally conductive, and cost-effective prosperity [7]. By considering all relevant priority, copper or graphite tool is most commonly used in EDM where copper is found superior to graphite as concerned with EDM performance [8]. However, ZrB_2-Cu alloy and ZrB_2-CuNi alloy electrodes were experimentally found as better alternatives compared to cupper [9,10]. Apart from that, literature reflecting the quality characteristic "electrode wear rate (EWR)" is one of the failure conditions, hampering the machining performance and leading to a defect production despite having a higher material removal rate

(MRR). Apart from electrode material, machining parameters also contribute a lot for the performance measures. Experimentally and statistically, machining parameters I and T_{on} are found to be most prominent parameters to deuterate EWR and parallelly improve MRR [11,12]. However, the deionization time of dielectric fluid during T_{off} duration has remarkable effect on electrode erosion rate as well as on the removal of material [13]. Review of EDM manifests that several efforts have been attempted by applying new technology to prevail over and improve the machining performance by bringing down the EWR minimum and maximize MRR [14].

Along with this, machined surface characteristic in terms of surface roughness Ra also has been essential to improve EDM performance. Therefore, Kiyak et al. attempted to analyze the effects of process parameter to machine tool steel with a cupper electrode and experimentally found I_P and T_{on} are the best contributors towards primary surface texture of a copper tool as well as for the tool steel [15]. Researcher also applied composite technology to develop electrode. Tsai et al. had worked out on a Cr/Cu composite-based tool. Based on their observation, they arrived at a conclusion that an increase in MRR recorded when electrode connected negative but was detrimental to the interests of surface finish [16]. Bhaumik M et al. attempted a cryotreated tungsten carbide electrode along with sic powder mixed flushing to improve EDM performance while machining of AISI304. Their observation reflected an improved machining efficiency due to a combined effect of cryotreated electrode with sic powder flushing [17]. When sic abrasive powder blended with dielectric fluid on EDM, it improved surface characteristics of H-11 die steel [18]. However, the gap voltage of EDM machine parameter had adverse effects on MRR while machining EN19 steel [19]. To improve surface quality, air and argon gas flushing attempted and observed uniform recast layer formation with absence of potholes and cracks on Ti_6Al_4V alloy [20]. A researcher team, Lee et al., had investigated the interests of surface crack and applied a full factorial design approach to get the significant effects of EDM parameters [21].

From the past work done on EDM, it has been observed that several efforts and techniques have been put together to bring down the EWR minimum and improve its productiveness. Owing to the fact that copper has a high electrical conduciveness property and zirconium has a high temperature strength with softening temperature to 300°C without hampering conductivity, a new electrode was set. AISI 202 stainless steel has wide applications in the automotive industry, process industry machinery, electrical machinery, petrochemical, chemical, and aviation fields [22] because of its corrosion resistance property. Despite of its use in making chemical containers, kitchen appliances, bolts, etc., an insufficient study was reported from literature. Hence, an experimental investigation was conducted to get important findings of EDM performance and its scope by creating quality discharge between a Cu-Cr-Zr alloy electrode and AISI 202 stainless-steel work. The performance potentials measures were evaluated by means of MRR, EWR, and R_a with respect to I, T_{on}, and T_{off} machining parameters. THe central composite design of response surface methodology (RSM) was used to design the experiment and predict optimal machining parameters to satisfy all responses parallelly: MRR, EWR, and R_a with equal importance. The EDM performance measures were interpreted with technical clarity by scanning electron microscope (SEM), atomic force microscopy (AFM), and x-ray diffraction (XRD) technology.

10.2 EXPERIMENTAL DETAILS

10.2.1 Electrode Details

For the experimental investigation, a copper-based alloy Cu-Cr-Zr (copper-chromium-zirconium) was selected as an electrode and whose elements are presented in Table 10.1 and its EDS image represented over Fig. 10.1 ensured no other compound present on it. Individual chemical constituents of Cu-Cr-Zr have their own contribution towards the alloy. The technicality of choosing such an electrode is due to of its high thermal and electrical conductive property as a copper element that is highly electrically conductive and zirconium adds strength to it with a softening temperature to 300°C without hampering conductivity.

Table 10.1 Electrode composition

Element	Copper	Chromium	Zirconium	Carbon	Oxygen
Wt. %	86.50	0.65	0.20	8.25	4.40

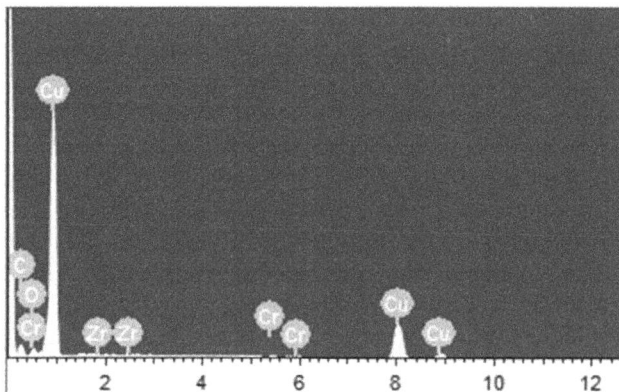

Figure 10.1 EDS figure of Cu-Cr-Zr.

Table 10.2 AISI 202 grade stainless-steel elements

Element	Carbon	Silicon	Manganese	Phosphorus	Sulfur	Nickel	Chromium	Iron
%es of Wt	≤0.15	≤1	7.50–10	≤0.060	≤0.030	4–6	17–19	68

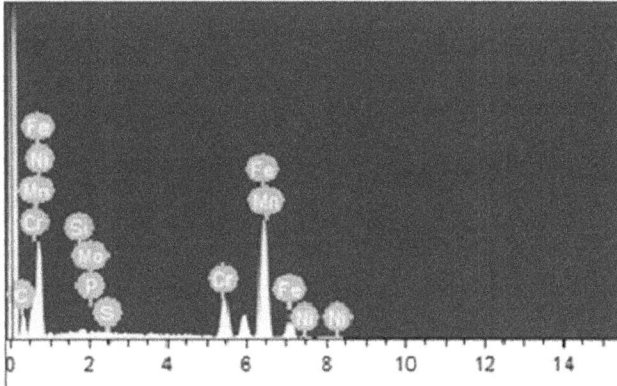

Figure 10.2 EDS figure of AISI 202 grade.

10.2.2 Workpiece Details

AISI 202 grade of stainless steel was taken as a workpiece whose chemical composition is shown in Table 10.2 and its EDS figure presented in Fig. 10.2, ensuring absence of any compounds. Because of its corrosion-resistance nature, stainless steel perceives a huge demand and a wide application in new product manufacturing and product development industries. In addition to this, it also can be used in medical instruments and utensils due to its hygienic properties.

10.2.3 Experimental Procedure

For the current research, a die-sinking EDM setup was used to machine AISI 202 grade stainless steel with commercial EDM oil-grade dielectric fluid medium. In die-sinking EDM, the machining zone was always kept submerged in a dielectric fluid medium so that accurate machining could be ensured. The basic architecture of a die-sinking EDM operation is replicated in Fig. 10.3. Since the workpiece needs to be melted, therefore a greater amount of thermal energy has been required to be generated over the workpiece. This condition can be ensured by connecting electrode negative and workpiece positive so that a greater amount of heat is liberated at the workpiece and less heat is liberated at the electrode.

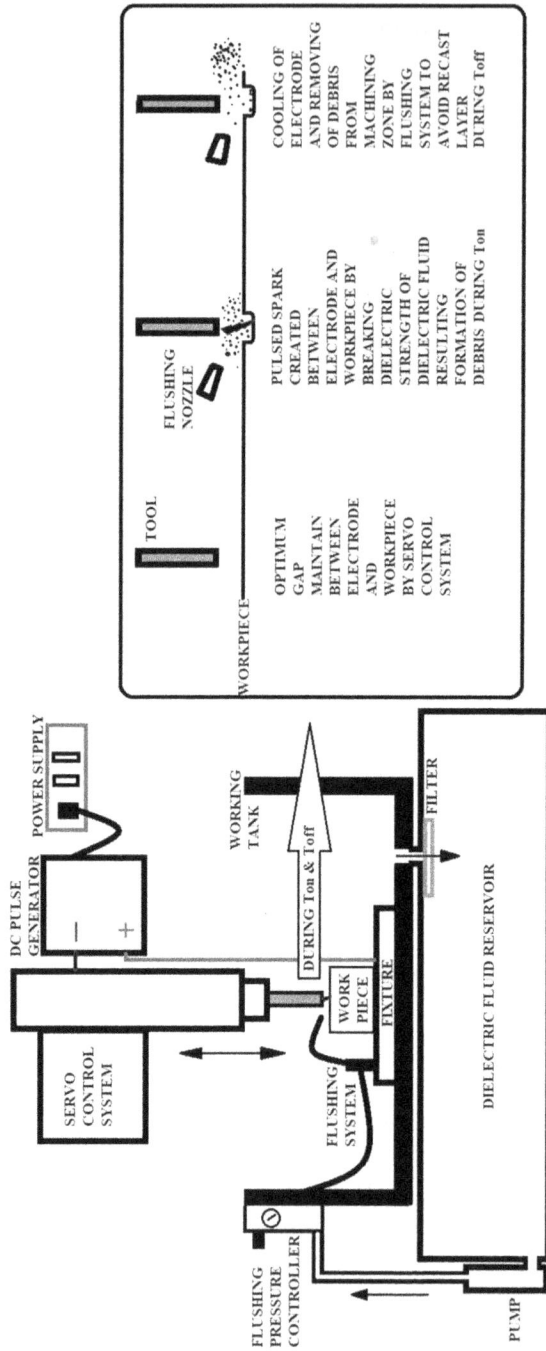

Figure 10.3 Schematic view of EDM operation.

Table 10.3 Controllable EDM inputs with its level

Machine Inputs	Pulse current	Pulse on-time	Pulse off-time
Technical tag	I in A	T_{on} in μs	T_{off} in μs
Low level	5	55	5
Medium level	7	65	7
High level	9	75	9

The process parameters responsible for the quality discharge that ensures EDM performance are I, T_{on}, and T_{off}. Therefore, the factors and its levels were selected such that it falls under the semi-finish operational zone reflected from literature presented in Table 10.3.

For smooth conduction of the experiment, first a stainless-steel sheet is cut into a number of pieces of specified dimensions, followed by grinding in order to make the surface parallel. After that, an electrode is prepared in a cylindrical shape with a diameter of 7 mm.

A central composite design of RSM is a mathematical combined with statistical technique used to generate a quadratic equation to express the response with the least number of experimental runs without damaging accuracy [23–25] and was applied to the model and analyzed the problem. It generates a regression equation that can be used to predict the response described in equation (1).

$$Y = \beta_0 + \sum_{i=1}^{n} \beta_i X_i + \sum_{i=1}^{n} \beta_{ii} X_i^2 + \sum_{i=1}^{n} \beta_{ij} X_i X_j \dots \dots \quad (10.1)$$

where Y is the response; β_0, β_i, β_{ii}, β_{ij} are the regression coefficients; and X_i (i varies from 1, 2, 3 … n) are the quantitative machining variables. A total of 20 experimental runs were conducted on economical aspects for the investigation presented in Table 10.4 with evaluated values of MRR, EWR, and average roughness. For statistical analysis, MINITAB 19 was used.

The performance characteristics of MRR from the workpiece and EWR from the electrode were calculated by using the weight comparison method formula described in equations (2) and (3):

$$MRR = \frac{\text{premachining workpiece weight} - \text{postmachining workpiece weight}}{\text{Time of machining}}$$

$$(10.2)$$

$$EWR = \frac{\text{premachining electrode weight} - \text{postmachining electrode weight}}{\text{Time of machining}}$$

$$(10.3)$$

Table 10.4 Design layout with experimental outcomes

S. No.	Coded levels			Levels value			Responses		
	X	Y	Z	I	T_{on}	T_{off}	EWR in mg/min	MRR in mg/min	R_a in μm
1	−1	1	1	5	75	9	0.000165	0.00155	1.92
2	0	−1	0	7	55	7	0.000204	0.01847	2.12
3	0	0	−1	7	65	5	0.000175	0.013095	2.14
4	−1	−1	−1	5	55	5	0.000163	0.00198	1.72
5	0	0	0	7	65	7	0.000179	0.012415	2.17
6	1	1	−1	9	75	5	0.000521	0.0342	3.33
7	0	1	0	7	75	7	0.000273	0.00669	2.03
8	−1	1	−1	5	75	5	0.000168	0.001495	1.71
9	0	0	0	7	65	7	0.000181	0.015095	2.19
10	1	0	0	9	65	7	0.000286	0.045095	3.12
11	0	0	0	7	65	7	0.000184	0.01382	2.21
12	−1	−1	1	5	55	9	0.000155	0.001205	1.87
13	1	1	1	9	75	9	0.000413	0.00264	3.61
14	0	0	0	7	65	7	0.000187	0.013985	2.25
15	1	−1	−1	9	55	5	0.000372	0.048175	2.50
16	0	0	1	7	65	9	0.000215	0.013375	2.53
17	0	0	0	7	65	7	0.000188	0.01385	2.33
18	−1	0	0	5	65	7	0.000162	0.001665	1.88
19	1	−1	1	9	55	9	0.000335	0.04203	2.72
20	0	0	0	7	65	7	0.000192	0.014535	2.41

The primary surface texture of the EDM-machined surface was measured in terms of average roughness R_a expressed by equation (4) more accurately by using the optical surface profiler (OSP) for which WYKO Veeco Metrology Group, 1100 Metrology Vision 32 was used in the VSI mode with a magnification range of 10.43 and scan area of 50 microns.

$$\text{Average Roughness } R_a = \frac{1}{Ls} \int_0^{Ls} |y(x)| dx \tag{10.4}$$

where y represents roughness peaks and valley heights of profile over a cut-off length Ls in the intended profile direction x. Furthermore, the post-machined surface was investigated by atomic force microscope (AFM) using NT–MDT NTEGRA. AFM is a powerful microscope used to generate high-resolution 3-D surface profile images and helps to manipulate matter at a nanoscale.

To understand the machining physics, scanning electron microscopy (SEM) using LEO 435VP SEM at magnification from 500x to 6.00 Kx with 6 nm VP and 4 nm HV was used. Furthermore, energy dispersive x-ray spectroscopy (EDS) coupled with SEM was used for the chemical characterization or elemental analysis of EDM-machined samples. The machined specimens were further analyzed by an x-ray diffractometer (XRD) (Bruker AXS Diffrakto meter D8) with Cu-Kα radiation at a wavelength of 0.154 nm, current of 30 mA, and voltage of 40 KV at room temperature. A 2θ scanning range of X-ray intensity employed at 10°–90° with a scanning rate of 2°/min [26]. Here, Bragg's law is applied to compute the crystallographic spacing of a matrix. X-ray radiation (1.548 for Cu) of wavelength λ and Scehrrer constant K is taken as 0.9.

10.3 RESULTS AND DISCUSSION

10.3.1 Statistical Analysis

Electrode wear rate or electrode erosion rate (EWR), material removal rate (MRR), and average roughness (R_a) are three dimensions of electric discharge machine (EDM) performance measures that define the machining behavior, feasibility, and scope. The highest MRR, least EWR, and R_a are the main industrial interests as they indicate better control and utilization of electric discharge and hence ensure better productivity. Here, the experiment was conducted to measure and analyze the EDM performance potential. All the precise data for each experiment brought together in one edition and was mathematically applied (explained in equations (2) and (3)) to get the values of EWR and MRR. The primary surface texture of each specimen was measured by using the optical profilometer and documented. The experimental design along with its performance measures are recorded in Table 10.4.

The regression model for EWR, MRR, and R_a are expressed in terms of uncoded controllable factors derived from a RSM statistical approach expressed in equations (5), (6), and (7).

$$EWR = 0.000185 + 0.0001111I + 0.000031T_{on} - 0.000012T_{off}$$
$$+ 0.000039I \times I + 0.000053T_{on} \times T_{on} + 0.000010\ T_{off} \times T_{off}$$
$$+ 0.000026I \times T_{on} - 0.000017\ I \times T_{off} - 0.000008T_{on} \times T_{off}.$$
$$(10.5)$$

$$MRR = 0.01463 + 0.01642I - 0.00652\ T_{on} - 0.00381\ T_{off} + 0.00774I \times I$$
$$- 0.00306\ T_{on} \times T_{on} - 0.00240\ T_{off} \times T_{off} - 0.00665I \times T_{on}$$
$$- 0.00462I \times T_{off} + 0.00308T_{on} \times T_{off}.$$
$$(10.6)$$

$$R_a = 2.2553 + 0.6180I + 0.1670T_{on} + 0.1250T_{off} + 0.2518I \times I$$
$$- 0.1732T_{on} \times T_{on} + 0.0868T_{off} \times T_{off} + 0.2100I \times T_{on}$$
$$+ 0.0175I \times T_{off} + 0.0150T_{on} \times T_{off}. \tag{10.7}$$

After getting the machining data, the optimization technique was executed to get optimized EDM process parameters setting combination that satisfied industrial interests of having the highest material removal rate, least electrode erosion rate, and excellent surface finish. Fig. 10.4 and Table 10.5

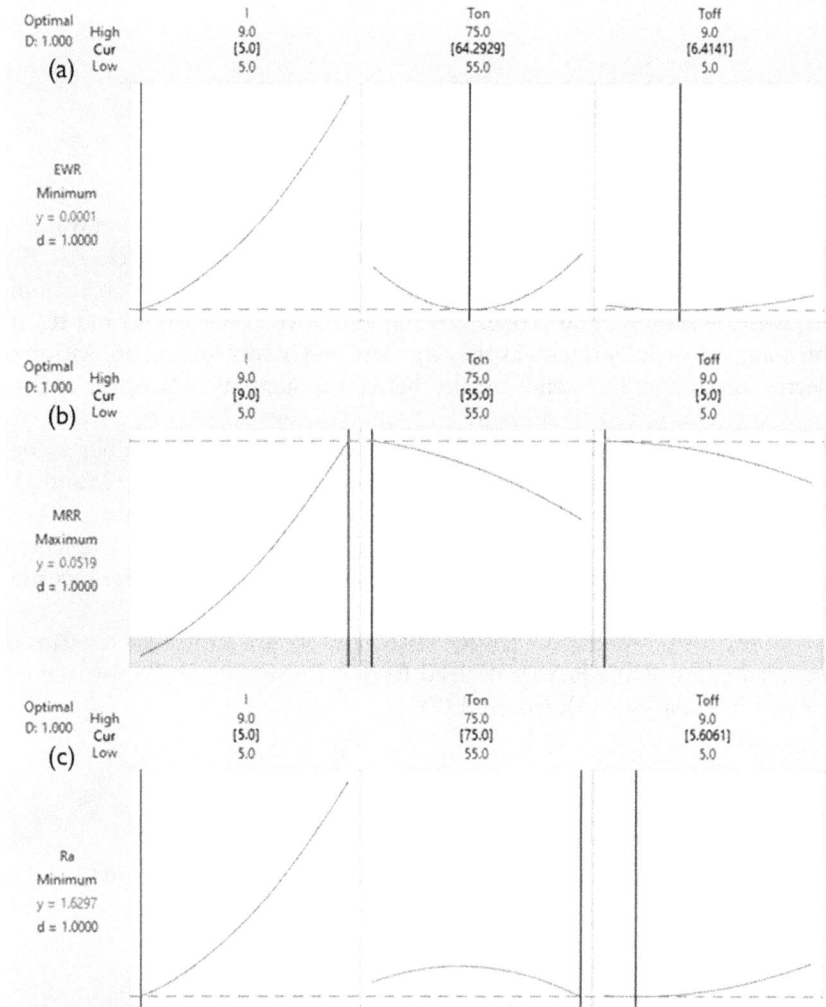

Figure 10.4 Optimized parameter (a) minimum EWR, (b) maximum MRR, and (c) minimum R_a.

Table 10.5 Obtained optimal parameters

Process Parameters	Value obtained		
	MRR	R_a	EWR
I in A	9	5	5
T_{on} in μs	55	75	64
T_{off} in μs	5	6	7

represent the obtained optimal EDM machining parameters to clinch the best responses.

10.3.2 EDM Machining Parameter Influence on EWR

Since electrode wear rate is a measure of attribution towards production potential and always the least EWR recommended to achieve higher production rate. Therefore, it is essential to study the physics behind EWR so that it can be minimized to the higher productivity. In this investigation, we investigate the effects of EDM performance characteristics: pulse current I, pulse on-time T_{on}, and pulse off-time T_{off} on electrode wear rate (EWR). By using statistical software, a 3-D surface has been plotted to interpret the silent feature of EDM machining parameters presented in Fig. 10.5. As pulse current increases from 4.5 A to 9.5 A, the EWR is found to increase, as shown in Fig. 10.5(a). This is quite natural that heat generation is proportional to the square of applied current and thereby an increase in energy discharge, causing more heat generation, resulting in more erosion. Fig. 10.5(a), 10.5(c) also reveal that pulse duration as an increase from 50 to approx. 60 μs, causing a reduction in EWR and thereafter found an increase in EWR. Therefore, a decrease with increase behavior pattern for T_{on} is observed. This phenomenon is appeared due to deposition of recast layer over the electrode. Due to high energy discharge, high temperature generated at electrode workpiece interface causing decomposition of dielectric fluid to carbon atoms deposited over the electrode surface, which enhances wear resistance with an increase in T_{on}. This phenomenon slightly reduces EWR. The deposition of recast layer over electrode surface can be seen in Fig. 10.10. However, high pulse current with high pulse on time caused detrimental effect to electrode and this happened due to higher energy discharge from electrode for a long duration of time. An AFM image of a post-machining electrode also is represented in Fig. 10.6. AFM findings presented in Fig. 10.6(a), Fig. 10.6(b), and Fig 10.6(c) indicate electrode surface peaks at currents of 5 A, 7 A, and 9 A, respectively, reveal that as the pulse current increases, surface picks get narrower and sharper. Rapid erosion of the electrode results from high-frequency, high-energy discharge and the principle of quenching leads to a change in hardness during T_{off}

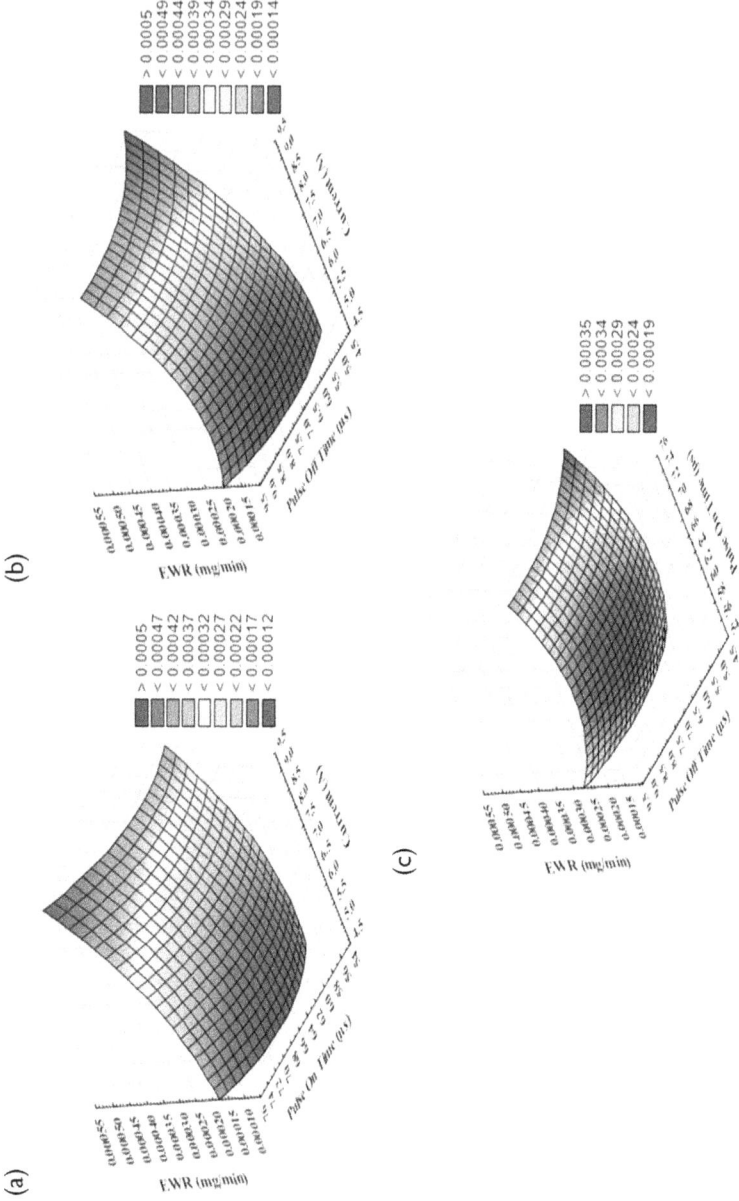

Figure 10.5 (a) Effects of current I and pulse on-time T_{on} on EWR, (b) effects of current I and pulse off-time T_{off} on EWR, (c) effects of pulse on-time T_{on} and off-time T_{off} on EWR.

Figure 10.6 Post-machining AFM images of electrode surface.

that could be the reason for finding such a surface texture. Literature surveyed also convey that the surface roughness depends on hardness [27].

Fig. 10.5(b) and 10.5(c) witnesses a decreasing-increasing behavior pattern of pulse off-time as it boosts from 4.5 to 9.5 μs. This is because as the pulse off-time T_{off} increases, the dielectric fluid finds enough time to chill decomposed carbon atoms formed from a dielectric fluid and coat the electrode surface, which slightly minimizes EWR. Heat generation dissipated due to thermal conductivity among electrode materials could be another reason to minimize EWR. The EDM process parameters T_{on} and T_{off} and contribution are presented in Fig. 10.5(c) and it shows the least electrode wear rate results from moderated T_{on} combined with moderated T_{off}.

10.3.3 EDM Machining Parameters' Influence on MRR

The material removal rate (MRR) is one of the major attributers towards EDM performance and it's a major challenge of EDM. Therefore, here the investigation was carried out to unfold the essence of EDM-controllable inputs: pulse current I, pulse off-time T_{off}, and pulse on time T_{on} towards improvement of MRR performance. By using statistical software, a 3-D surface plot was generated for analyzing the silent features of the machining parameters presented in Fig. 10.7.

MRR was found directly proportional to pulse current, clearly observed from Fig. 10.7. This is happening because of the heat generation, which is responsible for removing material by melting and vaporizing is directly proportional to the square of applied current. However, an interesting behavior can be observered from Fig. 10.7(a); that an increased pulse on-time combined with a low current contributes to an increased MRR, whereas an increased pulse on-time combined with a high current contributes to a decreased MRR. Since higher current combined with higher pulse on-time generates higher discharge thermal energy for a long duration, hence, causing more debris to form, which dissipates between spark gap disrupt the electrical sparks with plasma channel expansion [28,29] leads to decreased MRR and settles down over the machined surface, thereby producing a layer over the machined surface called a white layer or recast layer that may further hamper the machining process [30] could be another reason to decrease MRR. This recast layer or white layer presence is clearly visible in Fig. 10.11(b) and should be at a minimum to get defect-free job. But at a low-current setting with increased pulse duration generates less energy discharge for a long time duration which formed comparatively less debris doesn't hamper the transfer of thermal energy and hence resulted in an increased MRR. Similarly, from Fig. 10.7(b) it reveals that at high current with increased T_{off} decreases MRR. This may happen due to heat dissipation by dielectric, causing less heat transfer to the workpiece and the recast layer formation. Therefore, less MRR is observed in high T_{off} and T_{on} illustrated in Fig. 10.7(c).

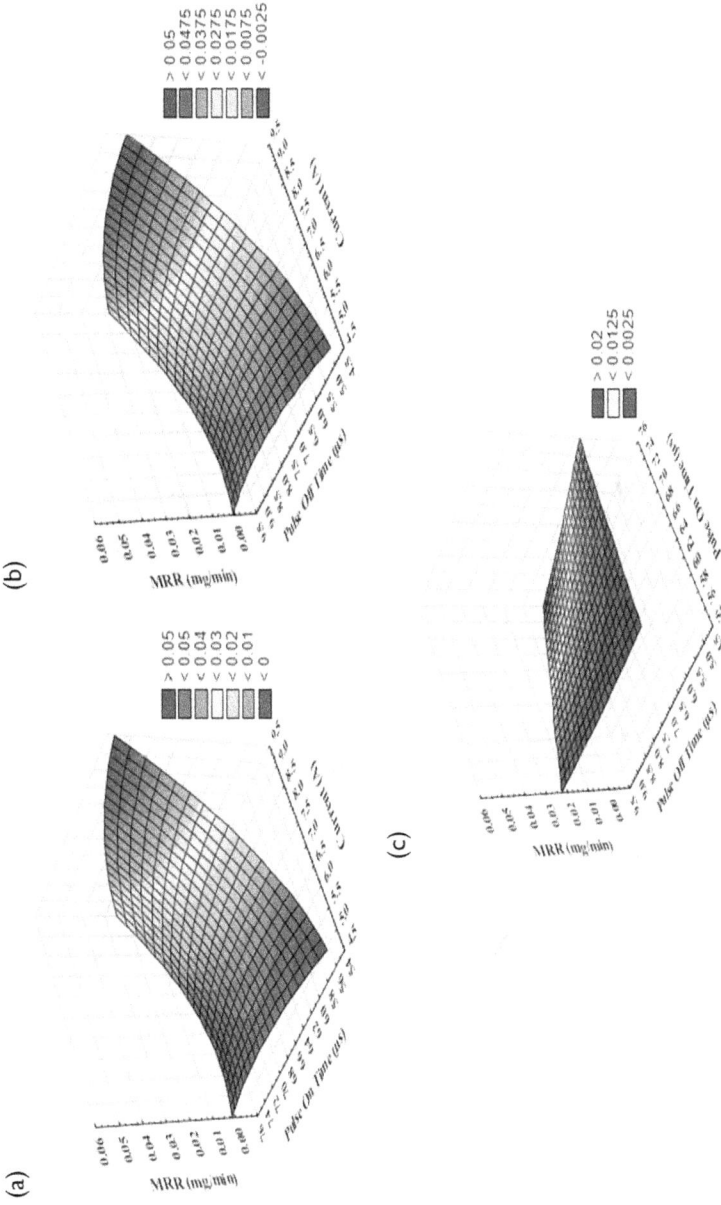

Figure 10.7 (a) Effects of current I and pulse on-time T_{on} on MRR, (b) effects of current I and pulse off-time T_{off} on MRR, (c) effects of pulse on-time T_{on} and pulse off-time T_{off} on MRR.

10.3.4 EDM Machining Parameters Influence on Surface Texture

Since the surface finish of work is a major aspect of manufacturing job, therefore it's necessary to analyze the process that governs this dynamic. Therefore, a study was made to analyze the influence parameters towards average surface roughness. A 3-D surface has been plotted based on the result obtained, as shown in Fig. 10.8. Surface roughness found an approximate constant from 4.5 A to 6.5 A and afterword it was proportional to pulse current along with low pulse off-time. Low-pulse current causing high-frequency, low-energy discharge creates smooth erosion of work material results and smaller crater with effective cooling of electrode and workpiece could be the reason for getting low roughness. But in high pulse on-time at 74 μs with an increase in pulse current from 4.5 A to 9.5 A increases the surface roughness. A higher discharge energy for long duration causing more impulsive force and resulting in deeper and larger erosion of work material could be the reason to increase surface roughness. The AFM image of post-EDM-machined surface taken for different current conditions also reveals the same as in Fig. 10.9. AFM findings presented in Fig. 10.9(a), Fig. 10.9(b), and Fig 10.9(c) show machined surface peaks and valleys at 5 A, 7 A, and 9 A, respectively. This analysis reflects longer and sharper peaks and valleys generation as the pulse current increases from 5 to 9 A. High discharge thermal energy along with a rapid chilling effect forms a recast layer over a machined surface that changes the hardness property could be the reason for getting such surface texture.

Pulse off-time T_{off} has a vital role while machining to cool the electrode after discharge and minimize workpiece heat-affected zone to retain the required shape leads to improvised EDM machining performance. For each current setting, as T_{off} increases from 4.5 to 9.5 μs, roughness found relatively increased, as illustrated in Fig. 10.8(b). This is due to a chilling effect, causing deposition of a recast layer whose property is different from the workpiece bulk property.

Mention figure showing pulse current contribution superior to that of other machining parameters and therefore needs to be optimized to obtained excellent quality of surface texture. This is due to stronger spark generating higher depth of penetration results in primary irregular surface texture. However, surface roughness also affected by T_{on} and T_{off} controlled by duty cycle is revealed in Fig. 10.8(c).

10.3.5 Study by Using SEM

EDM-machined surface and tool surface are investigated by a scanning electron microscope commonly known as a SEM. Since the machined surface undergoes various changes, such as getting heated by sparking and getting cooled by dielectric flushing, hence, to reveal a machined surface, a SEM was used for the detailed clarity represented in Fig. 10.10 and Fig. 10.11 for the electrode and workpiece at 5 A, 7 A, and 9 A, respectively.

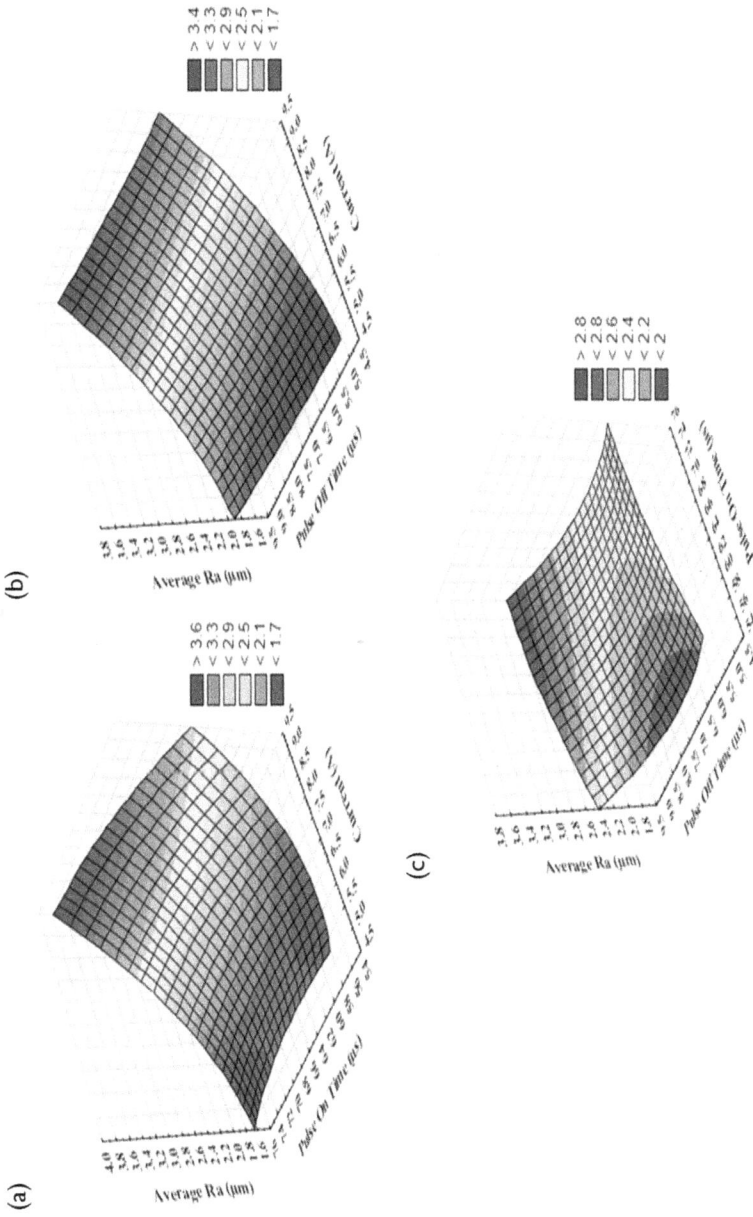

Figure 10.8 (a) Effects of current I and pulse on-time T_{on} cn R_a, (b) effects of current I and pulse off-time T_{off} on R_a, and (c) effects of pulse on-time T_{on} and pulse off-time T_{off} on R_a.

Figure 10.9 Post-machining AFM images of workpiece surface.

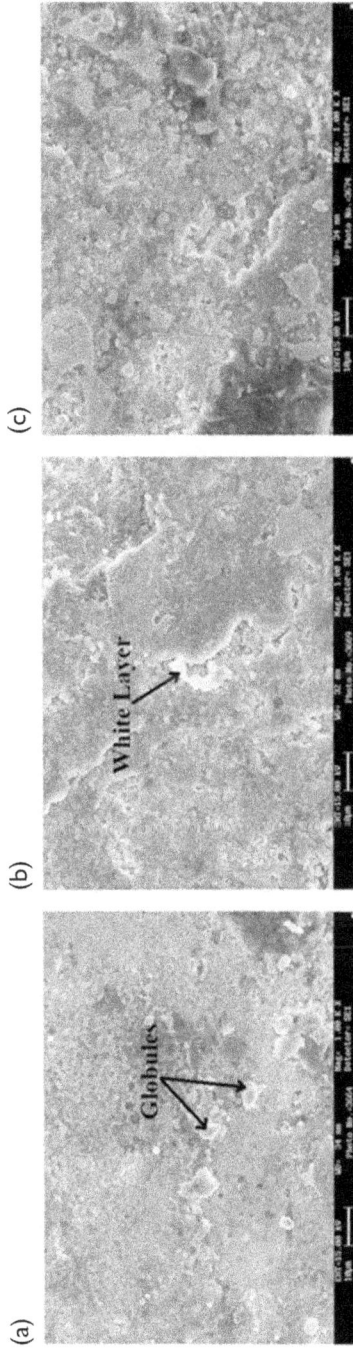

Figure 10.10 Post-machining SEM images of electrode surface.

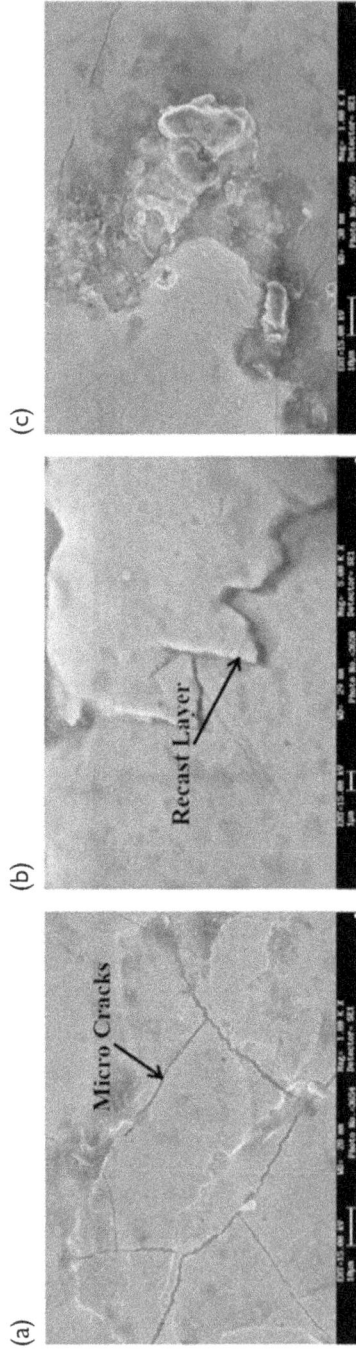

Figure 10.11 Post-machining SEM images of workpiece surface.

The presence of a white layer or recast layer on the electrode surface can be clearly seen in Fig. 10.10. More distribution of the white layer can be observed as we increased the current and this happened due to high energy discharge, causing more melting of metal, which migrated to the electrode.

Remarkable changes on a machined surface have been reported from a SEM figure for the workpiece. Micro-cracks in Fig. 10.11(a), recast layer in Fig. 10.11(b), pores, craters, and debris in Fig. 10.11(c) are clearly visible from the figure. High spark energy and subsequent quenching are responsible to form a re-solidify layer named a recast layer of molten material, which is harder than bulk material followed by heat-affected zone initiate cracks for such radial changes in machined surface. Micro-cracks are developed due to increased residual stresses resulting from non-homogeneities within a white layer [31]. THe microstructure of bulk material is not affected by EDM machining [32].

10.3.6 Study by Using X-Ray Diffraction Approach

Pre- and post-machining of the electrode and workpiece are studied by the x-ray diffraction approach at three different current settings. The XRD pattern of the pre-machined electrode surface texture is represented over Fig. 10.12(a). Fig. 10.12(b) represents the XRD pattern of the electrode at different current condition of 5 A, 7 A, and 9 A, respectively.

Electrode material has peaks with a higher intensity observed in Fig. 10.12(a) before machining and a noticeable decrease in peak intensity as the current setting increases, shown in Fig. 10.12(b). Since the EDM machining process involves melting and vaporizing the workpiece material, therefore the workpiece is encountered by many physical and temperature changes. These physical and temperature changes are responsible for changes in microstructure as well as changes in phase. This reflects changes in intensities in the XRD pattern since the grain size has a reverse relationship to the lattice strain.

Pre- and post-EDM machining of a workpiece at three different parameter settings is shown in Fig. 10.13. Intensity is found decreasing with an increase in current. However, an imperceptible shift and broadening of the peak is observed due to a very high temperature change. This high-temperature change results in a phase change. Migration of material between the electrode and workpiece is also noticed in the figure and justified in the above discussion.

10.3.7 Multi-Objective Optimization

From the above technical discussion, EDM is reflected as a complex process as MRR, EWR, and roughness are results of three distinct process combinations. To establish the industrial applicability, this is essential to integrate the write EDM controllable parameters: pulse current I, pulse off-time T_{off}, and pulse on-time T_{on} such that a highest MRR with least EWR and average surface roughness should be clinched. Therefore, multi-objective

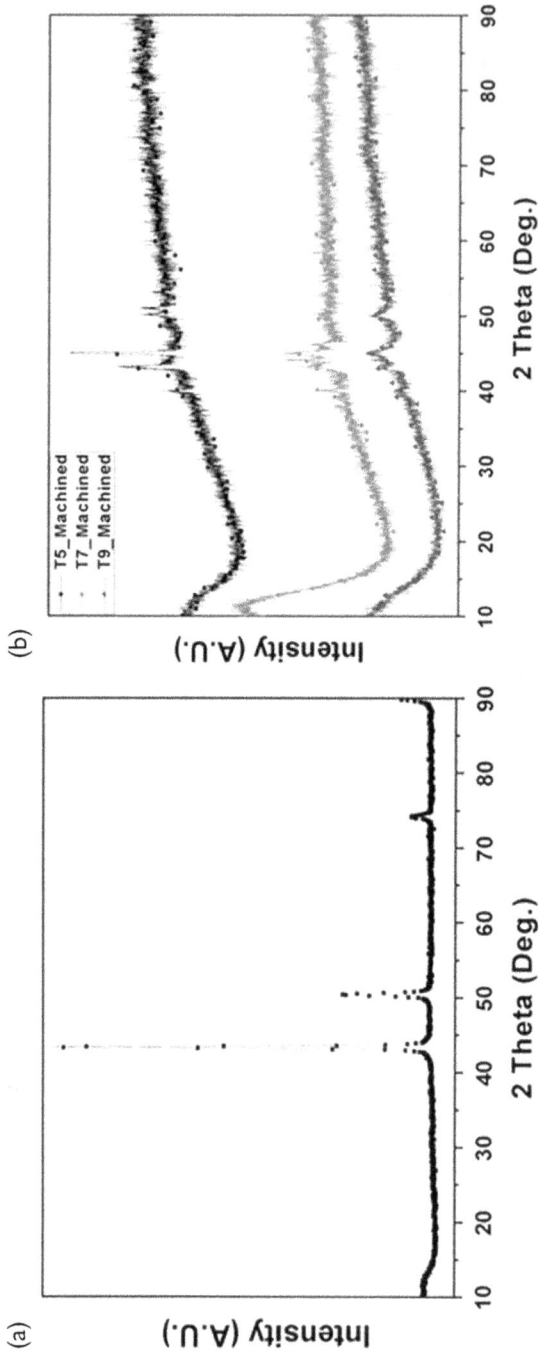

Figure 10.12 (a) Pre-machining XRD patterns of electrode and (b) post-machining XRD patterns of electrode at varying currents.

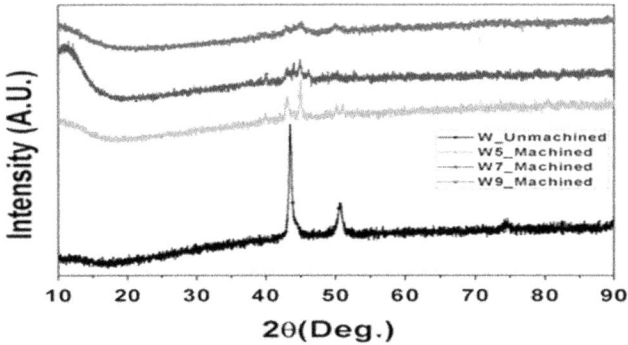

Figure 10.13 Pre- and post-machining XRD patterns.

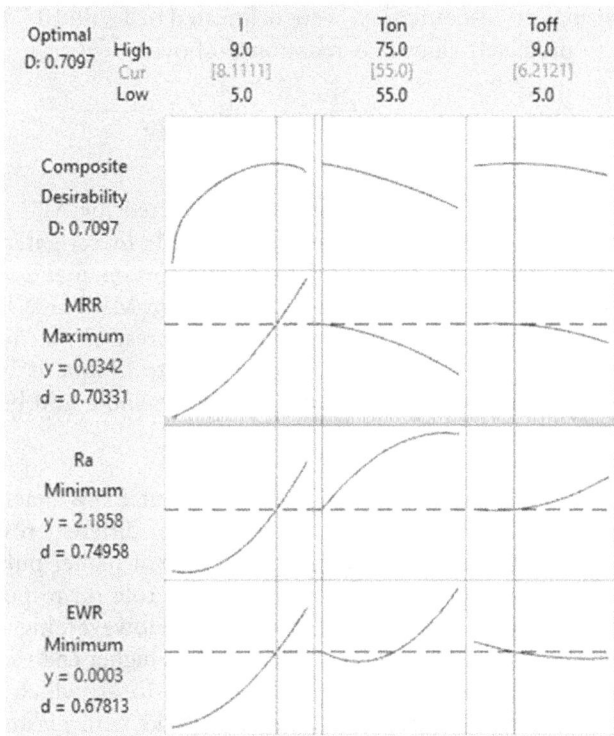

Figure 10.14 Optimized parameters with predicted value.

optimization is conducted by using a RSM optimizer tool in MINITAB 19 by giving equal importance to all responses. The obtained optimized parameters to satisfy the highest MRR with the least EWR and R_a with 0.709729 desirability are presented in Fig. 10.14. The predicted results are presented in Table 10.6.

Table 10.6 Optimal machining condition with their predicted response

I in A	T_{on} in μs	T_{off} in μs	MRR in mg/min	EWR in mg/min	R_a in μm	Desirability
8	55	6	0.0342395	0.0002728	2.18579	0.709729

Table 10.7 Confirmation test results

Sl No.	I in A	T_{on} in μs	T_{off} in μs	MRR in gm/min	EWR in gm/min	R_a in μm
1	8	55	6	0.02785	0.000234	2.11
2	8	55	6	0.02981	0.000223	1.98
3	8	55	6	0.0312	0.000251	1.85

The conformation test outcomes were delineated in Table 10.7 and found fairly close to predicted values, as mentioned above.

10.4 CONCLUSION

In this research, a precise experiment was conducted for AISI 202 grade stainless steel by a nontraditional machine EDM. Investigated was the micro-structural behavior of an EDM-machined specimen and scrutiny of the characteristics of the EDM input parameters on MRR, EWR, and R_a. For finding the best results to meet industrial interest of machining AISI 202 grade stainless steel, the RSM optimization technique is been used. The upshot of this investigation is discussed above and could be reflected as follows:

- Pulse current I was found the most prominent EDM machine parameter out of all. Pulse current affects a lot of different responses of EWR, MRR, and R_a. A combinational effect of higher pulse current setting and pulse on-time plays a significant role on response EWR due to high energy discharge thermal energy. However, higher current combined with higher pulse on-time generates higher energy discharge for a long duration, causing more debris to form, which dissipates between spark gap disrupt the electrical sparks with plasma channel expansion leads to decrease MRR.
- The micro-structures of a EDM processed surface studied by SEM and XRD technology and these show recast layer formation over a stainless-steel specimen and homogeneous dispersion as the peaks broaden with an increase in current [33,34]. The recast layer is the carbon layer deposition over the machined surface. This is happening due to quick quenching of melted material known as debris unable to be flushed out by dielectric fluid from the machining zone, leading

to hamper machining. Further study may be necessary on flushing pressure to improve machining.

- As the pulse current increases, surface peaks and valleys get narrower and sharper. Rapid erosion of the electrode and workpiece results from high-frequency discharge and the principle of quenching leads to a change in hardness during T_{off}, causing such machining architecture.
- Development of micro-cracks were detected over a machined surface by SEM. This physical change happened due to localized thermal stress (sudden change to high temperature during T_{on} and getting chilled during T_{off}). Some microstructure changes are also observed at the electrode side due to pulse current.
- EDM process parameters are optimized by RSM and found that the least erosion rate at pulse current 5 A, pulse off-time 7 µs, and pulse on-time 64 µs; found maximum MRR at pulse current 9 A, pulse off-time 5 µs, and pulse on-time 55 µs; found least average surface roughness at pulse current 5 A, pulse off-time 6 µs, and pulse on-time 75 µs, respectively.
- Multi-objective optimization performed by an RSM optimizer tool and the optimized parameters to satisfy maximize MRR, minimize EWR, and minimize average roughness parallelly are pulse current 8 A, pulse off-time 6 µs, and pulse on-time 55 µs, respectively. The predicted value for electrode wear rate, material removal rate, and average roughness are 0.0002728 gm/min, 0.0342395 mg/min, and 2.18579 µm, respectively. The conformation test conveys the experimental value fairly close to the predicted value.
- Hence, further study may be needed to improve machining by selecting a different dielectric fluid and finding the role of flushing pressure. Another multi-objective optimization algorithm needs to be implemented to predict more accurately. A power circuit needs to be studied for effective pulsed spark production.

ACKNOWLEDGEMENTS

This research work had financial support from IIT Roorkee, NIT Jamshedpur and Ministry of Human Resource and Development (MHRD), Government of India.

REFERENCES

[1] Hu CF, Zhou YC, Bao YW (2008) Material removal and surface damage in EDM of Ti$_3$Si$_2$ ceramic. *Ceramics International* 34: 537–541.
[2] Rajmohan T, Prabhu R, Subbarao G, Palanikumar K (2012) Optimization of machining parameters in electrical discharge machining (EDM) of 304

stainless steel. *Procedia Engineering International Conference on Modeling, Optimization and Computing* 38: 1030–1036.

[3] Muthuramalingam T, Mohan B (2014) A review on influence of electrical process parameters in EDM process. *Archives of Civil and Mechanical Engineering* 15(1): 87–94.

[4] Marafona J, Wykes C (2000) A new method of optimizing material removal rate using EDM with copper–tungsten electrode. *International Journal of Machine Tool and Manufacture* 40: 153–164.

[5] Ho KH, Newman ST (2003) State of the art electrical discharge machining (EDM). *International Journal of Machine Tool and Manufacture* 43: 1287–1300.

[6] Lopez-Esteban S, Gutierrez-Gonzalez CF, Mata-Osoro G, Pecharroman C, Diaz LA, Torrecillas R, Moya JS (2010) Electrical discharge machining of ceramics/semiconductor/metal nano composites. *Scripta Materialia* 63: 219–222.

[7] Li L, Wong YS, Fuh JYH., Lu L (2001) EDM performance of TiC/copper-based sintered electrodes. *Materials and Design* 22: 669–678.

[8] Haron CH, Ghani JA, Burhanuddin Y, Seong YK, Swee CY (2008) Copper and graphite electrode performance in electrical–discharge machining of XW42 tool steel. *Journal of Materials Processing Technology* 201: 570–573.

[9] Khanra AK, Sarkar BR, Bhattacharya B, Pathak LC, Godkhindi MM (2007) Performance of ZrB_2-Cu composite as an EDM electrode. *Journal of Materials Processing Technology* 183: 122–126.

[10] Czelusniak T, Amorim FL, Lohrengel A, Higa CF (2014) Development and application of copper-nickel zirconium diboride as EDM electrodes manufactured by selective laser sintering. *International Journal of Advanced Manufacturing Technology* 72: 5–8.

[11] Tang L, Guo YF (2013) Electrical discharge precision machining parameters optimization investigation on S-03 special stainless steel. *International Journal of Advanced Manufacturing Technology* 70: 5–8.

[12] Zarepour H, Tehrani AF, Karimi D, Amini S (2007) Statistical analysis on electrode wear in EDM of tool steel DIN 1.2714 used in forging dies. *Journal of Material Processing Technology* 187–188: 711–714.

[13] Fonseca J, Marafona JD (2013) The effect of deionization time on the electric discharge machining performance. *International Journal of Advanced Manufacturing Technology* 71: 1–4.

[14] Rajurkar KP, Sundaram MM, Malshe AP (2013) Review of electrochemical and electro-discharge machining. *CIRP Conference on Electro Physical and Chemical Machining (ISEM)* 6: 13–26.

[15] Kiyak M, Cakir O (2007) Examination of machining parameters on surface roughness in EDM of tool steel. *Journal of Material Processing Technology* 191: 141–144.

[16] Tsai HC, Yan BH, Huang FY (2003) EDM performance of Cr/Cu-based composite electrode. *International Journal of Machine Tools and Manufacture* 43: 245–252.

[17] Bhaumik M, Maity K (2018) Effect of deep cryotreated tungsten carbide electrode and Sic powder on EDM performance of AISI304. *Particulate Science and Technology* 37(8): 981–992. 10.1080/02726351. 2018.1487491

[18] Tripathy S, Tripathy DK (2017) Multi-response optimization of machining process parameters for powder mixed electro-discharge machining of H-11 die steel using grey relational analysis and topsis. *Machining Science and Technology* 19(1): 62–70.

[19] Surekha B, Lakshami TS, Jena H, Samal P (2021) Response surface modeling and application of fuzzy grey relational analysis to optimize the multi re-sponse characteristics of EN-19 machining using powder mixed EDM. *Australian Journal of Mechanical Engineering* 19(1): 19–29. 10.1080/14484846.2018.1564527

[20] Kong L, Liu Z, Qiu M, Wang W, Han Y, Bai S (2019) Machining char-acteristics of submerged gas flushing electrical discharge machining of Ti6Al4V alloy. *Journal of Manufacturing Process* 41: 188–196.

[21] Lee HT, Tai TY (2003) Relationship between EDM parameters and surface crack formation. *Journal of Material Processing Technology* 142: 676–683.

[22] Bharwal S, Vyas C (2014) Weldability issue of AISI 202 SS (stainless steel) grade with GTAW process compared to AISI 304 SS grade, *International Journal of Advanced Mechanical Engineering* 4(6): 695–700.

[23] Hewidy MS, Al-Tawel TA, El-Safty MF (2005) Modelling and machining parameters of wire electrical discharge machining of inconel 601 using RSM. *Journal of Materials Processing Technology* 169: 328–336.

[24] Assarzadeh S, Ghoreishi M (2013) Statistical modeling and optimization of process parameters in electro-discharge machining of cobalt-bonded tungsten carbide composite. *Procedia of the 17th CIRP Conference on Electro Physical and Chemical Machining* 6: 463–468.

[25] Samesh SH (2009) Study of parameters in electric discharge machining through response surface methodology approach. *Applied Mathematical Modeling* 33: 4397–4407.

[26] Pal K, Rajasekar R, Pal SK, Kim JK, Das CK (2010) Influence of fillers on NR/SBR blends containing ENR-Organoclay nanocomposites: Morphology and wear. *Journal of Nanoscience and Nanotechnology* 10: 3022–3033.

[27] Guu YH (2005) AFM surface imaging of AISI D2 tool steel machined by EDM process. *Applied Surface Science* 242: 245–250.

[28] Ranjith R, Tamilselvam P, Prakash T, Chinnasamy C. (2019) Examinations concerning the electric discharge machining of AZ91/5B4CP composites uti-lizing distinctive electrode materials. *Materials and Manufacturing Processes* 34(10): 1120–1128.

[29] Paswan K, Pramanik A, Chattopadhyaya S (2020) Machining perfor-mance of Inconel 718 using graphene nanofluid in EDM. *Materials and Manufacturing Processes*; 35(1): 33–35. 10.1080/10426914.2020.1711924

[30] Rahul DS, Biswal BB, Mahapatra SS (2017) A novel satisfaction function and distance-based approach for machining performance optimization during electro-discharge machining on super alloy Inconel 718. *Arabian Journal for Science and Engineering* 42(5): 1999–2020. 10.1007/s13369-017-2422-5

[31] Ekmkci B (2007) Residual stress and white layer in electric discharge ma-chining (EDM). *Applied Surface Science* 253: 9234–9240.

[32] Mannan KT, Krishnaiah A, Arikatla SP (2013) Surface characterization of electric discharge machined surface of high speed steel. *Advanced Materials Manufacturing and Characterization* 3: 161–167.

[33] Mandaloi G, Singh S, Kumar P, Pal K (2016) Effect of crystalline structure of AISI M2 steel using tungsten-thorium electrode through MRR, EWR and surface finish. *Measurement* 90: 74–84.

[34] Mandaloi G, Singh S, Kumar P, Pal K (2015) Effect on crystalline structure of AISI M2 steel using copper electrode through material removal rate, electrode wear rate and surface finish. *Measurement* 61: 305–319.

Chapter 11

A Novel Method for Fabricating Functionally Graded Materials by Vibration-Assisted Casting

Divyanand Kumar, Dinesh Kumar, and Anand Mukut Tigga

Department of Production & Industrial Engineering, National Institute of Technology Jamshedpur, Jharkhand, India

CONTENTS

11.1 INTRODUCTION

Modern advances in engineering and material processing have enabled the development of a new form of a heterogeneous composite material known as functionally graded materials (FGMs). These second-generation composites are gaining a lot of attraction in the scientific community since they are intended to function better in severe environments. FGMs are classed based on their graded structure, which has spatially variable properties in specific directions and is designed to enhance performance by distributing such properties. It may be a progressive alteration of chemical compositions, structure, grain size, degree of texturization, density, and other physical qualities across layers or throughout the volume (Cannillo et al., 2007; Jha et al., 2013; Kawasaki & Watanabe, 1997).

FGMs can be synthesized in lithographic shape or 2-D material and 3-D materials by restricting the growth of dimensions and varying compositions. A 2-D material or a layered material can be synthesized by limiting only one dimension, and so 1-D or wired material can be obtained by restricting dimensions in two directions. Hence, the most defining parameters to determine the material properties is not only size but also its dimensionality. The properties of 2-D graphene and 3-D graphite are vastly different from one another (Geim & Novoselov, 2009). 2-D materials are projected to significantly affect a broad range of applications, including electronics, gas storage and separation, potentially high-performance sensors, and inert coating techniques (Mas-Ballesté et al., 2011).

The elementary idea of moving objects by acoustic vibration was given by Ernst Chladni back in 1787 (Stöckmann, 2007). Extensive research was conducted on the aggregation and accumulation of sand particles in some specific regions onto nodal lines of a vibrating plate. The ultimate aim was to understand the distribution of the particles and pattern formation, also called the Chladni figures. The experiments included in the appendix of Faraday's 1831 article were designed to vary the density of the fluid enclosing the moving particle in this context. It has been reported that regardless of the fluid's nature, the velocity induced by the oscillatory motion of the fluid has a significant effect on the motion of the particles (Périnet et al., 2017).

Since then, the dominant view has been that particle movement away from nodal lines is irregular and random, which indicates that nodal line movement is uncontrollable on a Chladni plate (Arango & Reyes, 2016). Instead, the particle's motion should be highly regular to be statistically modeled, anticipated, and governed. By playing appropriately selected musical tunes, we can control the position of several reinforcing particles concurrently and independently using a solitary sonic actuation system (Q. Zhou et al., 2016). Their method enables the tracking, pattern modification, and sorting of numerous small objects made of various materials, including metals, dewdrops on solid materials, plant seeds, sugar balls, and electronic items.

This chapter concentrates on several essential aspects of the novel processing technique for fabricating functionally graded materials (FGMs) and 3-D materials using vibration-assisted methodologies. They include:

1. The effect of Faraday waves on fluidization properties of fine cohesive reinforcing particles under vertical vibration
2. The controlled manipulation of particles on a flat metallic surface
3. The outcome of mechanical vibration on reinforcing particles suspended in a fluidic medium

4. The effect of process parameters like vibrational frequency and amplitude
5. The pattern formation and cymatic visualization of dispersion of particles in a fluid
6. The displacement of particles subjected to vertical and horizontal vibrations in a fluid

11.2 THEORY OF THE PROPOSED FABRICATION TECHNIQUE

A 3-D material and FGMs can be synthesized by using a novel technique called vibration-assisted casting, which is based on a method that comprises the generation of streaming patterns in Faraday waves by imparting a vibration of a specific frequency and amplitude at the bottom of the molds. This phenomenon is now known as the Faraday instability. It occurs due to the periodic vibration caused by Faraday waves, which causes regular stationary waves to be generated on the liquid surface that oscillate at half the frequency of the forcing wave (subharmonic waves) (Périnet et al., 2017). The system can control the dispersion of the re-inforcing particles in a fluidic medium, as shown in Fig. 11.1, a function waveform generator that can generate waveforms (sine, square, pulse, and arbitrary). This wave is amplified by using an amplifier. This amplified wave can be imparted to the flat bottom of the mold filled with a molten fluid using mechanical vibration or by using an acoustic system. It can be

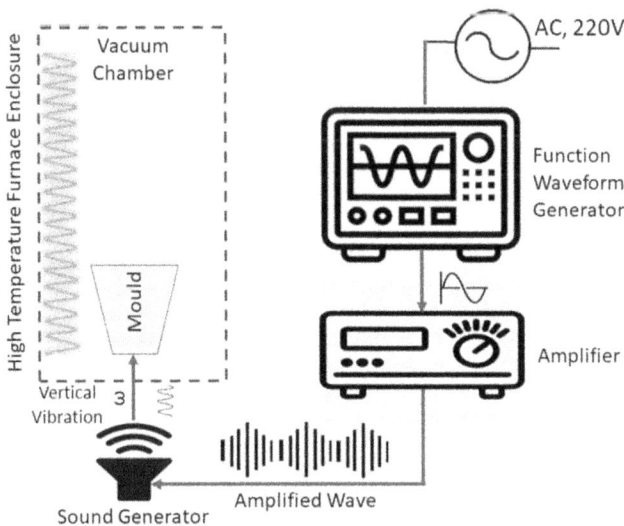

Figure 11.1 Schematic representation of the vibration-assisted casting methodology.

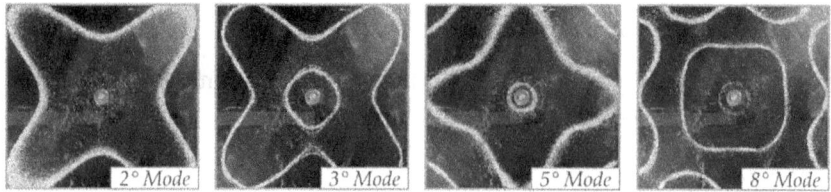

Figure 11.2 (Misseroni et al., 2016) The Hooke-Chladni-Faraday approach to represent the four eigenmodes of a square resilient plate with an open boundary.

achieved by creating ripples in the free surface of a liquid using vibration in the vertical direction.

Long before Michael Faraday's time, Robert Hooke and later Ernst Chladni devised an innovative approach for visualizing standing waves in elastodynamics (Arango & Reyes, 2016). A violin bow-like pattern was drawn on a metallic or glass plate to visualize the standing wave that existed in the plate. These researches have stimulated the interest of scientists, mathematicians, physicists, and musicians worldwide and gave a new dimension to cymatics, which deals with analysis of various methods of making sound and vibration visible. As an illustration, Fig. 11.2 shows instances of Chladni patterns for normal modes of an oscillating system of a square elastic plate having an open boundary. These patterns are influenced by the boundary conditions as well as any plate inhomogeneities, such as voids or inclusions, that may exist. Powder accumulates along the vibrating plate's nodal lines, providing a clear picture of the wavefront's position (Misseroni et al., 2016).

When it comes to Faraday waves, there are two distinct components to the velocity field. The first is an oscillatory component, whose importance for surface waves has been well recognized and whose linear and nonlinear causes have been extensively explored (Miles, 1990). This early understanding was made feasible in part by the fact that viscous effects aren't required to explain the oscillatory part accurately. On the other hand, viscosity is critical to describing the second part of the velocity field that is visible, particularly after a prolonged period of time (Society et al., 1953).

Specifically, the motion created by a vibrating fluid surface that oscillates is a highly intriguing subject. Numerous physical processes, including streaming flows and turbulence, contribute to contaminant scattering on the free surface of the fluid. Multiple mechanisms have been investigated in recent experimental studies (Clement et al., 1993; Geldart, 1973) but a clear assessment of the significant consequences remains lacking. Faraday waves have been suggested to deposit heavy particles and suspend light particles' templates to create particulate films (Geldart et al., 1984; Gutierrez and Aumaitre 2016). Once again, streaming patterns play a crucial part in the emergence of conventions. In another context, it has been shown that Faraday streaming flows

Table 11.1 Behaviour of fine cohesive particles in a fluidized bed subjected to mechanical vibration

Particle behaviours in a fluidized bed	Frequency (Hz)	Nature of vibration	Reference
Ring-shaped structure of floaters in liquid	46	Vertical	(Chen et al., 2014)
H-shaped structure of floaters in liquid	60	Vertical	
Geldart Group C powder: No agglomeration of particle	Less than 60	Vertical	(Lee et al., 2020)
Agglomeration of particle	60	Vertical	
Agglomeration and chemical bonding of particle		No Vibration	
Geldart Group C powder: The level of size segregation and the average size of agglomerates reduced	30	Vertical	(Xu & Zhu, 2005)
Growth of agglomerates	Greater than 50	Vertical	
Binary mixtures of particle effective segregation of particle	15	Vertical	(Sun et al., 2014)
Less efficient particle segregation		No Vibration	

are essential for the dynamics of wave patterns, particularly for their drift and mode interactions (Higuera & Knobloch, 2006; Maritín & Vega, 2006; Vega et al., 2001). Due to the scope and significance of its application domains, a comprehensive analysis of streaming flow is needed (analytical, numerical, and experimental under actual circumstances).

When it came to tracking the flow, particle image velocimetry (PIV) data was used in another experiment (Périnet 2017). The aluminum piece affixed to the top of the electromechanical shaker connects the trough to the shaker. The sinusoidal signal generated by the function waveform generator and bipolar amplifier is utilized to power the shaker. Accelerometers and lock-in amplifiers are employed to calibrate the system's acceleration precisely. The imaging system's second function channel uses a synthesized oscillation to serve as a synchronization signal for triggering the device at a programmable phase (Table 11.1).

11.2.1 Generation of Faraday Waves

When the forcing acceleration Γ exceeds a threshold Γ_0, which is determined by the forcing frequency $2f$, Faraday waves are introduced at the fluid's free surface, causing it to vibrate. Subharmonic waves have a natural frequency half that of the forcing frequency f (Périnet et al., 2017). For closed basins, through the dispersion relation, the wave number $k = 2\pi/\lambda$ and wavelength λ are associated with the metallic surface's natural frequencies $f_{m,n} = \omega_{m,n}/(2\pi)$.

11.2.2 Controlled Manipulation of Particles in Faraday Waves

Controlled manipulation of particles on a surface is critical in various applications, including material science and bioengineering. Controlled particle arrangement on a liquid surface would provide a convenient and customizable technique of engineering surface attributes like electrical and thermal conductivity, for example (Aubry & Singh, 2008; Grzybowski et al., 2000; Whitesides & Grzybowski, 2002). New ways to particle manipulation at a fluid surface have recently been developed. To control particle motion at the liquid-gas interface, they rely on the creation of surface waves (Francois et al., 2017). Whenever the liquid surface is vertically vibrated over a specific acceleration threshold, parametrically excited waves are generated (Umbanhowar et al., 1996). Faraday waves are modulational unstable waves that can easily be broken down into ensembles of localized oscillons or solitons (Shats et al., 2012). Oscillators can form spatially periodic patterns in viscous liquids, and such patterns have been proposed as metamaterials. Faraday wave patterns have recently been presented as micro-scale assembly templates. The current understanding of what can be done with such waves is based on various particle qualities like wettability and density (Falkovich et al., 2005).

11.3 GROUP C (GELDART POWDERS), FLUIDIZATION, AND AGGLOMERATION OF FINE PARTICLES

Fine particles are particularly appealing in the advanced materials, food-stuffs, and pharmaceutical sectors due to the increased surface-to-volume ratio and many more unique properties. However, as the size of these tiny granules gets smaller, managing them becomes considerably more challenging. Geldart classified particles with a diameter less than 30 μm as group C (cohesive) particles in 1973. Such particles are speculated to be inappropriate for fluidization due to their proclivity to aggregate due to interparticle solid interaction forces (Sanll et al., 2014; Wright and Saylor, 2003). However, the particle bed may display self-agglomerating fluidization due to the creation of stable and approximately single-sized agglomerates for nanoparticles when the interparticle force is substantially higher compared to gravity (Chirone et al., 1993). As a consequence, the fluidization behaviour of fine particles is highly impacted by the chartersistics (i.e. size, dispersion, and intensity) of the agglomerates produced during fluidization, as well as the observed agglomeration behaviours.

11.3.1 Fluidization Characteristics of Fine Cohesive Particles under Vibration

When a bed of fine particles acts like a liquid, gas, or fluid, this is called fluidization. Fluidization is a property of particles. Fine cohesive particle fluidization is inherently unstable. In fluidization research, the pressure drop is a key aspect in determining the fluidization phenomena in a fluidized bed.

When the interparticle interactions between most group C particles are insufficient, the agglomerates generated during fluidization are uncertain. They frequently exhibit significant size segregation, resulting in fractional fluidization or even de-fluidization. As a result, the fluidization tendency of tiny cohesive particles is strongly influenced by the characteristics (i.e. size, dispersion, and intensity) of the agglomerates formed during fluidization, as well as the agglomeration patterns observed during fluidization (Chirone et al., 1993; Pacek & Nienow, 1990; Xu & Zhu, 2005).

11.3.2 Techniques to Break Agglomerates

External factors like those of mechanical vibration, acoustic, and magnetic forces have the ability to fluidize small cohesive particles (Gupta & Mujumdar, 1980; Lee et al., 2018; van Ommen et al., 2012). The following limits apply to fluidization techniques for fine cohesive particles that use magnetic, acoustic, or electric fields: For 100 and 115 dB, magnetic particles are required, an extensive frequency range of 25 to 1000 Hz is needed, and frequencies of 1 kHz and greater are necessary to produce an electric field (Ammendola & Chirone, 2010; Kaliyaperumal et al., 2011; Valverde et al., 2009). By contrast, mechanical vibration fluidization is not restricted by particle type and is capable of fluidizing Geldart group C particles inside a vibration frequency band of 45 to 150 Hz (Barletta & Poletto, 2012; Liang et al., 2016; Wank et al., 2001).

11.3.2.1 Effect of Gas Velocity

A critical factor in the agglomeration of tiny particles undergoing fluidization is the fluidizing gas velocity, which is one of the most important factors to consider. Moreover, the impact of gas velocity on agglomeration remains an open question. On one hand, some have stated that the velocity of the fluidizing gas has minimal impact on the diameter of agglomerates (Chaouki et al., 1985); on the other hand (Iwadate & Horio, 1998) and (T. Zhou & Li, 1999) have presented models that suggest that increasing gas velocity should result in a reduction in agglomerate diameter. The dynamic self-agglomeration of microscopic particles undergoing fluidization is explained by the fact that cohesive forces promote agglomerate growth. In contrast, splitting forces like collisions among particles and agglomerates produced by the fluidizing gas, lead to agglomerate fracturing. Thus, increased gas velocity generates greater spliting forces that may also lead to smaller agglomeration size.

11.3.2.2 Effect of Mechanical Vibration

It has been demonstrated that mechanical vibration is an efficient technique for assisting the fluidization of tiny cohesive particles because it causes the splitting of channels and agglomerates (Wank et al., 2001). The mean size and level of size segregation of the agglomerates in the entire bed are substantially decreased due to the vibration process. Another factor that may have an impact on the agglomeration tendency of particles during fluidization in response to vibration (A) is the vibration intensity, that is directly proportional to the amplitude (A) or vibration frequency (f). It is required to express vibration intensity in terms of the vibration strength, Λ, which is interpreted as the proportion of the vibration acceleration to the gravity acceleration, or $\Lambda = A(2f)2/g$.

11.3.2.3 Theoretical Analysis

Size-segregation of the agglomerate is a phenomenon that frequently occurs during the fluidization of cohesive particles. The agglomerates form a layered structure that spans the length of the bed column: the tinier, quite stabilized agglomerates are located at the peak of the bed, while the more oversized, more loose agglomerates are found at the bottom (Xu & Zhu, 2006; T. Zhou & Li, 1999). Thus, it is only for top-bed stable agglomerates that modeling of agglomeration size is relevant in this case. To make the analysis more straight-forward, Xu and Zhu (2005) make the following assumptions: Specifically, (1) all agglomerates formed are identical in size (with an average diameter of d_a) and possess almost identical characteristics; (2) the wall factor is ignored; and (3) the van der Waals force outweighs almost all forms of interparticle cohesive interactions. Whenever the total energy resulting from collisions plus external vibration (if any) exceeds the total energy resulting from cohesive forces, the agglomeration tends to disrupt or shatter. Thus, at the agglomerate's breaking point, it is possible to achieve the following energy balance:

$$E_{coll} + E_{vib,eff} = E_{coh}$$

The letters "coll," "vib," "eff," and "coh" stand for "collision," "effective vibration," and "cohesion," respectively. When vibration is applied to the agglomerates, the agglomerate mean size and size segregation across the entire bed are considerably decreased. Both experimental and theoretical evidence indicate that increasing gas velocity results in a decrease in the size of agglomerations. Meanwhile, it has been discovered that vibration has a significantly more complex effect than previously thought. When the vibration intensity exceeds a threshold amount, the mean agglomerate size initially decreases and then gradually increases. This indicates that vibration has a two-fold effect on agglomeration during the fluidization of cohesive particles: it can aid in breaking agglomerates through extra vibrational energy and can facilitate agglomeration due to an increased contact possibility between the particles and/or agglomerates during the fluidization process. These two aspects of the effect compete with one another during the self-agglomeration mechanism in the particle fluidization as a result of vibration. Moreover, one of them becomes influential depending on the strength of the vibration applied.

Fluidization and product characteristics of nickel oxide (NiO) placed in the Geldart group C in a fluidized bed with vertical vibration, as shown in Fig. 11.2, was investigated by (Lee et al., 2018, 2020). The findings indicated that the pressure drop tendency stabilizes, and the least fluidization velocity falls as the vibration and reaction temperature increases. SEM and TEM examinations indicated agglomeration and chemical affinity in the absence of vibration. While, a vibration resulted in the lack of agglomeration and the appearance of a porous surface. According to the findings of X-ray EDS and mapping examinations of products, the oxygen concentrations and dispersion dropped significantly after fluidization under vibration, leaving just a trace of oxygen at the product surface. In this case, the surface gas velocity was 0.45 m/s, and the vibration frequency was 60 Hz, and agglomeration took place once again. It has also been demonstrated that vertical vibration relaxes stresses between small cohesive particles, preventing agglomeration and allowing them to move freely.

Fine particles readily agglomerate and cannot flow against gravity because interaction forces among particle-particle and particle-substrate are higher than gravity for smaller particles. In industry, vibration is frequently employed to overcome contact forces in powder handling processes. Significant research has been conducted on particles' behaviour on a vibrating substrate; although, most of these studies have used noncohesive particles with sizes greater than a few hundred micrometers (Bapat et al., 1986; Chirone et al., 1993; Jo Luo & Han, 1996). However, there are only a few reports on fine particle behaviour. Four distinct particles with median mass diameters ranging from 0.5 to 500 m were investigated (Kobayakawa & Matsusaka, 2013). A high-speed camera equipped with a zoom lens was employed to examine the interaction of particles on a 2-D vibrating plate as depicted in Fig. 11.3. In addition, a model based on gravity, adhesion, and drag force is used to investigate the particles' vertical motion in this study. Larger particles saltate more, while smaller ones show the tendency to agglomerate that saltate marginally, as observed by the high-speed digital camera. The values estimated by the model using gravity, adhesion, drag force, and restitution accorded well with the experimentally determined saltation levels of the particles. According to the model, the level of saltation of agglomerated particles are evaluated by the velocity and acceleration of the vibrating substrate.

11.4 DISPLACEMENT OF PARTICLES SUBJECTED TO VERTICAL AND HORIZONTAL VIBRATIONS IN A FLUID

The interaction of the particles held in oscillatory fluid flows due to the vertical and horizontal vibrations are critical to analyzing the dynamics of reinforcing medium in aluminum-based FGMs. Recently, there has been a

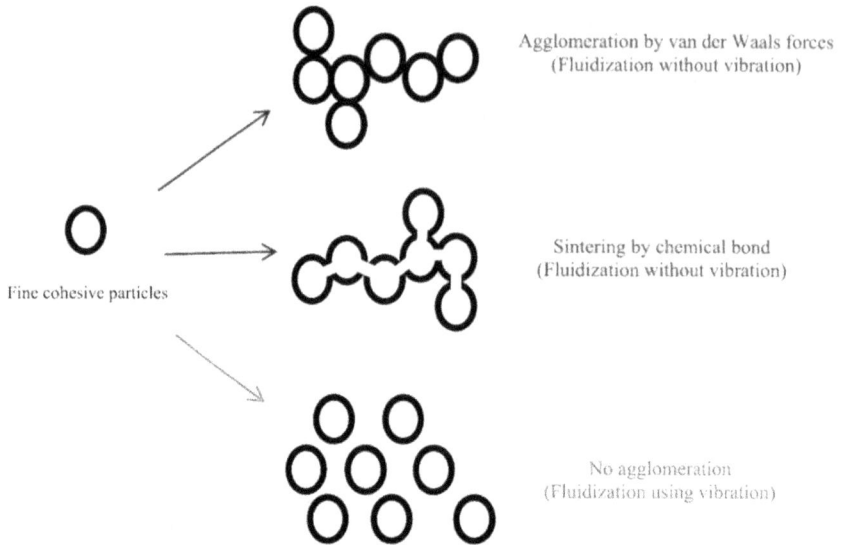

Figure 11.3 (Lee et al., 2020) Fluidization's impact on the agglomeration of tiny cohesive particles in the existance and non-existance of vibration.

surge of interest among the scientific community and industrial processes to study and how the dynamic behaviour of granular systems can be controlled effectively. The reinforcing ceramic particles such as B_4C, SiC, and Al_2O_3 can be considered spherical ones for determining the motion of suspended particles in a fluidic medium and thus facilitates creating a model for further analysis. There are several factors such as fluid viscosity (v), vibrational frequency (f), amplitude (a) and the dimensional acceleration (T), gravity force, buoyancy force, diffusion, acoustics radiation forces, and particle size, which researchers have reported studying the dynamics of particles in a fluid under the vibration.

The effect of vibration and fluid viscosity to analyze the interaction among a pair of metallic spheres placed in a liquid-filled container subjected to vertical vibration was investigated by (Klotsa et al., 2007). As illustrated in Fig. 11.4, the line connecting the centres of two spherical particles is perpendicular to the direction of oscillation (Fig. 11.5).

The gaps between the particles widen with the increase in viscosity and vice versa. Also, the sinusoidal motion of particles may be associated by the frequency (f), and by the dimensionless acceleration (*T*) of the liquid-filled container:

$$T = \frac{A\omega^2}{g}$$

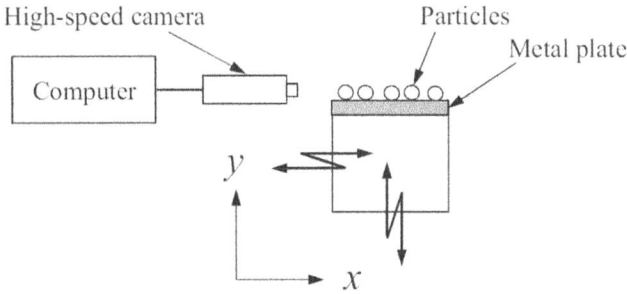

Figure 11.4 (Kobayakawa 2013) Schematic representation of the experimental setup equipped with high-speed camera to study the behaviour of the partcles on a vibrating plate.

Figure 11.5 Alignment of two spherical particles subjected to vertical vibration in an oscillatory fluidic system.

where
 A = amplitude of the vibrating cell
 $\omega = 2\pi f$ is the angular frequency
 g = acceleration due to gravity

Additionally (Nam et al., 2011) proposed a novel technique for gradient-size particle disintergration using a standing surface acoustic wave (SSAW) in deionized water. According to their findings, flat rectangular channels are more efficient for imparting SSAWs to particles. Additionally, three fundamental forces, namely viscous, diffusion, and acoustic radiation forces, are required to determine the particle's displacement from the fluidic channel's center plane.
Acoustic radiation force can be calculated using

$$Fr = -\left(\frac{\pi p_0^2 V_p \beta_m}{2\lambda}\right) \emptyset(\beta, \rho) \sin 2kx$$

where
 p_o = pressure amplitude
 V = volume
 λ = ultrasonic wavelength
 k = wave factor
 x = distance from an anti-pressure node to a pressure node
 ρ = density
 β = compressibility.

Subscripts p and m indicate particle and medium, respectively.

Also, the displacement of the particle, considering the three forces, i.e., hydrodynamic focusing, diffusion, and acoustic radiation force, can be inserted into a mathematical equation as:

$$d = \frac{w_f}{2} + \sqrt{Dt_w} + \frac{\lambda}{2\pi}\tan^{-1}(e^c)$$

where

w_f = width of the focused flow

t_w = working time

λ = SAW operating wavelength, and can be calculated by $f = \frac{C}{\lambda}$

The displacements of the particle are a function of particle size and working time of SSAW under the effect of acoustic force. Thus, the bigger particles show maximum displacement towards the wall, while smaller particles show the least displacements.

Further, Lyubimov et al. (2013) carried out the numerical investigation of the behaviour of a system of particles suspended in a fluid under high-frequency translational vibrations. Interaction forces fall out as the gap between particles increases and depends on particle pairs' orientation corresponding to the vibrational direction. Additionally, the velocity of a particle moving through an oscillating fluid can be characterized by

$$u = Re\,(Ue^{i\omega t})$$

The particles are clustered in layers perpendicular to the vibration direction under vibrations of linear polarizations.

Furthermore, Voth et al. (2002) observed the clustering and ordered crystalline patterns of the particles due to the attractive and repulsive interactions in a fluid subjected to vertical vibration. The concave-shaped bottom glass plate was employed to restrict the drifting of particles to the edges of the container due to inexact leveling or moderately nonlinear vibrational motion. The particle distribution was found to be dependent on acceleration given by $\left(T = \frac{S\omega^2}{g}\right)$. Hexagonal distribution was obtained for acceleration ranges of $2.8 < T < 3.0$.

In another experiment, the discharge rates and trajectories of the particles were examined in a hopper subjected to vertical vibrations. Wassgren et al. (2002) vibrated the hopper box sinusoidally, $z = a\sin\omega t$ where 'a' is the amplitude of vibration and ω is the radian frequency, and based on the dimensionless acceleration amplitude, the bed of particles exhibited a variety of flow patterns $\left(T = \frac{a\omega^2}{g}\right)$.

Also, Jaeger et al. (1996) reported that the temperature has no effect on the granular system and the movement of particles depends on the angle of repose, static friction and inelastic collisions. Further, Hassan et al. (2006) deduced that when a particle is suspended by thin wire in a fluidic container subjected to horizontal vibrations, the amplitude of the particle is linearly proportional to the amplitude of the fluidic cell.

11.5 CONCLUSIONS

The interaction of particles held in an oscillatory fluid flow due to vertical and horizontal vibrations is critical for understanding the dynamics of the reinforcing medium in aluminum-based FGMs. The reinforcing ceramic particles such as B_4C, SiC, and Al_2O_3 can be considered spherical ones for determining the motion of suspended particles in a fluidic medium, thus facilitating creating a model for further analysis. In view of the dynamics of particles in a fluid under vibration, several factors, such as fluid viscosity (v), vibrational frequency (f), amplitude (a) and dimensional acceleration (T), gravity force, buoyancy force, diffusion, acoustic radiation forces, and particle size have been investigated. When vibration is introduced to a fluidized bed, both the average size of the agglomerates and the extent of size segregation across the entire bed are drastically decreased. All this experimental and theoretical evidence indicates that increasing the gas velocity results in a reduction in the agglomerate size.

Whereas it has been discovered that mechanical vibration has a considerably more complicated effect, when the vibration intensity exceeds a threshold value, the mean agglomerate size decreases initially but gradually increases. Thus, vibration's effect on agglomeration during the fluidization of cohesive particles is bidirectional. It can aid in the disintegration of agglomerates through the addition of vibrational energy, and favors agglomeration through the increased associating possibility between particles or agglomerates. Additionally, vertical vibration alleviates forces between tiny cohesive particles, thereby preventing agglomeration and restoring them to an active fluidization state. Based on several studies, a novel processing technique to synthesize 3-D materials and functionally graded materials (FGMs) has been proposed based on the generation of Faraday waves on the free surface of the liquid using mechanical vibration. Thus, the distribution of fine particles can be controlled efficiently in a fluidic medium by adjusting the various process parameters such as vibrational frequency, amplitude, density, and average particle sizes.

Eventually, this technique will provide much-needed flexibility in controlling the dispersion of particles in fluids, and it will be possible to obtain numerous spatially varying patterns of reinforcing particles in the microstructure of FGMs and 3-D materials.

REFERENCES

Ammendola, P., & Chirone, R. (2010). Aeration and mixing behaviours of nano-sized powders under sound vibration. *Powder Technology*, *201*(1), 49–56. 10.1016/j.powtec.2010.03.002

Arango, J., & Reyes, C. (2016). Stochastic models for chladni figures. *Proceedings of the Edinburgh Mathematical Society*, *59*(2), 287–300. 10.1017/S0013 091515000139

Aubry, N., & Singh, P. (2008). Physics underlying controlled self-assembly of micro- and nanoparticles at a two-fluid interface using an electric field. *Physical Review E – Statistical, Nonlinear, and Soft Matter Physics*, *77*(5), 1–11. 10.1103/ PhysRevE.77.056302

Bapat, C. N., Sankar, S., & Popplewell, N. (1986). Repeated impacts on a sinusoidally vibrating table reappraised. *Journal of Sound and Vibration*, *108*(1), 99–115. 10.1016/S0022-460X(86)80314-5

Barletta, D., & Poletto, M. (2012). Aggregation phenomena in fluidization of co-hesive powders assisted by mechanical vibrations. *Powder Technology*, *225*, 93–100. 10.1016/j.powtec.2012.03.038

Cannillo, V., Lusvarghi, L., Siligardi, C., & Sola, A. (2007). Prediction of the elastic properties profile in glass-alumina functionally graded materials. *Journal of the European Ceramic Society*, *27*(6), 2393–2400. 10.1016/j.jeurceramsoc. 2006.09.009

Chaouki, J., Chavarie, C., Klvana, D., & Pajonk, G. (1985). Effect of interparticle forces on the hydrodynamic behaviour of fluidized aerogels. *Powder Technology*, *43*(2), 117–125. 10.1016/0032-5910(85)87003-0

Chen, P., Luo, Z., Güven, S., Tasoglu, S., Ganesan, A. V., Weng, A., & Demirci, U. (2014). Microscale assembly directed by liquid-based template. *Advanced Materials*, *26*(34), 5936–5941. 10.1002/adma.201402079

Chirone, R., Massimilla, L., & Russo, S. (1993). Bubble-free fluidization of a co-hesive powder in an acoustic field. *Chemical Engineering Science*, *48*(1), 41–52. 10.1016/0009-2509(93)80281-T

Clément, E., Luding, S., Blumen, A., Rajchenbach, J., & Duran, J. (1993). Fluidization, condensation and clusterization of a vibrating column of beads. *International Journal of Modern Physics B*, *07*(09n10), 1807–1827. 10.1142/ S0217979293002602

Falkovich, G., Weinberg, A., Denissenko, P., & Lukaschuk, S. (2005). Surface tension: Floater clustering in a standing wave. *Nature*, *435*(7045), 1045–1046. 10.1038/4351045a

Francois, N., Xia, H., Punzmann, H., Fontana, P. W., & Shats, M. (2017). Wave-based liquid-interface metamaterials. *Nature Communications*, *8*, 1–9. 10.103 8/ncomms14325

Geim, A. K., & Novoselov, K. S. (2009). The rise of graphene. *Nanoscience and Technology: A Collection of Reviews from Nature Journals*, 11–19. 10.1142/ 9789814287005_0002.

Geldart, D. (1973). Types of gas fluidization. *Powder Technology*, *7*(5), 285–292. 10.1016/0032-5910(73)80037-3

Geldart, D., Harnby, N., & Wong, A. C. (1984). Fluidization of cohesive powders. *Powder Technology*, *37*(1), 25–37. 10.1016/0032-5910(84)80003-0

Grzybowski, B. A., Stone, H. A., & Whitesides, G. M. (2000). Dynamic self-assembly of magnetized, millimetre-sized objects rotating at a liquid-air interface. *Nature, 405*(6790), 1033–1036. 10.1038/35016528

Gupta, R., & Mujumdar, A. S. (1980). Aerodynamics of a vibrated fluid bed. *The Canadian Journal of Chemical Engineering, 58*(3), 332–338. 10.1002/cjce.5450580309

Gutiérrez, P., & Aumaître, S. (2016). Clustering of floaters on the free surface of a turbulent flow: An experimental study. *European Journal of Mechanics, B/Fluids, 60*, 24–32. 10.1016/j.euromechflu.2016.06.009

Hassan, S., Lyubimova, T. P., Lyubimov, D. V., & Kawaji, M. (2006). Motion of a sphere suspended in a vibrating liquid-filled container. *Journal of Applied Mechanics, Transactions ASME, 73*(1), 72–78. 10.1115/1.1992516

Higuera, M., & Knobloch, E. (2006). Nearly inviscid faraday waves in slightly rectangular containers. *Progress of Theoretical Physics Supplement, 161*(161), 53–67. 10.1143/PTPS.161.53

Iwadate, Y., & Horio, M. (1998). Prediction of agglomerate sizes in bubbling fluidized beds of group C powders. *Powder Technology, 100*(2–3), 223–236. 10.1016/S0032-5910(98)00143-0

Jaeger, H. M., Nagel, S. R., & Behringer, R. P. (1996). Granular solids, liquids, and gases. *Reviews of Modern Physics, 68*(4), 1259–1273. 10.1103/RevModPhys.68.1259

Jha, D. K., Kant, T., & Singh, R. K. (2013). A critical review of recent research on functionally graded plates. *Composite Structures, 96*, 833–849. 10.1016/j.compstruct.2012.09.001

Jo Luo, A. C., & Han, R. P. S. (1996). The dynamics of a bouncing ball with a sinusoidally vibrating table revisited. *Nonlinear Dynamics, 10*, 1–18. https://link-springer-com.proxy.lib.ohio-state.edu/content/pdf/10.1007%2FBF00114795.pdf

Kaliyaperumal, S., Barghi, S., Zhu, J., Briens, L., & Rohani, S. (2011). Effects of acoustic vibration on nano and sub-micron powders fluidization. *Powder Technology, 210*(2), 143–149. 10.1016/j.powtec.2011.03.007

Kawasaki, A., & Watanabe, R. (1997). Concept and P/M fabrication of functionally gradient materials. *Ceramics International, 23*(1), 73–83. 10.1016/0272-8842(95)00143-3

Klotsa, D., Swift, M. R., Bowley, R. M., & King, P. J. (2007). Interaction of spheres in oscillatory fluid flows. *Physical Review E – Statistical, Nonlinear, and Soft Matter Physics, 76*(5), 1–8. 10.1103/PhysRevE.76.056314

Kobayakawa, M., & Matsusaka, S. (2013). Analysis of behaviour of small agglomerated particles on two-dimensional vibrating plate. *AIP Conference Proceedings, 1542*(June 2013), 991–994. 10.1063/1.4812100

Lee, J. R., Hasolli, N., Jeon, S. M., Lee, K. S., Kim, K. D., Kim, Y. H., Lee, K. Y., & Park, Y. O. (2018). Optimization fluidization characteristics conditions of nickel oxide for hydrogen reduction by fluidized bed reactor. *Korean Journal of Chemical Engineering, 35*(11), 2321–2326. 10.1007/s11814-018-0137-2

Lee, J. R., Lee, K. S., Park, Y. O., & Lee, K. Y. (2020). Fluidization characteristics of fine cohesive particles assisted by vertical vibration in a fluidized bed reactor. *Chemical Engineering Journal, 380*(August 2019), 122454. 10.1016/j.cej.2019.122454

Liang, X., Zhou, Y., Zou, L., Kong, J., Wang, J., & Zhou, T. (2016). Fluidization behaviour of binary iron-containing nanoparticle mixtures in a vibro-fluidized bed. *Powder Technology*, *304*, 101–107. 10.1016/j.powtec.2016.01.012

Lyubimov, D. V., Baydin, A. Y., & Lyubimova, T. P. (2013). Particle dynamics in a fluid under high frequency vibrations of linear polarization. *Microgravity Science and Technology*, *25*(2), 121–126. 10.1007/s12217-012-9336-3

Maritín, E., & Vega, J. M. (2006). The effect of surface contamination on the drift instability of standing Faraday waves. *Journal of Fluid Mechanics*, *546*(figure 1), 203–225. 10.1017/S0022112005007032

Mas-Ballesté, R., Gómez-Navarro, C., Gómez-Herrero, J., & Zamora, F. (2011). 2-D materials: To graphene and beyond. *Nanoscale*, *3*(1), 20–30. 10.1039/c0nr00323a

Miles, J. (1990). Parametrically forced surface waves. *Annual Review of Fluid Mechanics*, *22*(1), 143–165. 10.1146/annurev.fluid.22.1.143

Misseroni, D., Colquitt, D. J., Movchan, A. B., Movchan, N. V., & Jones, I. S. (2016). Cymatics for the cloaking of flexural vibrations in a structured plate. *Scientific Reports*, *6*(October 2015), 1–11. 10.1038/srep23929

Nam, J., Lee, Y., & Shin, S. (2011). Size-dependent microparticles separation through standing surface acoustic waves. *Microfluidics and Nanofluidics*, *11*(3), 317–326. 10.1007/s10404-011-0798-1

Pacek, A. W., & Nienow, A. W. (1990). Fluidisation of fine and very dense hard-metal powders. *Powder Technology*, *60*(2), 145–158. 10.1016/0032-5910(90)80139-P

Périnet, N., Gutiérrez, P., Urra, H., Mujica, N., & Gordillo, L. (2017). Streaming patterns in Faraday waves. *Journal of Fluid Mechanics*, *819*, 285–310. 10.1017/jfm.2017.166

SanlI, C., Lohse, D., & Van Der Meer, D. (2014). From antinode clusters to node clusters: The concentration-dependent transition of floaters on a standing Faraday wave. *Physical Review E – Statistical, Nonlinear, and Soft Matter Physics*, *89*(5), 1–9. 10.1103/PhysRevE.89.053011

Shats, M., Xia, H., & Punzmann, H. (2012). Parametrically excited water surface ripples as ensembles of oscillons. *Physical Review Letters*, *108*(3), 1–5. 10.1103/PhysRevLett.108.034502

Society, T. R., Transactions, P., Society, R., & Sciences, P. (1953). Mass transport in water waves. *Philosophical Transactions of the Royal Society of London. Series A, Mathematical and Physical Sciences*, *245*(903), 535–581. 10.1098/rsta.1953.0006

Stöckmann, H. J. (2007). Chladni meets Napoleon. *European Physical Journal: Special Topics*, *145*(1), 15–23. 10.1140/epjst/e2007-00144-5

Sun, L., Zhao, F., Zhang, Q., Li, D., & Lu, H. (2014). Numerical simulation of particle segregation in vibration fluidized beds. *Chemical Engineering and Technology*, *37*(12), 2109–2115. 10.1002/ceat.201400158

Umbanhowar, P. B., Melo, F., & Swinney, H. L. (1996). Localized excitations in a vertically vibrated granular layer. *Nature*, *382*(6594), 793–796. 10.1038/382793a0

Valverde, J. M., Espin, M. J., Quintanilla, M. A. S., & Castellanos, A. (2009). Electrofluidized bed of silica nanoparticles. *Journal of Electrostatics*, *67*(2–3), 439–444. 10.1016/j.elstat.2009.01.021

van Ommen, J. R., Valverde, J. M., & Pfeffer, R. (2012). Fluidization of nano-powders: A review. *Journal of Nanoparticle Research*, *14*(3): 1–29. 10.1007/s11051-012-0737-4

Vega, J. M., Knobloch, E., & Martel, C. (2001). Nearly inviscid Faraday waves in annular containers of moderately large aspect ratio. *Physica D: Nonlinear Phenomena*, *154*(3–4), 313–336. 10.1016/S0167-2789(01)00238-X

Voth, G. A., Bigger, B., Buckley, M. R., Losert, W., Brenner, M. P., Stone, H. A., & Gollub, J. P. (2002). Ordered clusters and dynamical states of particles in a vibrated fluid. 88(23): 234301. 10.1103/PhysRevLett.88.234301

Wank, J. R., George, S. M., & Weimer, A. W. (2001). Vibro-fluidization of fine boron nitride powder at low pressure. *Powder Technology*, *121*(2–3), 195–204. 10.1016/S0032-5910(01)00337-0

Wassgren, C. R., Hunt, M. L., Freese, P. J., Palamara, J., & Brennen, C. E. (2002). Effects of vertical vibration on hopper flows of granular material. *Physics of Fluids*, *14*(10), 3439–3448. 10.1063/1.1503354

Whitesides, G. M., & Grzybowski, B. (2002). Self-assembly at all scales. *Science*, *295*(5564), 2418–2421. 10.1126/science.1070821

Wright, P. H., & Saylor, J. R. (2003). Patterning of particulate films using Faraday waves. *Review of Scientific Instruments*, *74*(9), 4063–4070. 10.1063/1.1602936

Xu, C., & Zhu, J. (2005). Experimental and theoretical study on the agglomeration arising from fluidization of cohesive particles – Effects of mechanical vibration. *Chemical Engineering Science*, *60*(23), 6529–6541. 10.1016/j.ces.2005.05.062

Xu, C., & Zhu, J. (2006). Parametric study of fine particle fluidization under mechanical vibration. *Powder Technology*, *161*(2), 135–144. 10.1016/j.powtec.2005.10.002

Zhou, Q., Sariola, V., Latifi, K., & Liimatainen, V. (2016). Controlling the motion of multiple objects on a Chladni plate. *Nature Communications*, *7*, 1–10. 10.1038/ncomms12764

Zhou, T., & Li, H. (1999). Estimation of agglomerate size for cohesive particles during fluidization. *Powder Technology*, *101*(1), 57–62. 10.1016/S0032-5910(98)00148-X

Chapter 12

Processing of Composite Materials Using Microwave Energy

Radha Raman Mishra

Department of Mechanical Engineering, Birla Institute of Technology and Science Pilani, Pilani, India

CONTENTS

12.1 INTRODUCTION

Composite materials have been used in many engineering applications such as structures, aerospace, automobiles, and nuclear reactors in recent years. The industrial components manufactured in these applications involve complex heating characteristics of composite materials. Many of the manufacturing processes used to process composite materials undergo interaction with thermal energy and rely on effective heat transfer between the energy source and material. Non-uniform heating of the materials results in a temperature gradient across them, affecting the grain size, microstructure, and mechanical properties. Therefore, developing a manufacturing process that can offer energy and time saving with rapid and eco-friendly processing. Microwave energy can be a resource-efficient alternative that can address some of these challenges. In recent years, microwave energy has been exploited for the processing of composite materials, including metallic materials, which are difficult to process using microwaves [1–3]. Additionally, various combinations of materials mixing fibres, granules, and powders have been utilized to develop composite materials using microwave energy.

The selective microwave heating process, in which only a selected volume is exposed to microwave radiation, offers better properties in polymer composite materials (PMC) [4]. Moreover, conductive fibres, conductive particles, and a conductive coating on nanofibres improve bonding strength due to enhanced microwave absorption in the conductive phase. The literature is very little on natural fibre-based PMC processing using microwave energy. Microwave processing of ceramic matrix composites (CMC) has been reported with improved properties [5]. Enhanced densification rate, reduced porosity, improved sinterability, and better mechanical properties were reported with nano reinforcements in CMCs. Processing of metal matrix composites (MMCs) is challenging as metallic materials reflect microwaves at room temperature. However, decreasing the size of powder particles allow them to couple with microwave energy. Microwave sintering of MMCs enhances densification rate, reduces porosity, reduces oxidation, and improves mechanical properties than conventional sintering [6,7]. The smaller size or higher dielectric loss factor of reinforcements and matrix improves mechanical properties [8]. It has been observed that most of the literature on microwave processing of composites reports processing of materials; however, physics involved in the interaction of microwaves with matrix and reinforcements have been largely unexplored. In the present chapter, a brief overview of microwave interaction with engineering materials and microwave heating mechanisms involved in the processing of composite materials have been discussed. Challenges and opportunities associated with the processing of composite materials have been discussed.

12.2 MICROWAVE-MATERIAL INTERACTION PHENOMENA

Absorption of microwave energy in a material depends upon various factors such as penetration depth (D) of microwaves inside the material, electrical and thermal properties of the material, the strength of electric (E), and magnetic (M) fields. Fig. 12.1 illustrates the effects of E, H, D, and material characteristics on the power dissipated inside the different materials. Based on the microwave interaction characteristics, the materials can be segregated into three groups:

- **Category-1:** No loss of the strengths in E and H fields due to almost zero loss factor of the materials such as teflon, quartz, etc.
- **Category-2:** Loss of the strengths in E and H fields due to increasing dielectric loss factor (ε'') of materials. Most of the lost materials, such as water, SiC, etc., fall in the flat portion of the curve (P curve, Fig. 12.1).

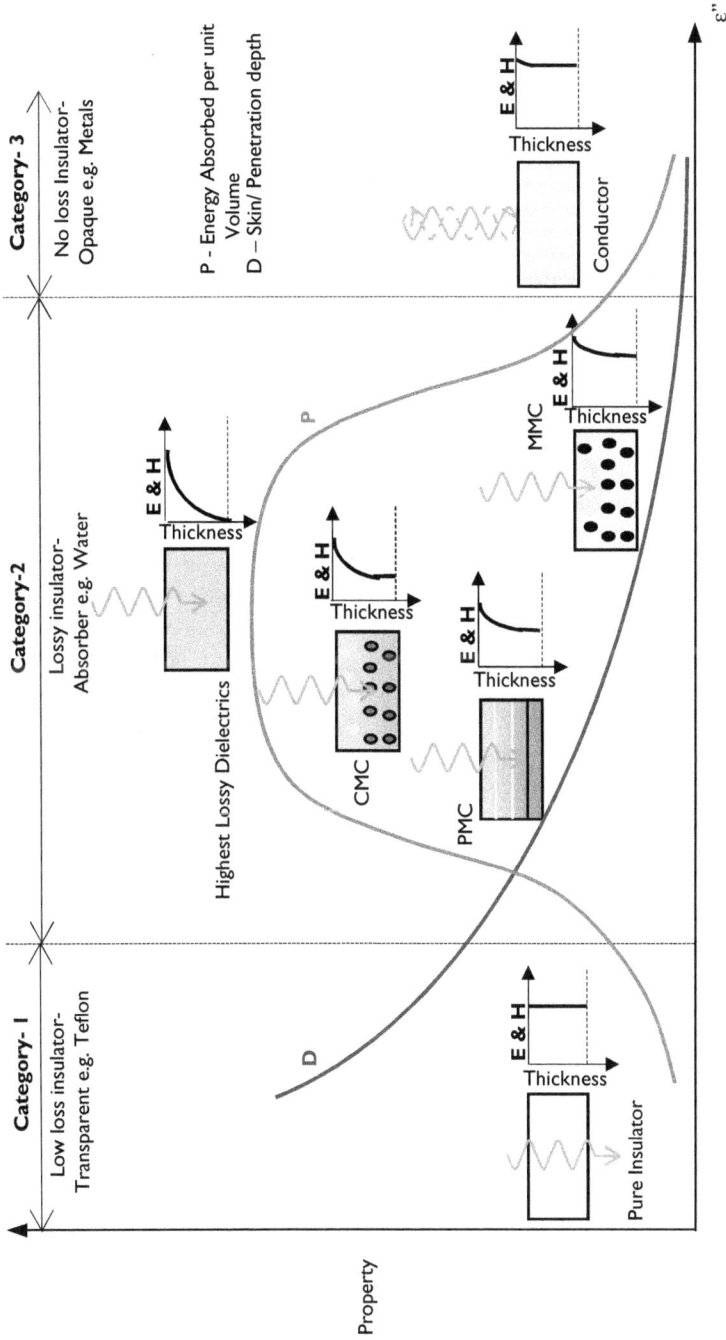

Figure 12.1 Microwave interactions with materials [2].

- **Category-3:** Negligible loss of the strength in E and H fields up to a few micrometer thicknesses of the materials such as metals, etc. The major amount of microwave energy gets reflected in the surfaces of these materials.

On the basis of microwave energy absorption characteristics, the materials can be classified into four major groups [9]. Fig. 12.2 illustrates responses possible by different materials once exposed to microwaves:

- **Transparent:** all incident microwaves get transmitted through these materials (Fig. 12.2, indicated as T) without any absorption/heating of the material (such as teflon, quartz, etc.).
- **Absorber:** materials (Fig. 12.2, indicated as A) that absorb total incident microwaves up on it and get heated (such as water, SiC, etc.) are known as the absorber.
- **Opaque:** almost all incident microwaves get reflected from the surfaces of these materials (Fig. 12.2, indicated as R) without or negligible energy absorption (such as all bulk metals).
- **Mixed absorbers:** materials (Fig. 12.2, indicated as M) that have multi-phases and at least one phase is microwave absorber (such as PMC, CMC, MMC).
- **Partial absorbers:** materials (Fig. 12.2, indicated as P) that absorb a portion of the incident microwaves with reflection and transmission of some parts of the incident microwave energy are known as partial absorbers (such as polymers and ceramics).

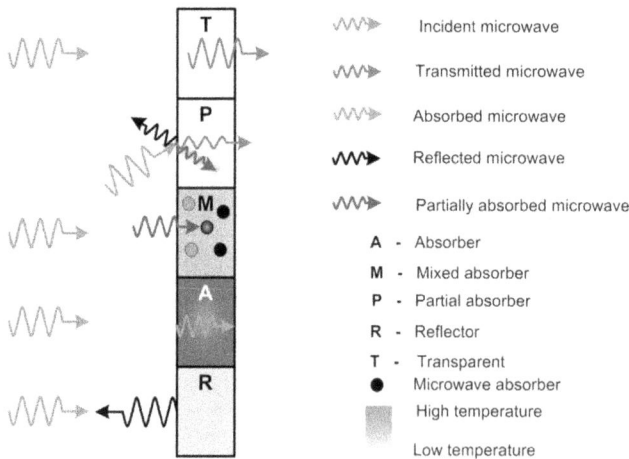

Figure 12.2 Microwave absorption characteristics of different materials [9].

12.3 MICROWAVE PROCESSING OF COMPOSITE MATERIALS

The composite materials are mixed absorber materials, and the properties of the constituents (matrix and reinforcements) affect the heating during microwave exposure. The constituent, which has a high dielectric loss factor, couples with microwaves rapidly and absorbs microwave energy. Subsequently, the constituent gets heated (selective heating) and becomes a heat source for the hybrid heating of the other constituent(s). The localized hybrid and selective heating improve the properties of composites compared to the composites developed using conventional heating methods. The presence of constituents with higher conductivity (such as CNT, Al, Cu, Ni, C powders, etc.) in composites affects properties such as dielectric constant, dielectric loss factor, loss tangent, and relative permittivity penetration depth and conductive properties [10]. The conductive reinforcements generally form the conductive networks after a certain volume percentage known as percolation threshold; consequently, significant improvement in the properties of PMCs/CMCs occurs [10]. Moreover, the size and magnetic properties of the reinforcements also affect the microwave heating characteristics of the composites [10]. In the following sections, microwave heating of the composite materials has been discussed.

12.3.1 Polymer Matrix Composite

Higher dielectric loss factor constituent (reinforcement/matrix) dominates over low dielectric loss materials while exposing the PMC to microwave energy. The PMCs containing low conductivity fibres such as fibre glass, aramid, natural, and absorption of the microwave energy depends upon the dielectric properties of the polymer matrix [1]. At the same time, high conductivity fibres (such as carbon fibre) are rapidly heated by microwave energy and dominate over a polymer matrix. The same concept has been illustrated in Fig. 12.3. The localized selective heating of carbon fibres improves mechanical properties of PMC due to stronger bonding at the interface of fibre and matrix; however, the low dielectric loss factor of fibers causes poor bonding of fibres with polymer matrix and poor mechanical properties in PMCs. An increase in temperature of the polymer matrix affects its chemical structure and dielectric properties during microwave exposure. In the thermoset polymer matrix, cross-linking starts with an increase in temperature during microwave heating. This results in the formation of the internal network structure and an increase in the viscosity of the matrix. Enhanced viscosity of thermoset polymer matrix at elevated temperature results in a decline in microwave absorption in the matrix due to restriction in dipoles orientations in the applied. On the other hand, the thermoplastic polymer matrix behaves as microwave transparent at room temperature, and it does not absorb microwave

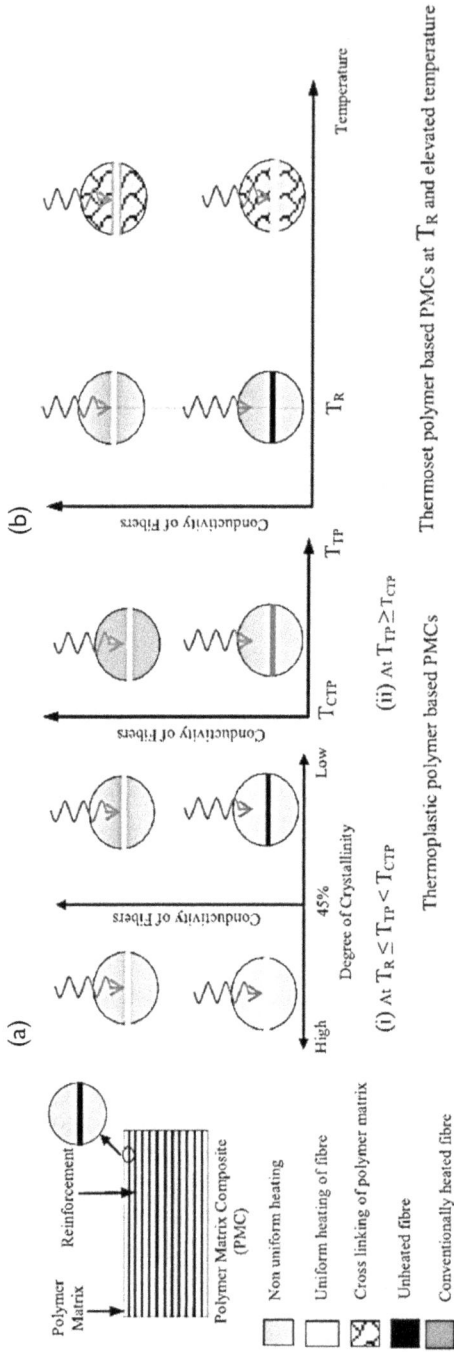

Figure 12.3 Microwave heating of PMCs [2].

energy until it attains a critical temperature. These effects are attributed to their low dielectric loss factor and degree of crystallinity. The degree of crystallinity of a thermoplastic adversely affects its dielectric loss factor. The degree of crystallinity above 45% pushes the thermoplastic matrix in the microwave transparent material domain [11]. The microwave energy interaction with thermoset and thermoplastic polymer matrix PMCs has been schematically illustrated in Fig. 12.3. Below the critical temperature, microwave energy absorption in thermoplastic polymers with a higher degree of crystallinity depends upon the microwave-absorbing characteristics of the reinforced fibres and heat is transferred by conduction to the thermoplastic matrix from the reinforcements.

12.3.2 Ceramic Matrix Composites

Generally, both the constituents (matrix and reinforcement) in the CMCs are ceramics. Microwave heating of the CMCs depends on the dielectric properties of matrix and reinforcement materials, the critical temperature of ceramics, and powder size [12]. Ceramics behave as transparent/lossy materials at room temperature depending upon their dielectric loss factors. Ceramics such as SiC and zirconia absorb all incident microwaves; whereas ceramics such as alumina, SiO_2, and Fe_3O_4 are transparent to microwaves at room temperature [13]. As the temperature of ceramics increases, microwave absorption gets enhanced due to rapid increase in the dielectric loss factor beyond a temperature known as the critical temperature (for example, above 800–900°C for alumina). The microwave hybrid heating technique is helpful in processing the CMCs compared to the conventional heating techniques and ensures lesser possibilities of thermal cracking due to more uniform heating. For the processing of CMCs, a suitable combination of matrix (low dielectric loss factor) and reinforcement (high dielectric loss factor) materials can be mixed to develop the green compact. On microwave exposure, the reinforcement acts as a localized susceptor in the low-loss factor ceramic matrix. The mechanism of heating of CMCs is schematically illustrated in Fig. 12.4. Conduction of heat from the reinforcement to the matrix continues until the matrix temperature reaches beyond the critical temperature. Subsequently, microwave hybrid heating of the matrix and direct microwave heating of the reinforcements results in a higher densification rate with uniform heating. Better mechanical properties can be achieved using the finer powder particles of matrix and reinforcement, as it offers volumetric heating.

12.3.3 Metal Matrix Composites

In the MMCs, usually, reinforcements and matrix are ceramic particles and metallic particles, respectively. Microwave heating of MMCs is affected by

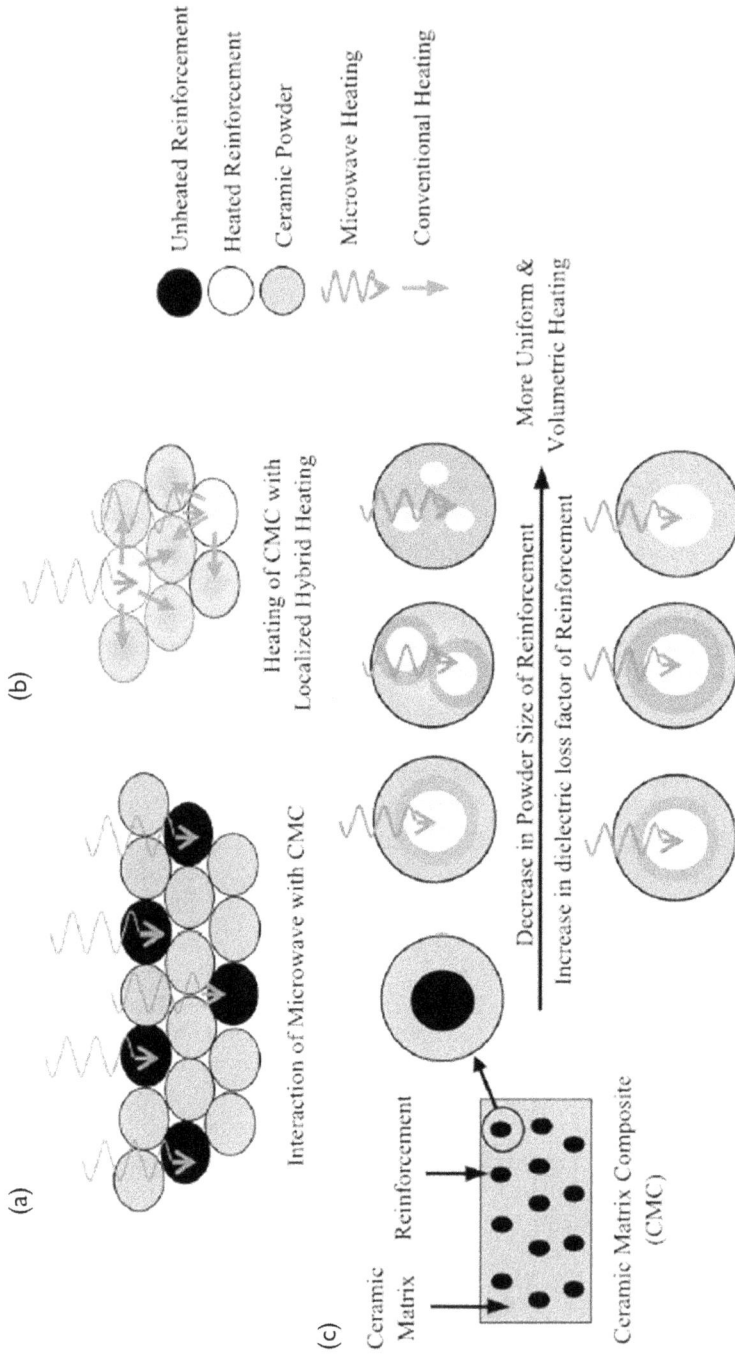

Figure 12.4 Microwave heating of CMCs [2].

powder size, properties of matrix and reinforcement materials, temperature, and processing frequency [14]. In microwave processing of MMCs,
the green compact (matrix and reinforcement) is exposed to microwave
energy. High dielectric loss factor ceramic reinforcement (such as, Sic) offers rapid heating of MMCs; whereas metallic particles absorb microwave
energy only when their sizes are comparable to the skin depth. In the case of
coarser metallic particles, the hybrid mode of microwave absorption is
slower [2]. The localized hybrid heating occurs in the MMCs with coarser
metal particles. On the other hand, micro- or nanosize metallic powders
such as the matrix provide more uniform heating of MMCs. The phenomenon of microwave heating of the MMCs is illustrated schematically in
Fig. 12.5.

12.4 CHALLENGES

Although microwave energy has the potential to address various issues in
the conventional manufacturing processes, there are challenges yet to
overcome for the effective use of microwaves for processing composite
materials. The challenges associated with microwave processing of composite materials can be categorized into the following two groups.

12.4.1 Challenges Inherent with Microwaves

Limited control over energy conversion during microwave processing of the
composite materials results in various challenges caused by the non-uniform
heating effects of microwave energy. Some relevant issues are below:

- Thermal damage of the processed composite materials
- Need of special tooling design to achieve rapid and uniform heating
- Real-time measurement of material properties during microwave
  exposure
- Limited flexibility and knowledge to carry the modelling of the
  process
- Lack of physical interpretations of microwave heating phenomena
- The poor repeatability of results and unavailability of experimental
  data
- Issues related to safety and health of people due to leakage

12.4.2 Challenges Inherent with Composite Materials

The anisotropic properties and heterogeneous distribution of reinforcements in composite materials cause issues during microwave heating. The
main challenges are:

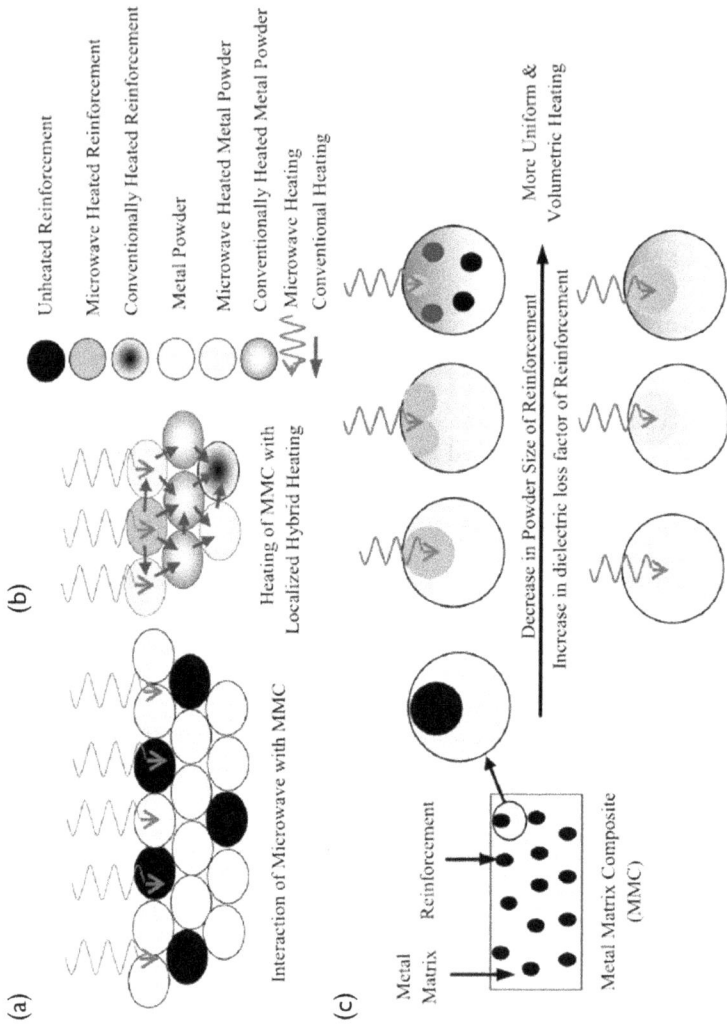

Figure 12.5 Microwave heating of MMCs [2].

- Thermal damage due to non-uniform heating of matrix and reinforcements
- Poor adhesion of matrix and reinforcement due to different microwave energy absorption rate
- Absorption of moisture by the PMCs
- Control of densification rate of MMCs and CMCs and heating of different constituents
- Elimination of oxidation of matrix phases in MMCs

12.5 OPPORTUNITIES

Microwave processing of composite materials is an emerging area of material processing. It offers various opportunities to exploit the potential of microwave energy due to inherent processing challenges provided by the conventional heating techniques for composite processing. A summary of possible future research directions has been outlined as below:

- Approaches are needed that can utilize microwave energy more effectively
- Innovations for tooling design
- Exploring the physical fundamentals behind the microwave heating
- Development of realistic models for theoretical studies
- Design of novel composite materials with enhanced properties

12.6 CONCLUSIONS

The potential of microwave energy in composite materials processing has been discussed. Processing of composite materials using microwave energy is gaining popularity over the conventional techniques due to advantages such as time reduction, energy-saving, eco-friendly processing, and improvement in mechanical properties of targeted parts. Microwave energy has been used for curing and sintering PMCs and CMCs, respectively, almost from the beginning; however, their use for sintering of MMCs has been realized very late due to the unexplored interaction of microwaves with metallic particles. Various research groups are popularly researching microwave energy across the globe; however, industrial adoption and technology transfer are yet to be completed. Exploring the physical fundamentals of microwave interaction with composites will enrich the domain knowledge and provide a better understanding to control the process and optimize the processes. The numerical simulations are tools to explore further the complex heating of composite materials during microwave exposure. There are immense opportunities in the area for possible improvements and innovations in the process.

REFERENCES

[1] Thostenson, E. T., & Chou, T. W. (1999). Microwave processing: fundamentals and applications. *Composites Part A: Applied Science and Manufacturing*, *30*(9), 1055–1071.

[2] Mishra, R. R., & Sharma, A. K. (2016). Microwave–material interaction phenomena: heating mechanisms, challenges and opportunities in material processing. *Composites Part A: Applied Science and Manufacturing*, *81*, 78–97.

[3] Mishra, R. R., & Sharma, A. K. (2016). A review of research trends in microwave processing of metal-based materials and opportunities in microwave metal casting. *Critical Reviews in Solid State and Materials Sciences*, *41*(3), 217–255.

[4] Thostenson, E. T., & Chou, T. W. (2001). Microwave and conventional curing of thick-section thermoset composite laminates: experiment and simulation. *Polymer composites*, *22*(2), 197–212.

[5] Rybakov, K. I., Olevsky, E. A., & Krikun, E. V. (2013). Microwave sintering: fundamentals and modeling. *Journal of the American Ceramic Society*, *96*(4), 1003–1020.

[6] Roy, R., Agrawal, D., Cheng, J., & Gedevanishvili, S. (1999). Full sintering of powdered-metal bodies in a microwave field. *Nature*, *399*(6737), 668–670.

[7] Gupta, M., & Ling, S. N. M. (2011). *Magnesium, magnesium alloys, and magnesium composites*. John Wiley & Sons.

[8] Tun, K. S., & Gupta, M. (2007). Improving mechanical properties of magnesium using nano-yttria reinforcement and microwave assisted powder metallurgy method. *Composites Science and Technology*, *67*(13), 2657–2664.

[9] Sharma, A. K., & Mishra, R. R. (2018). Role of particle size in microwave processing of metallic material systems. *Materials Science and Technology*, *34*(2), 123–137.

[10] Pramila Devi, D. S., Nair, A. B., Jabin, T., & Kutty, S. K. (2012). Mechanical, thermal, and microwave properties of conducting composites of polypyrrole/polypyrrole-coated short nylon fibers with acrylonitrile butadiene rubber. *Journal of Applied Polymer Science*, *126*(6), 1965–1976.

[11] Chen, M., Siochi, E. J., Ward, T. C., & McGrath, J. E. (1993). Basic ideas of microwave processing of polymers. *Polymer Engineering & Science*, *33*(17), 1092–1109.

[12] Clark, D. E., & Sutton, W. H. (1996). Microwave processing of materials. *Annual Review of Materials Science*, *26*(1), 299–331.

[13] Agrawal, D. K. (1998). Microwave processing of ceramics. *Current Opinion in Solid State and Materials Science*, *3*(5), 480–485.

[14] Das, S., Mukhopadhyay, A. K., Datta, S., & Basu, D. (2009). Prospects of microwave processing: An overview. *Bulletin of Materials Science*, *32*(1), 1–13.

Chapter 13

Hybrid Machining of Metal Matrix Composites

Sahil Sharma, Akshay Dvivedi, and Pradeep Kumar
Department of Mechanical and Industrial Engineering, Indian Institute of
Technology Roorkee, Roorkee, India

Farhan Ahmad Shamim
Mechanical Engineering Department, Aligarh Muslim University,
Aligarh, India

Tarlochan Singh
Department of Mechanical Engineering, Indian Institute of Technology
Bombay, Mumbai, India

CONTENTS

13.1 INTRODUCTION

Composites like Al/SiC are widely used in various modern high-tech applications such as defence, aerospace, automobile, etc. as structural materials [1–3]. These composites offer several remarkable advantages: corrosion resistance, wear resistance, high strength to weight ratio, and stiffness. The MMCs contrast from different composites in numerous ways like (a) the matrix phase in MMCs is generally metallic (either pure or alloy metal) opposite to a ceramic and polymer, (b) MMCs show better toughness and ductility compared to ceramic matrix composites (CMCs), and (c) the reinforced particles in MMCs are used to improve the strength and modulus, whereas reinforcement in CMCs provides better damage tolerance capabilities [4]. MMCs are fabricated by introducing the discontinuous reinforcing phase in the continuous metal matrix phase. For instance, composite Al/SiCp is prepared by introducing the hard silicon carbide particles in the aluminium matrix phase. The discontinuous SiC particles have high modulus (250 GPa), hardness, good wear, and impact resistance and high melting point (2700°C) [5]. These properties provide unique characteristics to Al/SiCp composites. However, processing of such composite is a tedious task, as the SiC possesses low fracture toughness. The various machining processes for processing the MMCs are illustrated in Fig. 13.1.

Literature reveals that turning, milling, and grinding are the major conventional processes that have been utilized to machine metal matrix composites. However, the utilisation of such processes for machining MMCs results in high tool wear, low MRR, and high total machining cost, whereas electrochemical machining LBM, EDM, and ECM are few non-conventional processes that have also been utilized to machine MMCs. These non-conventional processes produce better-machined surface characteristics than conventional techniques, yet these are inadequate to fulfil the demand of modern industries. Thus, a concept of hybrid machining came into existence in which combination of different conventional and non-conventional machining mechanisms and/or tool-energies are used in same machining zone simultaneously [6]. Some of the hybrid machining

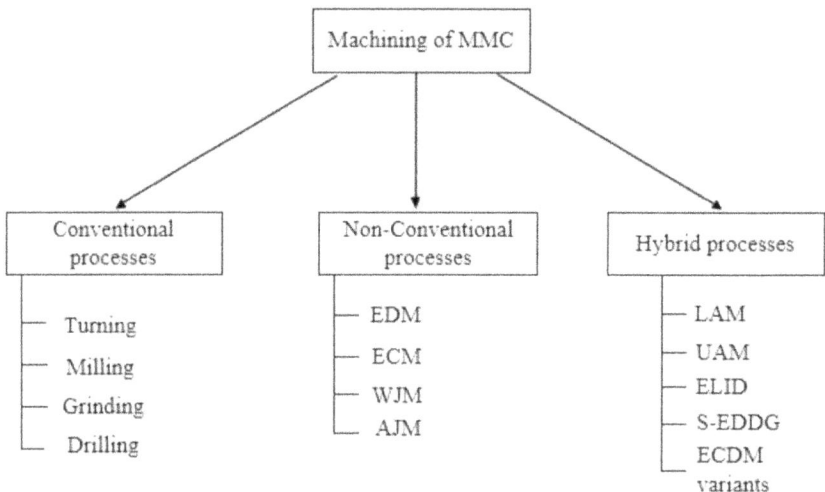

Figure 13.1 Machining methods to process MMCs.

processes included in this chapter are laser-assisted machining (LAM), ultrasonic-assisted machining (UAM), electrolytic in-process dressing (ELID) grinding, electrical discharge diamond grinding (EDDG), water jet guided laser (WJGL) drilling, rotary electrochemical discharge machining (R-ECDM), and near-dry wire electrochemical discharge machining (ND-WECDM), electrochemical discharge grinding (G-ECDM).

13.2 MACHINING OF MMCS BY CONVENTIONAL PROCESSES

Machining MMCs with conventional processes is generally carried out by turning, milling, grinding, and drilling. Among these processes, turning was the most utilized technique. PCD is usually preferred in the turning process because it provides high tool life and good surface quality [7]. Wu et al. [8] proposed a microstructure-based model for turning the particulate-reinforced MMC (PRMMC) with a diamond tool. As depicted in Fig. 13.2, when the tool and work material interact, compressive forces are generated near the tooltip, and tensile stresses distribute among the particles because of matrix deformation (Fig. 13.2a). With the tool advancement, particle fracture and particle-matrix debonding initiate when the tensile stresses reach the threshold value (Fig. 13.2b). With further advancement, the re-inforced grains located along the cutting path are dragged forward by the cutting edge, forcing the matrix material to bend and move along the shear plane in an upward direction and form a serrated chip (Fig. 13.2c–d). These chips move apart from the workpiece and connect with the next serrated chip. This process remained continuous until the machining is completed (Fig. 13.2e–f). Though turning is capable of machine MMC, the machined components consist of several surface defects: voids, micro-cracks, pits, matrix tearing, and protuberances.

13.3 MACHINING OF MMCS BY NON-CONVENTIONAL PROCESSES

Among all various processes, EDM is the most widely used process for processing the MMCs. The temperature approaches 12,000°C in the discharge plasma channel during EDM processing. This high temperature is sufficient to erode the tool and workpiece material by melting and evaporation. The MRR, SR, and TWR are the primary machining characteristics of the EDM process, which define machining time, surface quality, and tool life. Singh et al. [9] investigated the machinability of A6061/SiC(10%) composite by the EDM process. Process parameters such as pulse on/off-time, current and gap voltage were selected to investigate the effect on MRR, SR, and TWR. The obtained results show an increase in all the three output characteristics with the rise in pulse on time. The surface

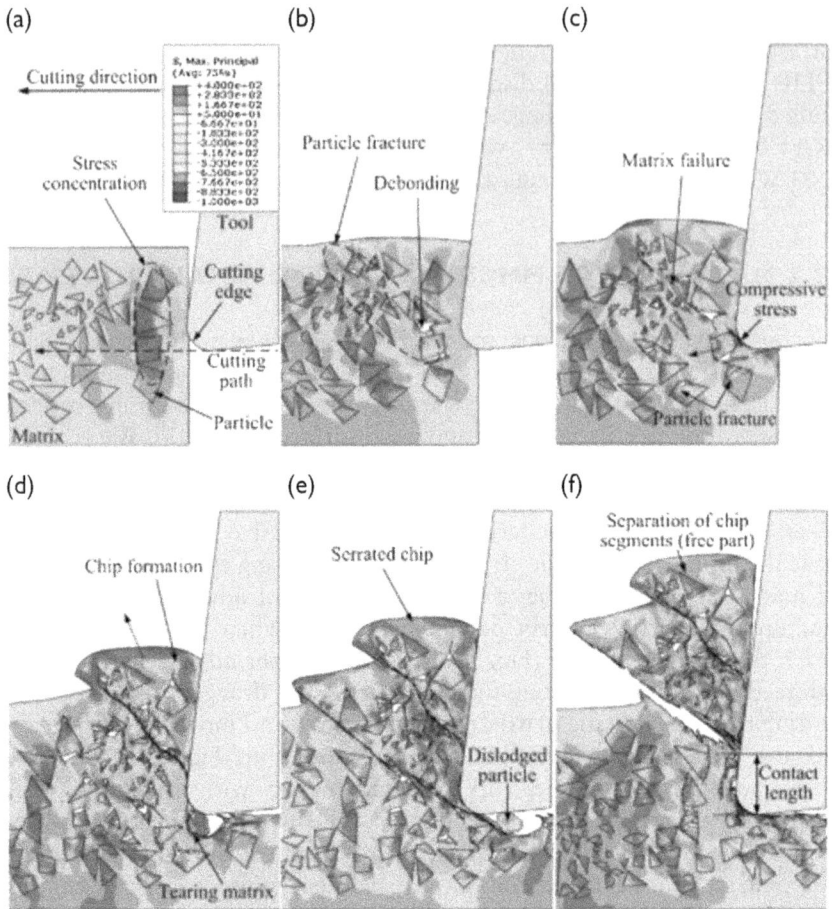

Figure 13.2 Principle mechanism of turning process of MMC [8].

integrity of EDMed components includes micro-pits due to spark penetration, cracks, formation of HAZ, and recast layer [10]. Powder-mixed EDM (PM-EDM) is an EDM process performed by mixing the powder particles in the dielectric medium. The mixed particles help to attain uniform spark discharges due to the bridging effect of the tool and workpiece [11]. A comparison between machined surfaces obtained by EDM and PMEDM techniques, as shown in Fig. 13.3, indicates that the PMEDM produces a smoother surface than EDM. Hu et al. [12] found out that PM-EDMed surfaces' SR was 31.5% lesser than EDMed surfaces.

Electrochemical machining is the second most used non-conventional process for processing composites. It works on the anodic dissolution of the work material. This process offers several benefits like high MRR, good precision, and surface quality of machined components. To investigate the effect of process parameters, Senthil Kumar et al. [13] machined the

(a)   (b)

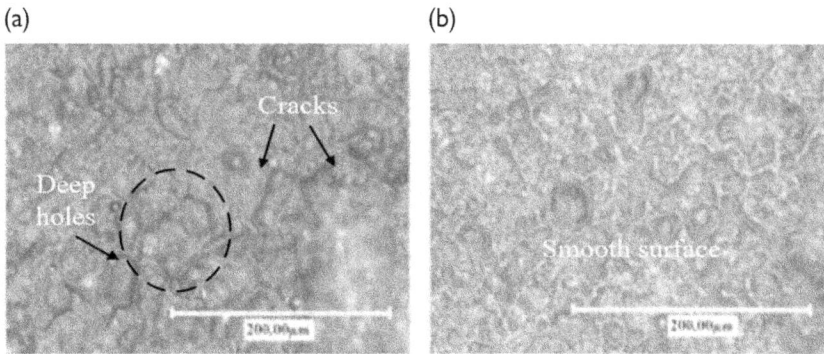

Figure 13.3 Micro-surface texture of Al/SiC 40% machined by **(a)** EDM, **(b)** PMEDM [12].

LM 25 Al/10%SiC composites by ECM. To optimize the process char-acteristics, response surface methodology (RSM) was employed. Results showed that MRR enhanced with an increase in applied voltage, electrolyte flow rate, and concentration. The optimized parameters were applied vol-tage 13.5 V, electrolyte flow rate 7.51 l/min, tool feed rate 1 mm/min, and electrolyte concentration 12.53 g/l. The obtained MRR and SR at the op-timized settings were 0.8773 g/min and 6.5667 μm, respectively.

Laser is a thermal-based non-conventional process suitable for machining the work materials with high MRR and narrow kerf. However, laser-processed components consist of surface defects, namely striation outlines, dross, and HAZ [14]. The laser beam can be used as a cutting tool or drilling to perform various machining actions. For instance, Biffi et al. [15] utilized a Nd:YAG laser beam to machine the truncated thread in A359/SiC(20%). Padhee et al. [16] fabricated drilled holes on Al/15% SiC matrix composites using a laser beam.

Abrasive water jet (AWJ) machining is a form of mechanical non-conventional process. In contrast to EDM and laser, AWJ induces no thermal damage to the workpiece and hence components without HAZ and recast layer can be obtained. In this process, a high-velocity water jet containing abrasive particles impacts the workpiece and removes the ma-terial by erosion action. Srinivas et al. [17] observed the AWJ cut surfaces and proposed a material removal mechanism. The jet stream of water and abrasives caused ductile fracturing of matrix material and fracturing and ploughing of SiC particles. Patel et al. [18] proved that the AWJ could also turn MMCs and provide better tool life, better-machined surface quality, and low tool wear than traditional turning operations.

13.4 HYBRID MACHINING

Hybrid machining is a technique in which more than two process mechan-isms and/or energy sources are used simultaneously at the same working station. The interaction of different energies takes place in the controlled manner during the process. The principle of using hybrid machining is to utilize the "1+1=3" effect, which is the sum of two individual processes that are more than their combined effect [19]. It is also beneficial to address the adverse effect of material removal processes when performed individually. For instance, EDM is a well-known thermal non-traditional process that can produce complex structures in conductive difficult-to-machine materials. Though the surface quality of EDMed components is good, it still suffers from low MRR. It consists of many surface defects such as recast layer, HAZ, micro-cracks, etc., that restrict the usage of this process for advanced modern applications.

On the other hand, ECM is an electrochemical process in which material removal occurs at the atomic level and provides excellent machined surface characteristics free from HAZ, recast layer, and micro-cracks. Combining these processes known as hybrid ED-ECM process, the advantages of both methods can be utilized at the same time. Some of the hybrid technologies are discussed below.

13.4.1 Laser-Assisted Machining (LAM)

LAM is a hybrid machining technology that has the potential to machine metal matrix composites at an overall low cost. It utilizes the combined process mechanism of laser machining and other unconventional and conventional processes such as turning, as shown in Fig. 13.4. This technique's basic principle works on decreasing the yield strength of the work material by heating at an elevated temperature. This reduces the cutting forces and thereby improves the machinability. Ideally, laser irradiation increases the temperate of the work material, especially at the shear deformation zone. As most of the heat is carried out by formed chips, no heat is diffused into work substrate and tool material. This process has the potential to machine metal matrix components with lower cutting forces and better surface finishing. Kawalec et al. [20] utilized the LAM for machining the Al/SiC to examine the tool wear and surface quality.

The composites were machined by CO_2 laser and universal lathe. Coated and uncoated sintered carbide and PCD insert were used to study the wedge

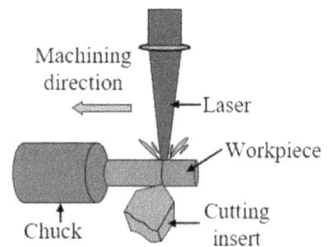

Figure 13.4 Schematic diagram of laser-assisted machining (LAM).

tool wear. The machining of the composite was conducted with constant cutting parameters: 0.04 mm/rev feed, 100 m/min cutting speed, and variable 0–1000 W laser power. Results showed that the abrasion wear of the tool flank was found to be the primary wear mechanism during laser-assisted conventional turning process. The uncoated sintered carbide inserts resulted in the lowest wedge wear than coated sintered and PCD inserts. The processing of material with 600 W laser power resulted in improved performance of the process.

Similarly, Kong et al. [21] demonstrated that the LAM of $Al/SiC_p/45\%$ MMC resulted in the machining of composite with 45% reduced machining time with higher MRR and tool life than conventional machining process. The maximum tool life for coated and uncoated tools was observed at 320°C, which was 2.31 and 1.93 times, respectively, higher than conventional machining. Moreover, the LAM process observed 40–50% cost saving/part with an additional cost of high-power diode laser and graphite particles.

Wang et al. [22] studied the LAM machining of $2024Al/SiC_p$ to investigate the surface characteristics of composite in terms of chip formation, surface integrity, cutting forces, and tool wear. Laser power and cutting depth were found the significant factors that affected the surface quality of the component. LAM resulted in a 38% reduction in flank wear and a 27% reduction in cutting forces than conventional machining. This study also revealed that the aluminium oxide nanoparticles of diameter 120 nm are generated in the cutting zone during the laser processing, which acts as a cavity filler and reduce the friction force between cutting tool and specimen.

13.4.2 Ultrasonic-Assisted Machining (UAM)

UAM is another type of hybrid technique in which ultrasonic vibrations are employed along with conventional machining processes. The application of vibration could decrease the effect of BUE, plastic deformation and tearing in cutting. It also restrains the flutter and hence stabilizes the cutting process [23]. The AFM graph of conventional and UAM highlights that the tool traces can easily be observed in conventional techniques (Fig. 13.5a), effectively removed by UAM. The surface machined by UAM is free from tool traces but consists of dense micro pits (Fig. 13.5b). This is because the tool cutter continuously squeezes and clashes the workpiece with high kinetic energy. Thus, ultrasonic-assisted machining effectively machines the MMC with smooth surface finishing. The UAM process has been employed to perform milling, turning, drilling, and grinding actions.

Xiang et al. [24] investigated the effect of UA-milling while machining the Al/SiC_p. The SiC content of volume of more than 40% makes the machining of material very difficult by the conventional process. The application of ultrasonic energy along with conventional milling effectivley machined the composite with higher surface characterisitcs and machining

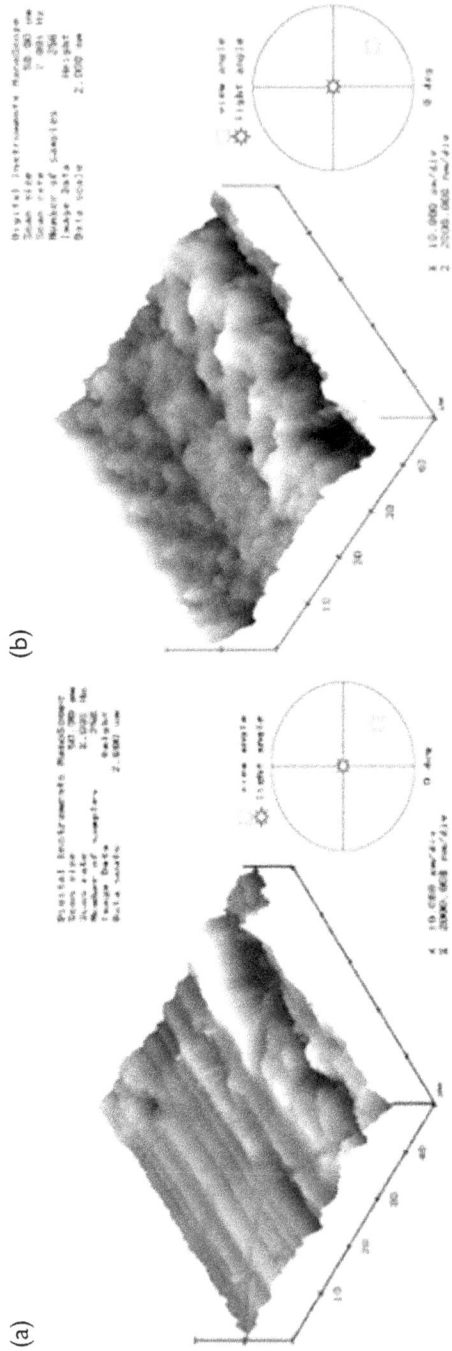

Figure 13.5 AFM graph of the composite under (a) conventional and (b) UAM process [23].

efficiency. It was observed that the cutting speed (≤290 m/min) of UAM was the major influencing parameter for reducing the surface roughness of the components. Moreover, the axial vibration applied to the milling cutter helped to reduce the torque and enhanced the rigidity of milling cutter. Zhong et al. [25] used this process for diamond turning of A359/SiC/20p MMC. Results showed that the UA-turning generates regular surface profile along cutting and vibration direction. In the case of drilling with ultrasonic energy, the cutting force and surface roughness decreased significantly by increasing machining precision and tool life. Xu et al. [26] used ultrasonic vibration along with drilling (UAD) to fabricate micro-holes in Al/SiC composite. The results obtained highlight that the UAD decreased the cutting forces (especially the torque by 30%) than common drilling and prolonged tool life. Kadivar et al. [27] investigated the characteristics of burr formed by UAD and conventional drilling of MMC. According to the results, burr dimensions in UAD were reduced (by 83%) as compared to traditional drilling. Thinner and smaller chip formation helped to generate high precision parts at low machining time. Ultrasonic grinding is another process that has been employed to machine MMCs. The three-dimensional grinding force can be reduced by vibration grinding than the normal grinding process for the same grinding parameters [28]. Zhou et al. [29] reported that with rotary ultrasonic grinding, the cutting forces and surface roughness can be reduced to 13.86% and 11.53%, respectively compared to grinding without ultrasonic.

Additionally, the surface obtained by rotary UAG was observed free from surface defects. Zhou et al. [30] proposed a single grinding force and end grinding model for predicting the grinding forces in UAG of Al/SiC$_p$ composites. The error percentage of the experimental and predicted model was found 11.7, 8.6, and 5% for tangential force, normal force, and axial force, respectively.

13.4.3 Electrolytic In-Process Dressing (ELID)

ELID is an electrochemical process that uninterruptedly dresses a metal bonded grinding wheel through in-situ electrolysis. It offers fresh ultra-fine abrasives grits to perform machining action and thus helps to attain superior surface quality during the processing of MMCs. The sequence of the ELID process mechanism is illustrated in Fig. 13.6.

Initially, a trued wheel is used before grinding. After truing, the abrasive grit of the wheel flattens (Fig. 13.6a) and thus requires pre-dressing. During pre-dressing, the bonding material comes out (Fig. 13.6b) and at the same time, an insulative oxidized layer formed on the wheel circumference (Fig. 13.6c). The oxidized layer decreases the wheel's electro-conductivity and thus prevents the excessive flow of bonding material. As the grinding begins, the grit and oxide layer wear out (Fig. 13.6d). Consequently, the wheel's conductivity increases that restart of dressing of wheel (Fig. 13.6e). This cycle repeats until the machining action is completed [31].

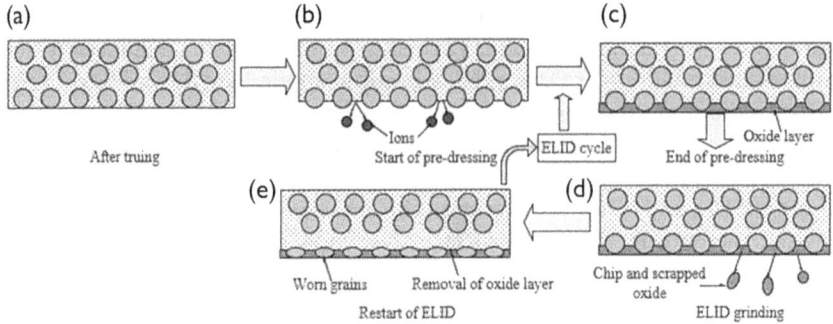

Figure 13.6 Schematic of ELID grinding mechanism.

Shanawaz et al. [32] employed ELID grinding process at different current duty ratios to machine the 2124 aluminium silicon carbide ($10\%SiC_p$) composite. Specimens ($125 \times 10 \times 10$ mm) were machined by the copper-bonded diamond wheel. Results show that the ELID grinding ease the machinability of composite and generate a smoother surface than the conventional grinding process. The minimum surface roughness of the material, i.e., 0.81 μm, was obtained at 60% duty cycle. Moreover, it was found that SR and MRR were improved with an increase in duty ratio. The higher value of duty ratio also resulted in decreasing the micro-hardness and normal cutting forces. Yu et al. [33] investigated the ELID while machining the SiCp/Al (56% volume fraction) composite. A cast-iron bonded diamond grinding wheel consisting of 5 μm abrasive grits was used to perform the machining action. The ELID process removed the reinforced particles by ductile fracture; as a result, smooth machined surfaces were obtained. These surfaces were found to consist lesser pits and cracks than wet grinding. The area of brittle fracture at the interface of the particle and matrix was also reduced.

13.4.4 Electrical Discharge Diamond Grinding (EDDG)

In the EDDG process, the removal of materials occurs due to synchronised interaction of (a) abrasion by diamond wheels abrasive grit and (b) erosion due to sparks produced between workpiece and bonding material of wheel. The synchronised interaction of abrasion and erosion by spark is shown in Fig. 13.7. The multitude discharges thermally soften the workpiece and

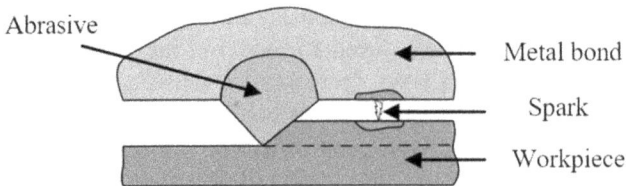

Figure 13.7 Workpiece and wheel interface in EDDG.

thereby reduces the grinding power. The combined action of abrasion and erosion makes the machinability of MMC very efficient with less specific grinding energy consumption. As the SiC particles have low fracture toughness, thus it has been found prone to cracking. Machining a composite consisting of SiC particles with EDM and conventional grinding is complicated as it leads to surface cracking or poor surface quality. To solve this problem, Agrawal et al. [34] performed the S-EDDG action to machine the Al-SiC composite to obtain good surface quality without compromising the MRR. The machining action were performed with the following range of parameters: cutting speed 1000–1400 rpm, pulse current 8–24 A, feed 20–40 μm, pulse on time (T_{on}) 50–150 μs, and duty factor (DF) 0.492–0.817. The surface obtained by S-EDDG was found free from the recast layer compared to processing by EDM mode. A five-time increment in MRR and decrement of surface roughness to almost half was observed when the cutting speed varied from 1000 to 1400 rpm. The wheel speed followed by current were the major influencing factors than other parameters. At the optimum settings i.e., cutting speed-1400 rpm, pulse current-24A, feed 20 μm, T_{on} 50 μs and DF- 0.817, the MRR improved significantly (up to 352%) at the expense of higher surface roughness from 3.01 to 4.61 μm.

In another work, Agrawal et al. [35] used the artificial neural network (ANN) technique for modelling and predicting the MRR and SR while machining of Al-10wt%SiC and Al-10wt%Al_2O_3 composites by EDDG. The prediction through presented modelling of S-EDDG demonstrated that MRR increased as the workpiece speed, wheel speed, pulse current, and depth of cut increased and reduced with an increased in duty ratio, whereas SR increased as the workpiece speed, duty ratio, depth of cut, and current increased while it decreased as the wheel speed increased.

13.4.5 Water Jet Guided (WJG) Laser Drilling

WJG laser drilling is a newly emerging hybrid technique for machining MMCs. In this process, a pressurized micro water jet stream is used as a laser beam guide. This technique can fabricate holes with good surface quality with no HAZ and no recast layer with a high level of hole circularity. Moreover, no change in microstructure with efficient removal of melted material is the additional benefit of this process. From the literature, it was observed that this technique is commonly used for processing hard monolithic materials. To study the feasibility of this process for processing the MMCs, Marimuthu et al. [36] employed the WJG laser drilling for fabricating the through holes in 2 mm thick Al/SiC material. The drilling action was performed with the following constant parameters: 200 ns pulse duration, 300 bar water jet pressure, and 8 mm nozzle workpiece offset. For the experimental work, the laser speed and total average power were varied from 5 mm/s to 25 mm/s and 30 W to 60 W, respectively. The results obtained by this experiment shows that in WJG laser drilling, the material is machined by cold ablation without leaving the traces of melted layer within the material. Both reinforcement particles and matrix were removed by the cold ablation phenomenon. Whereas, in conventional drilling, solid SiC particles were expelled along with melted aluminium matrix. The WJG laser-drilled holes were found highly consistent with minimal taper. It takes only 12 s to produce a fine quality hole of 0.8 mm over 2 mm thick MMC with 60 W power.

13.4.6 Rotary Electrochemical Discharge Machining (R-ECDM)

ECDM utilizes the combined mechanism of EDM and ECM for processing the conductive and non-conductive materials. Nowadays, variants of this process have been developed for various machining operations like die-sinking, milling, and drilling [37]. Singh et al. [38] developed rotary mode electrochemical discharge drilling (RM-ECDD) for MMCs. The rotary tool electrode assists flushing of debris and electrolyte replenishment in the vicinity of tool-workpiece vicinity. The machining sequence of the rotary mode-ECDD process is illustrated in Fig. 13.8.

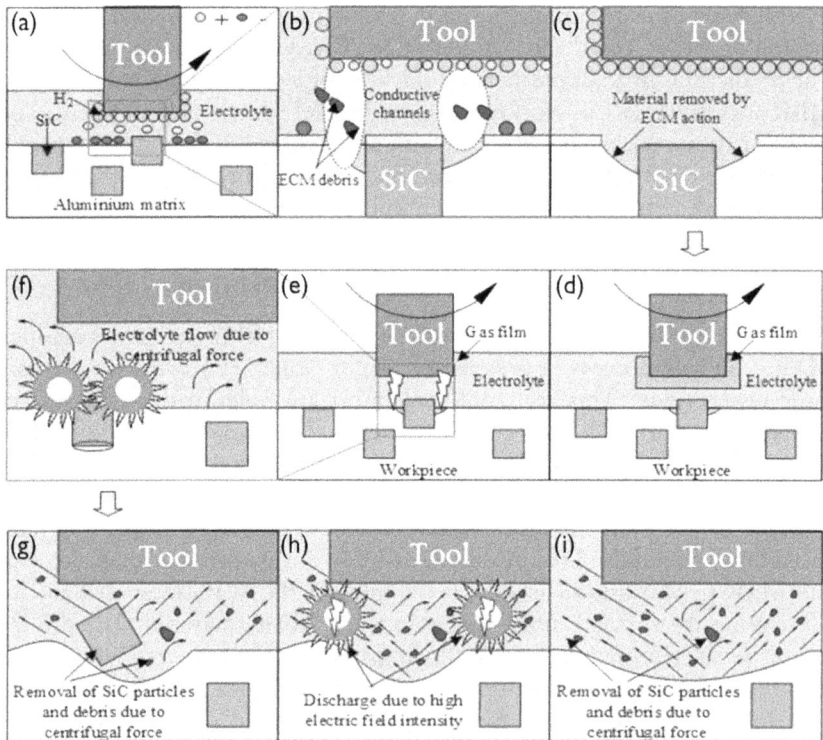

Figure 13.8 Sequence of material removal mechanism of RM-ECDD.

In this process, a rotary tool attached with negative terminal of power supply, placed nearly 1–3 mm above the workpiece in an electrolyte bath. As the voltage is switched on, the process of electrolysis starts and thereby, the electrons migrate towards the tool electrode (Fig. 13.8a). These electrons produce the hydrogen (H_2) gas bubbles around the tool circumference (Fig. 13.8b–c). As the voltage increases, a more significant number of H_2 bubbles formed and coalesced with each other. The coalescence of these H_2 gas bubbles results in an insulating gas film around the tool electrode (Fig. 13.8d). This insulating film offers high electric resistance and increases the electric field strength across the electrodes [39]. With further advancement of the voltage, the electric field becomes so strong that it breaks the gas film and generates a series of sparks on work material (Fig. 13.8e.). These high-temperature sparks instantly melt and evaporate the contacted work material and thus produce the desired pattern on the work material. During the processing of metal matrix composites by the RM-ECDD process, the anodic dissolution locally softens the matrix phase of composites. This anodic dissolution begins from the matrix and reinforced particle interface due to strong electric field intensity at the matrix–reinforcement interface and expelled out the reinforcement particle from the machining area (Fig. 13.8f–i). The anodic dissolution contributes to the material removal in the RM-ECDD process.

It is important to note that in RM-ECDD, a rotary tool electrode helps in electrolyte replenishment and evacuation of debris from the machining zone. It also helps to make the gas film thinner formed around the tool electrode [40]. The thin gas film modifies the characteristics of discharges and hence improves the machining performance. The obtained results from this study showed that the RM-ECDD process fabricated a 23% higher aspect ratio hole than conventional ECDD.

13.4.7 Near Dry Wire Electrochemical Discharge Machining (ND-WECDM)

In ND-WECDM, removal of work material occurs due to simultaneous action of sparking, anodic dissolution and thermal spalling. In this process, pressurised electrolyte mist flows across the wire tool (cathode) and work material (anode). Since a small amount of electrolyte is consumed during the process, the process is termed "near-dry." This process has the advantage of machining the MMCs with minimal tool breakage and controlled anodic dissolution than wire EDM and ECM, respectively. The machining mechanism of ND-WECDM of MMCs is depicted in Fig. 13.9. Initially, when the voltage is applied through the IEG, the electrochemical reactions start (Fig. 13.9a), resulting in the generation and accumulation of hydrogen gas bubbles on the cathode (Fig. 13.9b). Electrochemical reactions continue over the period (i.e. pulse on-time) during which more hydrogen gas bubbles generate (Fig. 13.9c). These gas

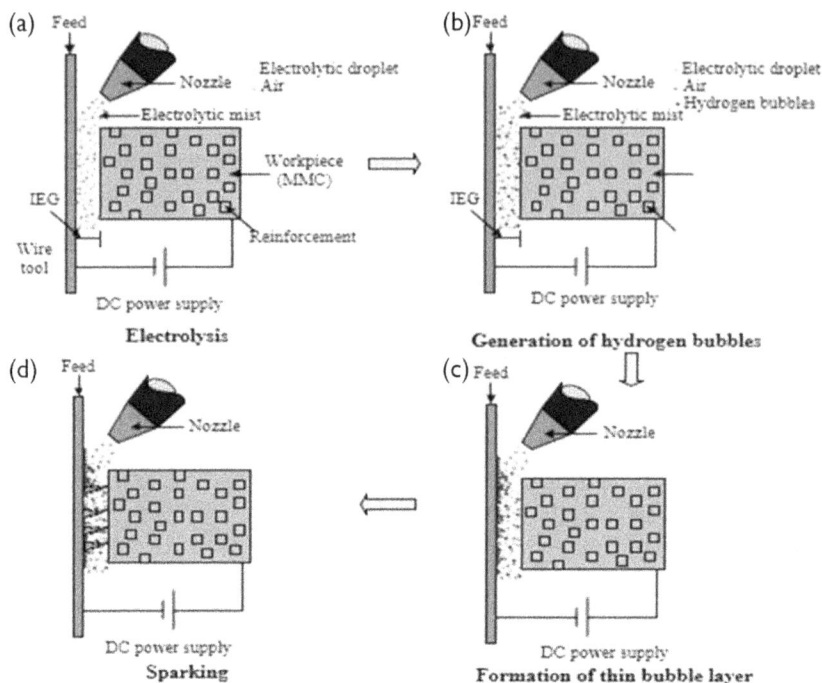

Figure 13.9 Schematic of material removal mechanism during ND-WECDM of MMC.

bubbles accumulate near the tool surface and increase the electrical resistance and voltage between the tool and work. As a result, a high
electric field is created between the electrodes, resulting in spark discharges (Fig. 13.9d).

The thermal energy liberated by these discharges produces a crater on
the work surface by melting and vaporisation and increases the electrolyte temperature. Consequently, the electrochemical dissolution of the
work material facilitates the elimination of reinforcement and thus
contributes to improving MRR. The expansion and contraction due to
heating and cooling cycles result in thermal stresses. As a result, the
physical bond between the matrix and reinforcement weakens, and reinforcements are expelled from the work material by thermal stresses.
This phenomenon is termed "thermal spalling." Thus, the combined
action of sparking, chemical dissolution, and thermal spalling contribute
to material removal. Shamim [41] fabricated the slit in 10 mm thick
Al6063/SiC material using this method. A brass wire of 0.25 mm diameter and a mixture of 10 wt.%/Vol. NaOH with air was employed as a
tool and electrolyte, respectively. Results showed that this technique can
produce micro-slits of good surface quality. The machined slit with dimensions is shown in Fig. 13.10.

(a)   (b)

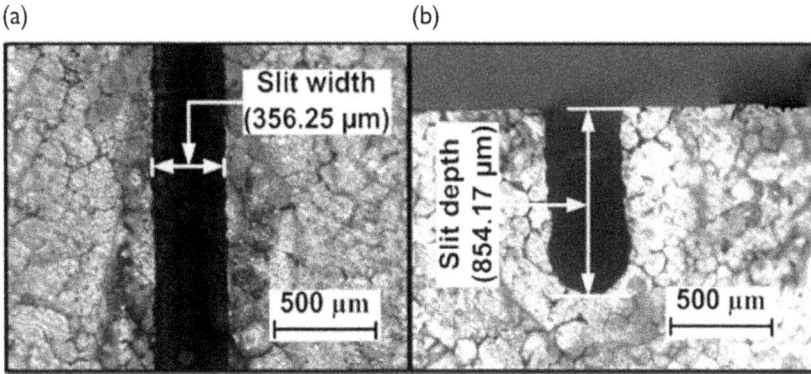

Figure 13.10 Stereo zoom microscope image showing machined slit produced by ND-WECDM process (a) width of the machined slit and (b) depth of the machined slit [41].

13.4.8 Electrochemical Discharge Grinding (G-ECDM)

ECDM gives a 30–40% higher material removal rate (MRR) than EDM and ECM processes. However, the sludge formation and accumulation of salt in the tool-workpiece vicinity restricts the bubble formation. As a result, discontinuation of ECDM action can take place. Additionally, during the processing of MMCs, a non-conductive passivation layer generally develops on the work surface that ceases the current flow and subsequently stops the electrolysis action. To overcome these limitations, Liu et al. [42] developed a grinding-aided electrochemical discharge machining (G-ECDM) process. This technique utilizes the simultaneous interaction of mechanical grinding, spark erosion and chemical etching, as illustrated in Fig. 13.11.

In this process, the abrasion imparted by the grinding action removes the non-conductive passivating layer and the centrifugal forces due to tool rotation helps to remove the debris from the tool-workpiece vicinity. The advantages of the G-ECDM process include high machining efficiency, reduction in surface/subsurface defects, and improved surface finish.

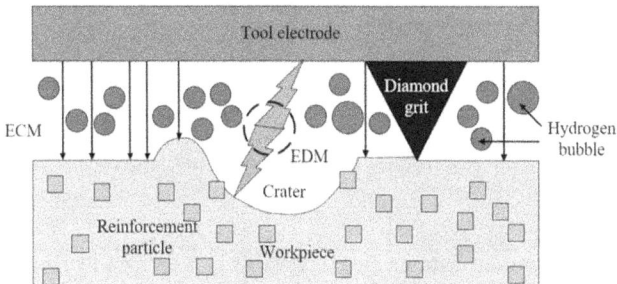

Figure 13.11 Schematic diagram depicting the simultaneous interaction of grinding, EDM, and ECM in G-ECDM.

(a)   (b)

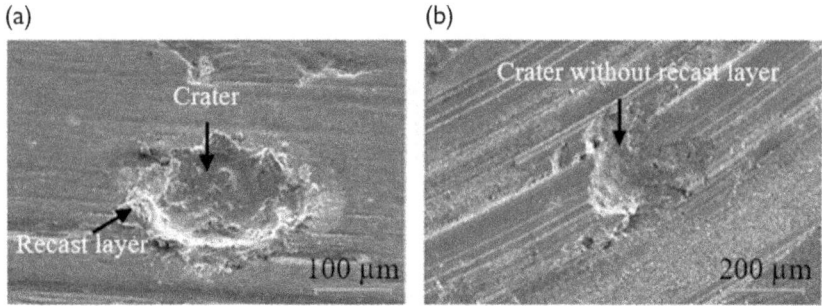

Figure 13.12 SEM image of 10 ALO material machined (**a**) with ECDM and (**b**) G-ECDM [42].

Liu et al. [42] conducted a comparative study to investigate the machining characteristics of ECDM and G-ECDM by using 10ALO as work material. The craters fabricated by ECDM and G-ECDM process on work material are shown in Fig. 13.12a–b.

From this figure, it can be evident that the fabricated crater by G-ECDM was free from re-solidified material (Fig. 13.12b) as opposed to the ECDMed crater (Fig. 13.12a). Additionally, the surface roughness of the G-ECDMed workpiece was found nearly ten times lower than the ECDMed component. Jha et al. [43] used the grinding assisted rotary disc electrochemical discharge machining (GA-RDECDM) process to develop micro slits in Al-6063/SiCp composites using NaNO$_3$ as an electrolyte medium. It is found that the triplex action of the abrasion (grinding), sparking (EDM), and chemical etching (ECM) resulted in the removal of wavy wall edges and thus produced slits with straight walls. The recommended optimized parametric settings were applied voltage-99 V, 30 rpm disk speed, 17 wt.%/vol. electrolyte concentration, and 3 ms pulse on-time.

13.5 CONCLUSIONS

This chapter presents an overview of the hybrid machining processes utilized for the subtractive processing of MMCs. The inferences drawn from this study are as follows:

- The laser assistance conventional machining reduces the material strength during processing and thereby decreases the tool wear and turning time by 38% and 45%, respectively, compared to traditional machining of MMCs.
- The ultrasonic assistance in conventional cutting lowers the effect of built-up edges (BUE), plastic deformation, tearing, and restrains flutters, which results in more reliable machining conditions. This technique can

be utilized to machine MMCs consisting of more than 40% SiC particles, which is impossible to machine by conventional processes.

- ELID grinding removes the reinforced particles from MMCs by ductile mode. Moreover, it reduces the area of brittle fracture of SiC particles at the particle-matrix interface. Thereby, it helps to attain an excellent machined surface than conventional grinding.
- Simultaneously interaction of sparks with abrasive grits in EDDG soften the work material and thereby reduces the cutting forces and grinding power.
- Water-jet-guided laser drilling removes the reinforced particles by the cold ablation phenomenon.
- Electrochemical discharge machining (ECDM) exhibits a nearly 30–40% higher machining rate compared to mere ECM and EDM processes while machining the MMCs.
- In RM-ECDD, a rotary tool electrode helps to take away the sludge from the tool and workpiece vicinity. Additionally, it also makes the gas film thinner, consequently improves the discharge characteristics and machining performance.
- Low tool breakage, minor electrolyte consumption, and controlled anodic dissolution are the few advantages offered by the near dry ECDM process compared to EDM and ECM processing of MMCs.
- G-ECDM process decreases the surface roughness by ten times compared to conventional ECDM, due to incorporating additional abrasion action by the tool.

REFERENCES

[1] F. Hakami, A. Pramanik, and A. K. Basak, "Tool wear and surface quality of metal matrix composites due to machining: A review," *Proc. Inst. Mech. Eng. Part B J. Eng. Manuf.*, vol. 231, no. 5, pp. 739–752, 2017, DOI: 10.11 77/0954405416667402

[2] W. H. Hunt; Kelly Anthony Zweben Carl, Editors, "Metal matrix composites," in *Comprehensive Composite Materials*. Elsevier, 2000, pp. 57–66.

[3] A. Pramanik, A. K. Basak, G. Littlefair, A. R. Dixit, and S. Chattopadhyaya, *Stress in the Interfaces of Metal Matrix Composites (MMCs) in Thermal and Tensile Loading*. Elsevier Ltd., 2019.

[4] J. Fan and J. Njuguna, "An introduction to lightweight composite materials and their use in transport structures," in *Lightweight Composite Structures in Transport: Design, Manufacturing, Analysis and Performance*. Elsevier Ltd., 2016, pp. 3–34.

[5] J. D. Selvam, I. Dinaharan, and R. S. Rai, *Matrix and Reinforcement Materials for Metal Matrix Composites*. Elsevier Ltd., 2021.

[6] B. Lauwers, F. Klocke, A. Klink, A. E. Tekkaya, R. Neugebauer, and D. McIntosh, "Hybrid processes in manufacturing," *CIRP Ann. – Manuf. Technol.*, vol. 63, no. 2, pp. 561–583, 2014, DOI: 10.1016/j.cirp.2014.05.003

[7] Y. Sahin, "The effects of various multilayer ceramic coatings on the wear of carbide cutting tools when machining metal matrix composites," *Surf. Coatings Technol.*, vol. 199, no. 1, pp. 112–117, 2005, DOI: 10.1016/j.surfcoat.2005.01.048

[8] Q. Wu, W. Xu, and L. Zhang, "Machining of particulate-reinforced metal matrix composites: An investigation into the chip formation and subsurface damage," *Journal of Materials Processing Technology*, vol. 274, pp. 116315, 2019, DOI: 10.1016/j.jmatprotec.2019.116315

[9] S. Singh, I. Singh, and A. Dvivedi, "Multi objective optimisation in drilling of Al6063/10% SiC metal matrix composite based on grey relational analysis," *Proc. Inst. Mech. Eng. Part B J. Eng. Manuf.*, vol. 227, no. 12, pp. 1767–1776, 2013, DOI: 10.1177/0954405413494383

[10] M. Ramulu, G. Paul, and J. Patel, "EDM surface effects on the fatigue strength of a 15 vol% SiCp/Al metal matrix composite material," *Compos. Struct.*, vol. 54, no. 1, pp. 79–86, 2001, DOI: 10.1016/S0263-8223(01)00072-1

[11] H. K. Kansal, S. Singh, and P. Kumar, "Effect of silicon powder mixed EDM on machining rate of AISI D2 die steel," *J. Manuf. Process.*, vol. 9, no. 1, pp. 13–22, 2007, DOI: 10.1016/S1526-6125(07)70104-4

[12] F. Q. Hu *et al.*, "Surface properties of SiCp/Al composite by powder-mixed EDM," *Procedia CIRP*, vol. 6, pp. 101–106, 2013, DOI: 10.1016/j.procir.2013.03.036

[13] C. Senthilkumar, G. Ganesan, and R. Karthikeyan, "Study of electrochemical machining characteristics of Al/SiCp composites," *Int. J. Adv. Manuf. Technol.*, vol. 43, no. 3–4, pp. 256–263, 2009, DOI: 10.1007/s00170-008-1704-1

[14] F. Müller and J. Monaghan, "Non-conventional machining of particle reinforced metal matrix composite," *Int. J. Mach. Tools Manuf.*, vol. 40, no. 9, pp. 1351–1366, 2000, DOI: 10.1016/S0890-6955(99)00121-2

[15] C. A. Biffi, E. Capello, and B. Previtali, "Laser and lathe thread cutting of aluminium metal matrix composite," *Int. J. Mach. Mach. Mater.*, vol. 6, no. 3–4, pp. 250–269, 2009, DOI: 10.1504/IJMMM.2009.027327

[16] S. Padhee, S. Pani, and S. S. Mahapatra, "A parametric study on laser drilling of Al/SiC p metal matrix composite," *Proc. Inst. Mech. Eng. Part B J. Eng. Manuf.*, vol. 226, no. 1, pp. 76–91, 2012, DOI: 10.1177/0954405411415939

[17] D. S. Srinivas and N. R. Babu, "Role of garnet and silicon carbide abrasives in abrasive waterjet cutting of aluminum-silicon carbide particulate metal matrix composites," *Int. J. Appl. Res. Mech. Eng.*, no. August 2015, pp. 109–122, 2011, DOI: 10.47893/ijarme.2011.1022

[18] R. Patel, M. Engineering, and E. Bengaluru, "Abrasive water jet turning of aluminum-silicon carbide metal matrix composites," *Department of Mechanical Engineering, BMS College of Engineering*, vol. 6061, pp. 412–415, 2017.

[19] X. Luo, Y. Cai, and S. Z. Chavoshi, "Introduction to hybrid machining technology," in *Hybrid Machining*, Elsevier Ltd., 2018, pp. 1–20.

[20] M. Kawalec, D. Przestacki, K. Bartkowiak, and M. Jankowiak, "Laser assisted machining of aluminium composite reinforced by SiC particle," *ICALEO 2008 – 27th Int. Congr. Appl. Lasers Electro-Optics, Congr. Proc.*, vol. 1906, pp. 895–900, 2008, DOI: 10.2351/1.5061278

[21] X. Kong, L. Yang, H. Zhang, G. Chi, and Y. Wang, "Optimisation of surface roughness in laser-assisted machining of metal matrix composites using Taguchi method," *Int. J. Adv. Manuf. Technol.*, vol. 89, no. 1–4, pp. 529–542, 2017, DOI: 10.1007/s00170-016-9115-1

[22] Z. Wang, J. Xu, H. Yu, Z. Yu, Y. Li, and Q. Du, "Process characteristics of laser-assisted micro machining of SiCp/2024Al composites," *Int. J. Adv. Manuf. Technol.*, vol. 94, no. 9–12, pp. 3679–3690, 2018, DOI: 10.1007/s00170-017-1071-x

[23] B. Zhao, C. S. Liu, X. S. Zhu, and K. W. Xu, "Research on the vibration cutting performance of particle reinforced metallic matrix composites SiCp/Al," in *Journal of Materials Processing Technology*, vol. 129, no. 1–3, pp. 380–384, 2002, DOI: 10.1016/S0924-0136(02)00696-9

[24] D. Xiang, X. Zhi, G. Yue, G. Gao, and B. Zhao, "Study on surface quality of Al/SiCp composites with ultrasonic vibration high speed milling," *Appl. Mech. Mater.*, vol. 42, pp. 363–366, 2011, DOI: 10.4028/www.scientific.net/AMM.42.363

[25] Z. W. Zhong and G. Lin, "Ultrasonic assisted turning of an aluminium-based metal matrix composite reinforced with SiC particles," *Int. J. Adv. Manuf. Technol.*, vol. 27, no. 11–12, pp. 1077–1081, Feb. 2006, DOI: 10.1007/s00170-004-2320-3

[26] X. Xu, Y. Mo, C. Liu, and B. Zhao, "Drilling force of SiC particle reinforced Aluminum-matrix composites with ultrasonic vibration," in *Key Engineering Materials*, vol. 416, pp. 243–247, 2009, DOI: 10.4028/www.scientific.net/KEM.416.243

[27] M. A. Kadivar, R. Yousefi, J. Akbari, A. Rahi, and S. M. Nikouei, "Burr size reduction in drilling of Al/SiC metal matrix composite by ultrasonic assistance," *Adv. Mater. Res.*, vol. 410, pp. 279–282, 2012, DOI: 10.4028/www.scientific.net/AMR.410.279

[28] D. H. Xiang et al., "Study on grinding force of high volume fraction sicp/al composites with rotary ultrasonic vibration grinding," *Adv. Mater. Res.*, vol. 1027, pp. 48–51, 2014, DOI: 10.4028/www.scientific.net/AMR.1027.48

[29] M. Zhou, M. Wang, and G. Dong, "Experimental investigation on rotary ultrasonic face grinding of SiCp/Al composites," *Mater. Manuf. Process.*, vol. 31, no. 5, pp. 673–678, 2016, DOI: 10.1080/10426914.2015.1025962

[30] M. Zhou and W. Zheng, "A model for grinding forces prediction in ultrasonic vibration-assisted grinding of SiCp/Al composites," *Int. J. Adv. Manuf. Technol.*, vol. 87, no. 9–12, pp. 3211–3224, 2016, DOI: 10.1007/s00170-016-8726-x

[31] N. Itoh, A. Nemoto, T. Katoh, and H. Ohmori, "Eco-friendly ELID grinding using metal-free electro-conductive resinoid bonded wheel," *JSME Int. J., Ser. C Mech. Syst. Mach. Elem. Manuf.*, vol. 47, no. 1, pp. 72–78, 2004, DOI: 10.1299/jsmec.47.72

[32] A. M. Shanawaz, S. Sundaram, U. T. S. Pillai, and P. Babu Aurtherson, "Grinding of aluminium silicon carbide metal matrix composite materials by electrolytic in-process dressing grinding," *Int. J. Adv. Manuf. Technol.*, vol. 57, no. 1–4, pp. 143–150, Nov. 2011, DOI: 10.1007/s00170-011-3288-4

[33] X. Yu, S. Huang, and L. Xu, "ELID grinding characteristics of SiCp/Al composites," *Int. J. Adv. Manuf. Technol.*, vol. 86, no. 5–8, pp. 1165–1171, 2016, DOI: 10.1007/s00170-015-8235-3

[34] S. S. Agrawal and V. Yadava, "Development and experimental study of surface-electrical discharge diamond grinding of Al–10 wt%SiC composite," *J. Inst. Eng. Ser. C*, vol. 97, no. 1, pp. 1–9, 2016, DOI: 10.1007/s40032-015-0183-z

[35] S. S. Agrawal and V. Yadava, "Modeling and prediction of material removal rate and surface roughness in surface-electrical discharge diamond grinding process of metal matrix composites," *Mater. Manuf. Process.*, vol. 28, no. 4, pp. 381–389, 2013, DOI: 10.1080/10426914.2013.763678

[36] S. Marimuthu, J. Dunleavey, Y. Liu, B. Smith, A. Kiely, and M. Antar, "Water-jet guided laser drilling of SiC reinforced aluminium metal matrix composites," *J. Compos. Mater.*, vol. 53, no. 26–27, pp. 3787–3796, 2019, DOI: 10.1177/0021998319848062

[37] T. Singh and A. Dvivedi, "Developments in electrochemical discharge machining: A review on electrochemical discharge machining, process variants and their hybrid methods," *Int. J. Mach. Tools Manuf.*, vol. 105, pp. 1–13, Jun. 2016, DOI: 10.1016/j.ijmachtools.2016.03.004

[38] T. Singh, R. K. Arya, and A. Dvivedi, "Experimental investigations into rotary mode electrochemical discharge drilling (RM-ECDD) of metal matrix composites," *Mach. Sci. Technol.*, vol. 24, no. 2, pp. 195–226, 2020, DOI: 10.1080/10910344.2019.1636270

[39] M. Goud, A. K. Sharma, and C. Jawalkar, "A review on material removal mechanism in electrochemical discharge machining (ECDM) and possibilities to enhance the material removal rate," *Precis. Eng.*, vol. 45, pp. 1–17, 2016, DOI: 10.1016/j.precisioneng.2016.01.007

[40] Z. P. Zheng, H. C. Su, F. Y. Huang, and B. H. Yan, "The tool geometrical shape and pulse-off time of pulse voltage effects in a Pyrex glass electrochemical discharge microdrilling process," *J. Micromechanics Microengineering*, vol. 17, no. 2, pp. 265–272, 2007, DOI: 10.1088/0960-1317/17/2/012

[41] F. A. Shamim, "Development and parametric investigation of near-dry wire ECDM process," [PhD thesis, IIT Roorkee, 2021].

[42] J. W. Liu, T. M. Yue, and Z. N. Guo, "Grinding-aided electrochemical discharge machining of particulate reinforced metal matrix composites," *Int. J. Adv. Manuf. Technol.*, vol. 68, no. 9–12, pp. 2349–2357, 2013, DOI: 10.1007/s00170-013-4846-8

[43] N. K. Jha, T. Singh, A. Dvivedi, and S. Rajesha, "Experimental investigations into triplex hybrid process of GA-RDECDM during subtractive processing of MMCs," *Mater. Manuf. Process.*, vol. 34, no. 3, pp. 243–255, 2019, DOI: 10.1080/10426914.2018.1512126

Chapter 14

Parametric Study of Process Parameters on MRR in Electrical Discharge Machining of Cu-Based SMA

Ranjit Singh, Ravi Pratap Singh, and Rajeev Trehan

Department of Industrial & Production Engineering

Dr. B.R. Ambedkar National Institute of Technology, Jalandhar, India

CONTENTS

14.1 INTRODUCTION

The need for smart materials in today's technologically advanced age is rising due to their exceptional and unique characteristics and uses in a wide range of areas, from biomedical to manufacturing. Shape memory alloys (SMAs) are a major kind of novel material that has the ability to revert to its original shape when subjected to changes in magnetic and temperature conditions [1]. The shape memory effect (SME) refers to the action of reverting to its original form. The SMAs first were identified in 1932 by Arne Olander [2]. There are three major types of SMAs, which are based on Ni-Ti, Fe-based, and Cu-based materials. Cu-based SMAs are a significant subcategory of SMAs having copper as major element in proportion of others. Cu-based SMAs have been known as low-cost alloys instead of NiTi alloys since they are conveniently processed using liquid and powder metallurgy techniques. The main Cu-based alloys are Cu-Al and Cu-Zn, with an additional alloy element added due to changes in the temperature or morphology of the process. Cu-based alloys have a wide temperature range, low hysteresis, a high super elastic effect and a large damping coefficient [3–6]. From such a viewpoint, the transition temperatures have been shown to be particularly susceptible to alloy composition [6–10]. It decays when overheated into the equilibrium

DOI: 10.1201/9781003327370-14

255

stages, resulting in martensite stabilization. All of these characteristics make this alloy suitable for a wide range of applications in the biomedical, automotive, and aerospace industries, among others [11,12]. The machining of such specialised materials is a task for scientists, researchers and industrialists to satisfy demand in the current scenario [12–15]. The intricate machining of such alloys is a challenge with the traditional machining methods [16–18]. So for preciseness and accuracy in machining intricate and complex shapes non-conventional machining operations have taken place over the traditional machining methods [19–22]. Non-conventional machining operations results into complex machining shapes on such advanced materials to cope with the needs of the current industry [23–27]. Electrical discharge machining (EDM) is a non-traditional machining technique that uses an electric spark to remove material from the work surface and achieve the required shape [28–32]. The electrical discharge causes a high temperature at the tool-work contact, melting and vaporizing the material. This process is used to promote highly precise machining, complicated shape of cuts, and substrate conditions in modern production [33–35]. One of the advantages of the EDM method is that an appropriate space between the working substance and the tool (electrode) is maintained. A residual stress-free material has been obtained after processing the material in the EDM process [36]. The EDM system's working concept is useful in a variety of ways, including micro-EDM, die-sink EDM, wire cut EDM, dry EDM, and powder mixed EDM to name a few [37–44]. EDM techniques are suitable for both large- and micro-scale machining regions because of their wide range of variables [45–48].

The current study looked at process factors for material removal rate while machining Cu-based SMA, such as Cu-Zn-Ni, in an EDM. The pilot study has been presented and discussed in this paper. The chemistry of MRR has been discussed and variation in the material removal while performing experiments at different range of parameters. The pictorial graphs has been presented for explaining the variation in MRR values while performing experiments.

14.2 MATERIALS AND METHODS

The Cu-based SMA i.e. Cu-Zn-Ni has been chosen for performing experiments. The electrode of copper has been selected as a tool for performing EDM operations. The electrode has diameter of 10 mm. The workpiece is in the shape of a round disc with dimensions of 15 mm in height and 100 mm in circumference. T_{on} time, T_{off} time, gap voltage, and peak current are the process parameters that are considered in the EDM process. The response parameter calculated with the variance of the method parameters is MRR.

Experiments have been done on the Oscar max CNC die-sink EDM system. The parameters chosen are T_{on} time, T_{off} time, gap voltage, and peak current. Material removal rate (MRR) has been selected as the response variable for investigating the impacts of various process parameters. The schematic view of CNC die-sinking EDM is shown in Fig. 14.1.

The set of chosen method parameters has been shown in Table 14.1 (Fig. 14.2).

Table 14.2 represents the MRR values at different setting of T_{on} time. The maximum and lowest material removal rate (MRR) values are 594.9725 and 540.9562 at T_{on} time of 30 μs and 400 μs.

Figure 14.1 Die-sinking EDM setup (Source: CIHT Maqsudan Jalandhar).

Table 14.1 Process parameter range

Parameters of the process	Range
Pulse on-time (T_{on})	30, 45, 60, 90, 120, 150, 250, 300, 400
Pulse off-time (T_{off})	20, 45, 60, 90, 120, 200, 250, 300
Peak current	8, 14, 20, 24, 30, 40, 50
Gap voltage	30, 40, 50, 60, 70, 80

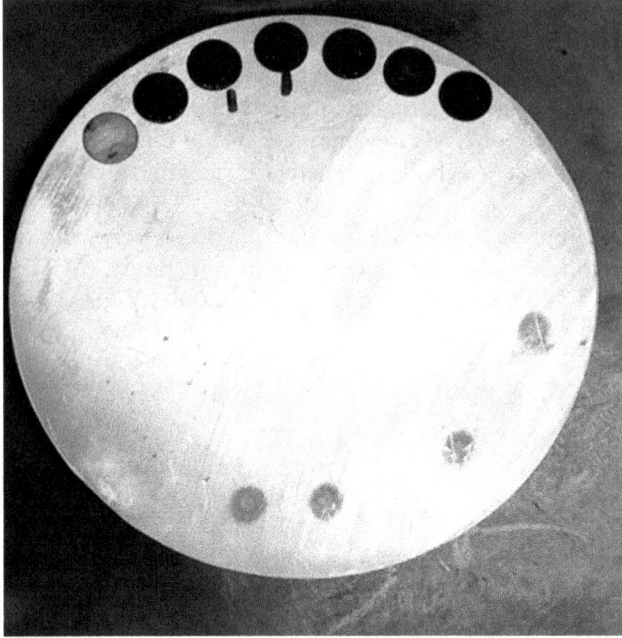

Figure 14.2 Cu-Zn-Ni shape memory alloy.

Table 14.2 Pulse on-time variation table for Cu-based SMA

Sr. No.	T_{on} time (μs)	Diameter in actual	Depth in actual	Time in minutes	Tool weight intial	Tool weight final	Intial wt-final wt.	MRR
1	30	10.03	6.85	10	29.27	27.01	2.26	540.9562
2	45	10.05	6.95	10	27.01	24.51	2.5	551.0444
3	60	10.08	6.99	10	35.27	32.18	3.09	557.5296
4	90	10.12	7.03	10	32.18	28.63	3.55	565.179
5	120	10.15	7.05	10	33.21	29.39	3.82	570.1523
6	150	10.19	7.08	10	29.39	25.48	3.91	577.1003
7	250	10.21	7.12	10	32.16	28.11	4.05	582.6411
8	300	10.23	7.18	10	28.11	23.86	4.25	589.8551
9	400	10.26	7.2	10	27.42	23.05	4.37	594.9725

MRR values at various pulse off-times are shown in Table 14.3. The minimum and maximum MRR value is 555.9608 and 588.5119 at T_{off} time of 300 μs and 20 μs.

Table 14.4 represents the MRR values at various peak current settings. The minimum and maximum values of MRR are 546.4842 mm^3/min and 599.783 mm^3/min for peak current value of 8 and 50 A.

Table 14.3 Pulse off-time variation table

Sr. No.	T_{off} time (μs)	Diameter in actual	Depth in actual	Time in minutes	Tool weight intial	Tool weight final	Int wt-final wt.	MRR
1	20	10.19	7.22	10	35.43	31.55	3.88	588.5119
2	45	10.17	7.17	10	31.55	28.03	3.52	582.1444
3	60	10.14	7.15	10	34.16	30.95	3.21	577.1007
4	90	10.12	7.11	10	30.95	27.93	3.02	571.6106
5	120	10.09	7.08	10	32.41	29.76	2.65	565.8291
6	200	10.07	7.06	10	29.76	27.57	2.19	561.9961
7	250	10.05	7.05	10	27.57	25.68	1.89	558.9731
8	300	10.03	7.04	10	33.17	31.72	1.45	555.9608

Table 14.4 Peak current variation table

Sr. No.	Peak current (I_p)	Diameter in actual	Depth in actual	Time in minutes	Tool weight intial	Tool weight final	Int. wt-final wt.	MRR
1	8	10.03	6.92	10	32.62	32.17	0.45	546.4842
2	14	10.07	6.98	10	32.17	31.59	0.58	555.6279
3	20	10.09	7.03	10	30.65	29.27	1.38	561.8331
4	24	10.15	7.11	10	29.27	27.01	2.26	575.0046
5	30	10.2	7.17	10	35.74	32.24	3.5	585.5839
6	40	10.24	7.19	10	32.24	27.97	4.27	591.832
7	50	10.28	7.23	10	27.97	23.05	4.92	599.783

Table 14.5 Gap voltage variation table

Sr. No.	Gap voltage (V_g)	Diameter in actual	Depth in actual	Time in minutes	Tool weight intial	Tool weight final	Int. wt-final wt.	MRR
1	30	10.15	7.13	10	26.42	22.91	3.51	576.6221
2	40	10.11	7.1	10	22.91	20.21	2.7	569.6791
3	50	10.09	7.09	10	32.24	30.08	2.16	566.6283
4	60	10.07	7.06	10	30.08	28.19	1.89	561.9961
5	70	10.05	7.03	10	34.46	33.21	1.25	557.3873
6	80	10.03	6.98	10	33.21	32.37	0.84	551.2225

Table 14.5 represents the values of MRR at different gap voltage values. MRR has a maximum and lowest value of 576.6221 mm^3/min and 551.2225 mm^3/min for gap voltages of 30 V and 80 V, respectively.

14.3 RESULTS AND DISCUSSION

The tests are conducted at CIHT Jalandhar using a CNC EDM machine. The effect of different process parameters has been studied while

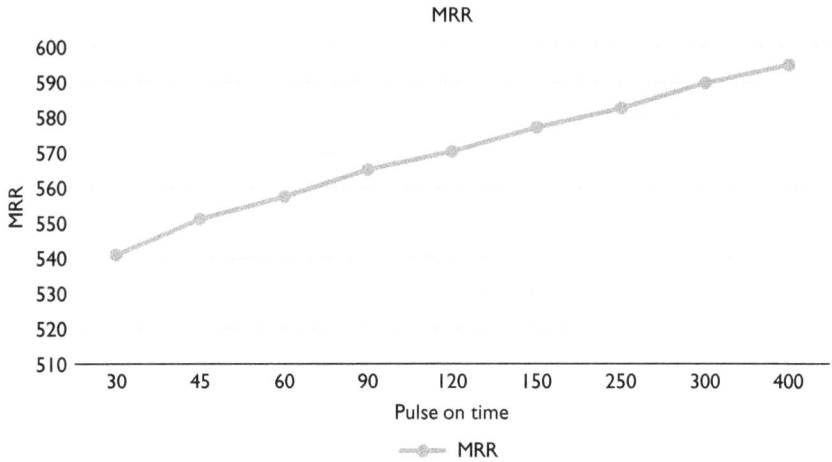

Figure 14.3 Pulse on-time v/s MRR.

processing of Cu-Zn-Ni SMA in EDM process. The effects have been observed and presented in graphical form. The graphs represent the relationship between the parameters. The behavior of graph represents the particular parameter effect on MRR.

Fig. 14.3 illustrates the effect of T_{on} on the MRR. The graph demonstrates an increase in MRR and surge in pulse on-time. The increase in T_{on} time leads in a shorter time interval between pulse occurrences, which contributes to the production of more sparks. The huge spark creation results in the development of a high temperature, which causes the workpiece to be removed, melted, and vaporized. MRR has a maximum and a minimum value of 594.9725 mm^3/min and 540.9562 mm^3/min for a T_{on} time of 30 s and 400 s, respectively. As the T_{on} time increases simultaneously, there is a linear-like increase in the MRR value, as depicted in the Fig. 14.3.

Fig. 14.4 shows the visual representation of the impact of pulse off-time. With the increase in pulse off-time, the value of MRR decreases. This occurs because an increase in T_{off} time increases the lag time between pulses, reducing spark generation. As a consequence of decreased spark generation, temperature rises less, and as temperature rises, thermal energy production at the tool-workpiece contact decreases, resulting in less material removal from the workpiece. The minimum and maximum MRR value is 555.9608 and 588.5119 at T_{off} time of 300 μs and 20 μs.

Fig. 14.5 graphically depicts the effect of peak current on MRR. The increase in peak current causes an increase in MRR. The increase in peak current causes a significant temperature to develop, resulting in

MRR

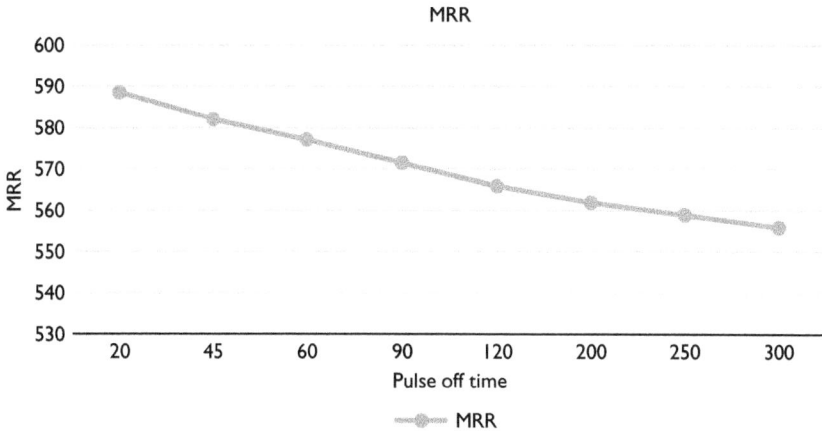

Figure 14.4 Pulse off-time v/s MRR.

MRR

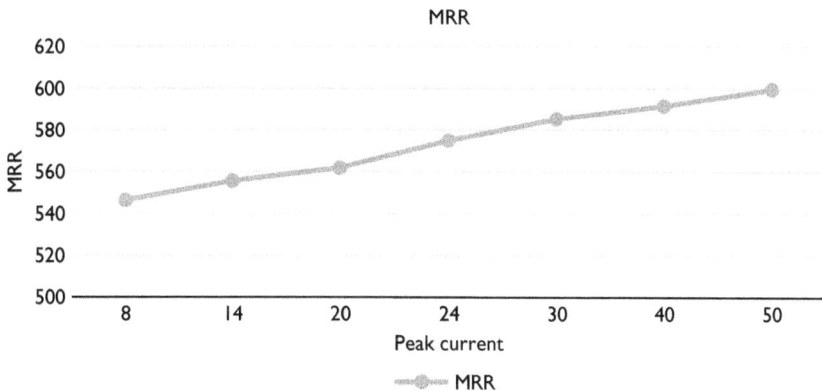

Figure 14.5 Peak current v/s MRR.

considerable thermal energy and the removal of the workpiece through melting and vaporization. At peak currents of 8 and 50 A, the lowest and highest MRR values are 546.4842 mm^3/min and 599.783 mm^3/min, respectively.

Fig. 14.6 depicts the effect of the voltage gap on the MRR visually. The substance removal rate decreases as the gap voltage increases. Longer discharge waiting time results in lower temperature, which results in a reduction in MRR. A smaller gap voltage implies shorter discharge waiting time, which means more spark generation, which leads to greater temperature formation, which leads to higher MRR. At gap voltages of 30 V and 80 V, the maximum and lowest values of MRR are 576.6221 mm^3/min and 551.2225 mm^3/min, respectively.

MRR

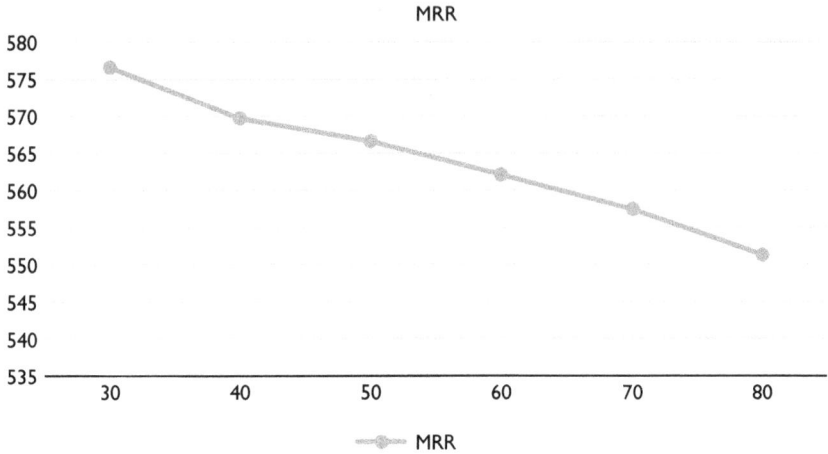

Figure 14.6 Gap voltage v/s MRR.

14.4 CONCLUSIONS

After experimenting with Cu-Zn-Ni SMA in a CNC EDM machine, the following interferences were discovered:

1. T_{on} time and peak current are two key factors that have a significant effect on the MRR. MRR rises as T_{on} time increases because there is no lag between matching pulses, resulting in greater discharge generation, which leads to higher temperature formation, and therefore more MRR.
2. With the increase in T_{off} time, the value of MRR decreases. This occurs because increasing the pulse off time increases the lag time between pulses, limiting the spark's growth. Due to less spark formation, there is less rise in temperature and, due to less temperature, there is less thermal energy formation at tool workpiece interface, which results in less material removal from the workpiece.
3. The increase in peak current causes an increase in MRR. The increase in peak current causes a significant temperature rise, which results in a big amount of thermal energy, which causes the workpiece to be removed by melting and vaporization.
4. The rate of material elimination decreases as the voltage difference rises. A large gap voltage indicates a longer discharge waiting time, which results in lower temperature and a lower MRR. Smaller gap voltages result in shorter discharge waiting time, resulting in greater spark generation, which leads to increased temperature formation, and therefore increased MRR.
5. MRR's maximum and lowest values are 594.9725 mm^3/min and 540.9562 mm^3/min at T_{on} time of 30 μs and 400 μs. The minimum

and maximum value of MRR is 555.9608 mm^3/min and 588.5119 mm^3/min at T_{off} time of 300 µs and 20 µs.

6. At peak currents of 8 and 50 A, MRR has a minimum and maximum value of 546.4842 mm^3/ min and 599.783 mm^3/min, respectively. At gap voltages of 30 V and 80 V, MRR reaches maximum and lowest values of 576.6221 mm^3/min and 551.2225 mm^3/min, respectively.

REFERENCES

[1] Jani, J.M., Leary, M., Subic, A. and Gibson, M.A. (2014), "A review of shape memory alloy research, applications and opportunities", *Mater Design (1980–2015)*, Vol. 56, pp. 1078–1113.

[2] Sun, L., Huang, W.M., Ding, Z., Zhao, Y., Wang, C.C., Purnawali, H. and Tang, C. (2012), "Stimulus-responsive shape memory materials: A review", *Mater Design*, Vol. 33, pp. 577–640.

[3] Ölander, A. (1932), "An electrochemical investigation of solid cadmium-gold alloys". *J Amer Chem Soc*, Vol. 54, No. 10, pp. 3819–3833.

[4] Vernon, L.B. and Vernon, H.M. (1941), "US Patent and Trademark Office". *Washington DC, US Patent No. 2*, Vol. 234, pp. 993.

[5] Gao, X.Y. and Huang, W.M. (2002), "Transformation start stress in non-textured shape memory alloys". *Smart Mater Str*, Vol. 11, No. 2, pp. 256–268.

[6] Bil, C., Massey, K. and Abdullah, E.J. (2013), "Wing morphing control with shape memory alloy actuators". *J Intel Mater Sys Str*, Vol. 24, No. 7, pp. 879–898.

[7] Van Humbeeck, J. (1999), "Non-medical applications of shape memory alloys". *Mater Sci Eng A*, Vol. 273, pp. 134–148.

[8] Singh, R.P. and Singhal, S. (2016), "Rotary ultrasonic machining: A review". *Mater Manuf Process*, Vol. 31, No. 14, pp. 1795–1824.

[9] Singh, R.P. and Singhal, S. (2017), "Investigation of machining characteristics in rotary ultrasonic machining of alumina ceramic". *Mater Manuf Process*, Vol. 32, No. 3, pp. 309–326.

[10] Singh, R.P., Tyagi, M. and Kataria, R. (2019); Sachdeva, A. Kumar P. Yadav O.P., editors, "Selection of the optimum hole quality conditions in manufacturing environment using MCDM approach: A case study", In *Operations Management and Systems Engineering*. Springer, Singapore, pp. 133–152.

[11] Singh, R.P. and Singhal, S. (2017), "Rotary ultrasonic machining of macor ceramic: An experimental investigation and microstructure analysis". *Mater Manuf Process*, Vol. 32, No. 9, pp. 927–939.

[12] Singh, R.P. and Singhal, S. (2018), "Experimental investigation of machining characteristics in rotary ultrasonic machining of quartz ceramic". *Proceedings of the Institution of Mechanical Engineers, Part L: Journal of Materials: Design and Applications*, Vol. 232, No. 10, pp. 870–889.

[13] Kheirikhah, M.M., Rabiee, S. and Edalat, M.E. (2010), "A review of shape memory alloy actuators in robotics". In *Rob Socc World Cup*. Springer, Berlin, Heidelberg, pp. 206–217.

[14] Sreekumar Mntsmzmmr, Nagarajan, T., Singaperumal, M., Zoppi, M. and Molfino, R. (2007), "Critical review of current trends in shape memory alloy actuators for intelligent robots". *Ind Rob An Int J*, Vol. 34, No. 4, pp. 285–294.

[15] Wang, W. and Ahn, S.H. (2017), "Shape memory alloy-based soft gripper with variable stiffness for compliant and effective grasping". *Soft Rob*, Vol. 4, No. 4, pp. 379–389.

[16] Petrini, L. and Migliavacca, F. (2011), "Biomedical applications of shape memory alloys". *J Metal*, Vol. 2011.

[17] Singh, R.P., Kataria, R., Kumar, J. and Verma, J. (2018), "Multi-response optimization of machining characteristics in ultrasonic machining of WC-Co composite through Taguchi method and grey-fuzzy logic". *AIMS Mater Sci*, Vol. 5, pp. 75–92.

[18] Singh, R.P., Kumar, J., Kataria, R. and Singhal, S. (2015), "Investigation of the machinability of commercially pure titanium in ultrasonic machining using graph theory and matrix method". *J Eng Res*, Vol. 4, No. 3, pp. 1–20.

[19] Singh, R.P. and Singhal, S. (2018), "An experimental study on rotary ultrasonic machining of macor ceramic". *Proceedings of the Institution of Mechanical Engineers, Part B: Journal of Engineering Manufacture*, Vol. 232, No. 7, pp. 1221–1234. (2018c)

[20] Singh, R.P. and Singhal, S. (2018), "Rotary ultrasonic machining of alumina ceramic: Experimental study and optimization of machining responses". *Proceedings of the Institution of Mechanical Engineers, Part L: Journal of Materials: Design and Applications*, Vol. 232, No. 12, pp. 967–986.

[21] Tyagi, M., Panchal, D., Singh, R.P. and Sachdeva, A. (2019), "Modeling and analysis of critical success factors for implementing the IT-based supply-chain performance system". In *Operations Management and Systems Engineering*. Springer, Singapore, pp. 51–67.

[22] Singh, R.P., Kataria, R. and Singhal, S. (2019), "Decision-making in real-life industrial environment through graph theory approach". In *Computer Architecture in Industrial, Biomechanical and Biomedical Engineering*. IntechOpen. DOI: 10.5772/intechopen.82011.

[23] Mantovani D. (2000), "Shape memory alloys: Properties and biomedical applications". *J Min Metals Mater Soc*, Vol. 52, No. 10, pp. 36–44.

[24] Wong, Y., Kong, J., Widjaja, L.K. and Venkatraman, S.S. (2014), "Biomedical applications of shape-memory polymers: How practically useful are they?" *Sci China Chem*, Vol. 57, No. 4, pp. 476–489.

[25] Duerig, T., Pelton, A. and Stöckel, D. (1999), "An overview of nitinol medical applications". *Mater Sci Eng A*, Vol. 273, pp. 149–160.

[26] Lieva, V.L. and Carla, H. (2004), "Smart clothing: A new life". *Int J Cloth Sci Tech*, Vol. 16, No. 1–2, pp. 63–72.

[27] Hewidy, M.S., El-Taweel, T.A. and El-Safty, M.F. (2005), "Modelling the machining parameters of wire electrical discharge machining of Inconel 601 using RSM". *J Mater Process Tech*, Vol. 169, No. 2, pp. 328–336.

[28] Yu, Z., Jun, T. and Masanori, K. (2004), "Dry electrical discharge machining of cemented carbide". *J Mater Process Tech*, Vol. 149, No. 1–3, pp. 353–357.

[29] Singh, R., Singh, R.P. and Trehan, R. (2020), "State of the art in processing of shape memory alloys with electrical discharge machining: A review". Proceedings of the Institution of Mechanical Engineers, Part B: Journal of Engineering Manufacture, Vol. 235, No. 3, pp. 333–366.

[30] Singh, R., Singh, R.P., Tyagi, M. and Kataria, R. (2019), "Investigation of dimensional deviation in wire EDM of M42 HSS using cryogenically treated brass wire". Mater Today: Proc, Vol. 25, pp. 679–685.

[31] Singh, R.P. and Singhal, S. (2018), "Experimental study on rotary ultrasonic machining of alumina ceramic: Microstructure analysis and multi-response optimization". *J Mater: Des Applications*, Vol. 232, No. 12, pp. 967–986.

[32] Gupta, A.K. and Singh, R.P. (2022), "Application of TLBO to optimize cutting variables for face milling of aluminium alloy Al-8090". In Proceedings of the International Conference on Industrial and Manufacturing Systems (CIMS-2020) (pp. 1–13). Springer, Cham.

[33] Chiang, K.T. and Chang, F.P. (2006), "Optimization of the WEDM process of particle-reinforced material with multiple performance characteristics using grey relational analysis". *J Mater Process Tech*, Vol. 180, No. 1–3, pp. 96–101.

[34] Lee, S.H. and Li, X.P. (2001), "Study of the effect of machining parameters on the machining characteristics in electrical discharge machining of tungsten carbide". *J Mater Process Tech*, Vol. 115, No. 3, pp. 344–358.

[35] Singh, R., Singh, R.P. and Trehan, R. (2022), "Investigation of machining rate and tool wear in processing of Fe-based-SMA through sinking EDM". In Proceedings of the International Conference on Industrial and Manufacturing Systems (CIMS-2020) (pp. 439–451). Springer, Cham.

[36] Ambade, S., Tembhurkar, C., Patil, A.P., Pantawane, P. and Singh, R.P. (2021), "Shielded metal arc welding of AISI 409M ferritic stainless steel: Study on mechanical, intergranular corrosion properties and microstructure analysis". *World J Eng*, Vol. 19, No. 3, pp. 266–273.

[37] Singh, R., Singh, R.P. and Trehan, R. (2021), "Parametric investigation of tool wear rate in EDM of Fe-based shape memory alloy: Microstructural analysis and optimization using genetic algorithm". *World J Eng*, Vol. 19, No. 3, pp. 418–428.

[38] Butola, R., Yuvaraj, N., Singh, R.P., Tyagi, L. and Khan, F. (2021), "Evaluation of microhardness and wear properties of Al 6063 composite reinforced with yttrium oxide using stir casting process". *World J Eng*, Vol. 19, No. 3, pp. 361–367.

[39] Ho, K.H., Newman, S.T., Rahimifard, S. and Allen, R.D. (2004), "State of the art in wire electrical discharge machining (WEDM)". *Int J Mach Tool Manuf*, Vol. 44, No. 12–13, pp. 1247–1259.

[40] Abbas, N.M., Solomon, D.G. and Bahari, M.F. (2007), "A review on current research trends in electrical discharge machining (EDM)". *Int J Mach Tool Manuf*, Vol. 47, No. 7–8, pp. 1214–1228.

[41] Singh, S., Maheshwari, S. and Pandey, P.C. (2004), "Some investigations into the electric discharge machining of hardened tool steel using different electrode materials". *J Mater Process Tech*, Vol. 149, No. 1–3, pp. 272–277.

[42] Datt, M. and Singh, D. (2015), "Optimization of WEDM parameters using Taguchi and ANOVA method". *Int J Curr Eng Tech*, Vol. 5, No. 6, pp. 3843–3847.

[43] Singh, R. and Singla, V.K. (2017), "Surface characterization of M42 Hss treated with cryogenic and non-cryogenated brass wire in WEDM process". *IMRF Biann Peer Rev Int Res J*, Vol. 5, pp. 28–32.

[44] Singh, R. and Singla, V.K. (2018), "Parametric modeling for wire electrical discharge machining of M42 HSS using untreated and cryogenated treated brass wire by using RSM". *Int J Mech Prod Eng*, Vol. 5, pp. 63–67.

[45] Kataria, R., Singh, R.P., Alkawaz, M.H. and Jha, K. (2021), "Optimization and neural modelling of infiltration rate in ultrasonic machining". *OPSEARCH*, Vol. 59, No. 1, pp. 1–20.

[46] Sahu, S., Ukey, P.D., Kumar, N., Singh, R.P. and Ansari, M.Z. (2021), "Three dimensional modelling of aluminum foam through computed tomography scan technique". *World J Eng*, Vol. 19, No. 3, pp. 340–345.

[47] Singh, R.P., Kataria, R. and Kumar, J. (2021), "Machining of WC-Co composite using ultrasonic drilling: Optimisation and mathematical modelling". *Adv Mater Process Technol*, Vol. 7, No. 2, pp. 317–332.

[48] Sharma, R.C., Dabra, V., Singh, G., Kumar, R., Singh, R.P. and Sharma, S. (2021), "Multi-response optimization while machining of stainless steel 316L using intelligent approach of grey theory and grey-TLBO". *World J Eng*, Vol. 19, No. 3, pp. 329–339.

Chapter 15

2-D Based Nanostructures and Their Machining Challenges

K. Santhosh Kumar

Department of Production and Industrial Engineering, National Institute of Technology Jamshedpur, Jharkhand, India

Subhash Singh

Department of Mechanical and Automation Engineering, Indira Gandhi Delhi Technical University for Women, New Delhi, India

CONTENTS

DOI: 10.1201/9781003327370-15

15.1 INTRODUCTION

In 1959, Nobel Prize–winner Richard Feynman introduced the concept of micromachines, manipulation and controlling compounds at undersized scale. Since then, miniaturization became the decisive rage in advanced science and technology [1]. Afterwards, the tremendous efforts of Gleiter and his companion on synthetization of granular ultrafine materials through in-situ consolidation of infinitesimal atomic clusters, the terrain of nanostructured materials turned out to be a crucial vogue in material science [2]. Nanotechnology is an area of research and innovations implicated with materials and devices for applications in human life. The term "nano" is a Greek word that means "extremely fine" [3]. Nanotechnology involves materials downsized to less than 100 nm, and connected with quantum/electron behavior. At this size, the material properties are revealingly peculiar in many aspects than that of its raw state. The nanostructures will form by means of molecular interaction, clogging, accumulation, aggregation, etc. In the case of biochemistry, intercellular components, DNA and RNA, are treated as nanostructures [4].

The domain of nanotechnology involves with the other interesting fields like applied physics, chemistry, biotechnology, bio-science, biochemistry, and molecular science. The queer physicochemical characteristics attributed to its freakish surface morphology, quantum size effect, and higher aspect ratio of 2-D nanostructures made them the trending area of research. High aspect ratio, eminent stability, inexpensive synthesis, intercalate morphology, molecular interactions, and superior physio-mechanical/chemical properties were the additional features of 2-D nanostructures [5,6]. Attributed to the outstanding features of 2-D nanomaterials set the stage, approaching a vast array of assured applications in numerous fields. Medical, energy storage, chemistry, engineering strategies, environmental studies, biochemistry, manufacturing and infrastructure, electronics and communication, consumer goods, instrumentation, sports, textiles, aerospace, automobile, defence, coatings, adhesives and sealants, biocides, etc., are the popular application sectors of nanomaterials and their structures. [7–11].

15.2 DEFINITION

In a nanostructure, when any one dimension is kept in nanometric range and other dimensions remains large, then those nanostructures are called as two-dimensional nanostructures (2-D nanostructures). In such kinds of structures, movements of electrons are restricted in one direction and spared to move in other two directions bound to quantum mechanics [12]. 2-D nanostructures generally exhibit sheet like structures in single layer manner or multi-layer manner [13]. This grade can be crystalline/amorphous and is made up of metallic, ceramic, and polymeric materials. Depending upon the restriction of movement of the electrons, these nanostructures have been categorized into four types, and are as follows [14,15]:

- 0-D – Zero-Dimensional: Electron movement restricted in three dimensions.
- 1-D One-Dimensional: Electron movement restricted in two dimensions.
- 2-D – Two-Dimensional: Electron movement restricted in one dimension.
- 3-D – Three-Dimensional: No restriction in electron movement.

The classifications, typical method(s) of synthesis, designations, and a few examples of various nanomaterials are tabulated in Table 15.1.

15.2.1 Advantages

- Vast scale of production.
- Superior particle inter-spacing.
- Enhanced flexibility in size, shape of particles, and morphologies.
- Excellent size distribution, higher purity, and reproducibility.
- High surface areas and uniformity and of the products.
- Compatible with flexible devices.
- Cost effective, easy to perform, efficient, vigorous, and almost green technique.
- Generally, no need of protecting surfactant or extra linker molecule.

Table 15.1 Classifications of nanomaterials

S. No.	Classification	Typical Method(s) of Synthesis	Designation	Example
I	0D	Sol-gel	Atom clusters and assemblies	Nanoparticles
2	I D	Vapour deposition Electro deposition	Modulated multilayers	Nanorods, nanotubes
3	2-D	Chemical vapour deposition Gas condensation	Ultrafine grained overlayers	Thin nanosheets
4	3D	Mechanical alloying	Bulky	Nanocomposites

15.2.2 Disadvantages

- High-energy consumption in some processes.
- Difficult-to-predict behaviour.
- Long reaction time and expensive reactors.
- In few preparative methods, equipment is highly expensive.
- Requires more time and steps.
- Safety concern particles with broad size distribution.
- Low out-of-plane electronic and ionic conductivity.
- Re-stacking.

15.3 CLASSIFICATIONS

Numerous procedures are there to produce and arrange nanoparticle and nanostructures, in which certain progenitors are introduced into liquidus, solidus, and gaseous states for proper arrangement of nanoparticles [16]. The arrangement of the nanoparticles is bound to chemical reaction or physical compaction between the material and precursors or among the particles. In spite of the sundry procedures, the production of nanomaterials is clubbed into two major groups: bottom-up and top-down approaches [17,18]. Fig. 15.1 and Fig. 15.2 represent the concept of top-down and bottom-up approaches and classifications of 2-D nanostructures available in different forms and shapes.

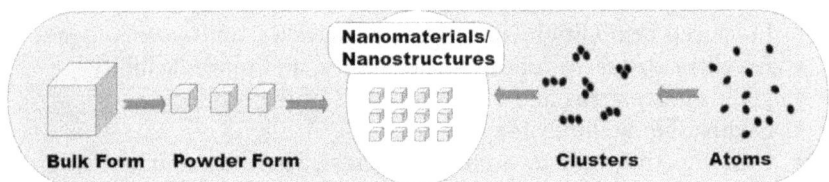

Figure 15.1 Delineated view of top-down and bottom-up approaches.

2D0 - Fullerene films

2D1 - Nanostraw, PhC, Fibers Films

2D2 - Tiling mosaic, layered films

2D00 - Heterofilms, fullereeno powders

2D Nanostructures

2D10 - Films of pods, fullereno-fibers

2D11 - Films of fibers/Nanotubes, PhC waveguides

2D20 - Fullereno-plate films

2D21 – Bridges, fiber-layer films

2D22 – MOS-structures

2D210 - Fullerene-fiber-layer films

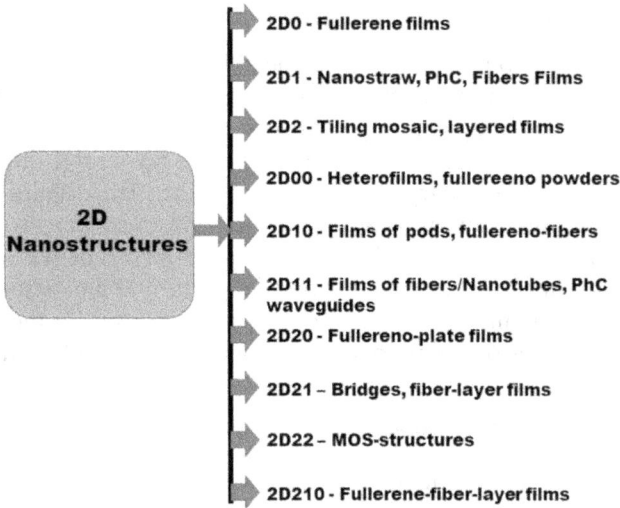

Figure 15.2 Classification of 2-D nanostructures.

15.4 PROPERTIES

On the whole, proper arrangement of atoms and their systematic alignment decides the properties of a structure/material. Every material exhibits additional characteristics when their size narrow downed to nano size (less than 100 nm). Size of the grain/constitutive phase, structure/interface, and chemical composition of the material are the three dominant factors which define the properties of 2-D nanostructures. Few conditions reported that, any one of the factors will deceive the respective properties of 2-D nanostructures [19].

15.4.1 Physical Properties

15.4.1.1 Diffusion and Sinterability

Diffusion paths are considered to be quicker in case of polycrystalline compounds, on account of their broad atomic alignment, grain boundaries, and interfaces. For polycrystalline materials, the activation energy of atoms during diffusion across the grains is double the activation energy of diffusion through grain boundary. Such increment in the diffusion energy will affect the mechanical properties substantially like, creep, superplasticity, efficient doping ability, and metal alloy synthesis even at low temperatures, which was critical in other systems. Along with their features, this enhanced diffusion also promotes limit of solubility, intermetallic phase transformation, and sintering capacity. The higher diffusibility of nanocrystalline materials drives the diffusion alongside bringing the formation of metastable phases proportionately at low temperatures [19,20].

15.4.1.2 Thermal Expansion Coefficient

Low-temperature phase transition metallic nanoparticles and semi-conductors are in contrast to their mass state. Whenever a fall is noticed in a phase transition temperature, it indicates a variation in energy from surface to volume as a role of grain size. The most favourable diffusion rate through the grain boundary is possible only when the disparity is in its thermal properties. Few antagonistic reports informed that a noticeable drop in thermal expansion coefficient of the material can be seen due to the effect of its grain sizes. But it is anticipated that the 2-D nanostructures will have a greater thermal expansion coefficient, as proven by the evolution of grain boundaries on a variety of metals and nanocrystalline alloys [20].

15.4.1.3 Electrical Resistance

In concern with the compounds of nanocrystalline and microcrystalline, the specific electrical resistivity rises with temperature. Despite that, at a standard temperature, electrical resistance improves and accelerates with the increase in size of the grains. Because of the entropy effect at the atomic level and configuration of grain boundaries, forces the electrons to scatter while crossing the grain boundaries. This scattering of the electrons extends the electrical resistance of the compounding material. Thereby, the specific electrical resistivity of the material can be altered easily by controlling the grain size. It is obvious that as the grain size decreases, or in other words the area of grain boundaries increases, electron dispersion, and consequently, electrical resistance, increases. Then, by controlling the grain size, it is possible to change the specific electrical resistance [19–21].

15.4.1.4 Solubility of Alloying Elements

Nanocrystalline materials revealed high solubility limits than the familiar polycrystalline materials. Attributable to factors like wider grain bounsdary, and lower concentration of atoms along the grain boundaries, dissolution of alloying elements in nanocrystalline materials is greater. Thusly, by rising the solubility limits of alloying elements, a new alloy with unique properties can be produced [20].

15.4.1.5 Magnetic Properties

Basically, magnetic properties of a material can be strengthened by rising the surface-area-to-volume-ratio. Generally, magnetic 2-D nanostructures corroborated errant behaviours, owing to their particle size and transmittance charge capacity. Magnetic properties of a 2-D nanostructured material differ from the bulk form, and this is due to the increment in surface energy results in enhancing sufficient energy for intuitive change of magnetic field in polar direction, such that the material adjusts itself into advanced paramagnetic state. This condition of the advancement in the paramagnetic state of a 2-D nanostructured material is referred to as a super-magnetic property [22]. To be specific, in view of when the mass state of a material is downsized to the nano range, paramagnetic energy develops into a super-magnetic energy. Ferromagnetic properties of a material are under control with the size of the grains, constitutional phase, and intermolecular distance. Consequently, the saturated magnetic properties (MPs) and curie temperature in case of nanocrystalline materials are significantly low contrasted with microcrystalline materials.

15.4.1.6 Corrosion Resistance

Based on several studies on corrosion resistance of crystalline materials, it's been reported that corrosion is an ever-existing factor, and corrosion rate of nanocrystalline material is more than that of microcrystalline materials. Due to the wide grain boundaries, the molecules with intensive energy are free to move and readily active in corrosion reactions; hence, nanocrystalline material possess greater corrosion rates. As the nanostructures are homogeneous structures, they are more active in corrosive reactions. Experimental investigations conducted on several nanocrystalline 2-D nanostructures showed high local corrosion resistance compared with traditional structures. Average grain size and corrosion resistance are inversely proportional for nanocrystalline structures, but this tendency will alter the nature and porosity of the materials [20–22].

15.4.2 Mechanical Properties

By decreasing the size of the nanomaterials, their mechanical properties can be enhanced greatly. Usually, in a crystalline structure the high-energy content defects dislocations, impurities, and micro-torsions are supposed to be eliminated. Because these high-energy content defects cause a slip in the grain sizes at the stress field, it results in expansion of atoms and dislocations in the stress field. Several investigations on 2-D nanostructures reported that grain size is single enough to decide the mechanical properties in comparison with traditional materials. Secondly, impurities, size/shape/distribution of grains and density of the crystalline defects are the influenceable factors for mechanical properties of 2-D nanostructured materials [22].

15.4.2.1 Elastic Properties

At the beginning, evaluation was done to find out the elastic constant of nanocrystalline materials on specimens fabricated through powder metallurgy method. Corresponding reports of these investigations specify that nanocrystalline materials exhibited superior elastic properties over the conventional microcrystalline materials. Nevertheless, experimental investigations evidently intimated that the existence of voids and cracks in samplings could have a noteworthy effect of drop in elastic modulus of nanocrystalline materials. It's difficult to alter the Young's modulus of a nanocrystalline material when it possesses insubstantial porosity, even with the decrement in the grain size [23]. This tendency is anticipated and reliable up to the grain size of 5 nm, below this limit atoms in the grain boundary will be substantial and may drastically differ their elastic properties. Therefore, in many cases, the elastic properties of nanocrystalline materials with grain size about 10 nm, is almost equal to regular microcrystalline materials.

15.4.2.2 Plastic Properties

In a polycrystalline material, size of the grain defines the yield strength and hardness of that structure. And the relation between the grain size and the yield strength can be represented by an empirical Hall–Petch equation and expressed as follows:

$$Yield\ Strength\ (\sigma_y) = \sigma_0 + kd^{-1/2}$$

$$Hardness\ (H) = H_0 + k'd^{-1/2}$$

where
  σ_y = yield stress
  σ_0 = yield stress for a single dislocation
  k = grain boundaries' constant
  d = grain diameter
  H = hardness, and
  H_0 = hardness for a single dislocation

The above expressions can be explicated by focusing on production, motion, and concentration of the dislocation behind the grains. The assessment of the above-mentioned results declared that, regardless to the productive method for all the cases, hardness and strength enhances by following the Hall-Petch equation and is true up to the critical range (15–25 nm). Wholly, whenever the grain size goes beneath the critical range, a reduction in the Hall-Petch coefficient will be noticed. In nanocrystalline structures, grain boundary slip or grain disorientation is the basic phenomenon to lead the plastic deformation and are affected by the critical ranged nanocrystalline materials [24].

15.4.2.3 Ductility and Toughness

The obtained results of nanocrystalline substances conducted to determine the ductility were miscellaneous and case sensitive to the factors like testing methods, surface texture, flaws, and porosity. However, the nanocrystalline structures with routine grain sizes exhibited reduced ductility while the nanocrystalline materials with lower grain sizes showed brittle nature [25]. These reflections were accredited to intensified diffusional creep providing the plasticity at critical temperatures at which traditional grain-sized compounds were doomed to fail in the elastic regimen.

15.4.2.4 Superplasticity

Certain polycrystalline materials at a particular temperature and strain rate demonstrated a greater plastic deformation (about 100–1000%) without any fracture or necking propagation. Such kinds of behaviour are called super-plasticity, which characteristically arises in substances with ultra-fine grains and above their melting temperatures [26]. The rate of deformation that promotes the superplasticity nature is inversely proportionate to the square of the grain sizes. The thermal stability of 2-D nanostructures and deficit of grain growth should be contemplated for benefit of their superplastic behaviour.

15.4.3 Thermal Properties

The thermal stability is crucial for the amalgamation of nanocrystalline precipitates unaffecting the morphological texture. Some typical physical characteristics of nanostructures are insightful to the thermal attributes such as heat capacitance, thermoelectric power, thermal conduction, and thermal expansion. The 2-D nanostructures are more consistent in terms of their thermal facets. The extensive interfaces within the 2-D nanostructures are the unique feature than the traditional materials. If the interfaces of the nanostructures are decreased, then their free energy capacity will be wea-kened, resulting in grain growth and phase particles. The above-mentioned grain growth in a nanostructure may initiate from several factors suchlike coaxial grains, impurities, grain slip, and voids [19–26].

15.5 PREPARATION METHODS

Preparative methods of nanomaterials are composed of two fundamental approaches, namely, top-down and bottom-up approaches. When a bulk structure breaking down into nanosized substances to form a nanomaterial, then it is called a top-down approach [2,5]. This approach requires severe and extreme conditions to prepare a nanostructure. Laser ablation, arc dis-charge, mechanochemical, ball milling, electrochemical, etc., are the different 2-D nanostructure preparative methods under the top-down method, whereas the bottom-up method depends on molecular-level precursor that advances agglomeration by means of physicochemical activities [5,6]. Reduced extreme conditions are enough to produce desired product in case of bottom-up method. These physiochemical processes involve polymerization, condensation, and pyrolysis in order to create a 2-D nanostructure. Betwixt the two methods, the bottom-up approach is the most popularly employed technique, as it advantageous in few aspects like inexpensive, expedite in process, and can perform mass scale production.

15.5.1 Top-down Method

In fact, the top-down technique is a destructive mechanism, which comprises removal of blocks from substrate, cutting, and sizing to the atomic planes. In the case of the top-down method, large-scale/bulk material is reduced to nonorange substances through chemical, physical, and mechanical processes. As this process entails the production of nanoparticles by means of mechanical and physical techniques, it is also termed a mechanical-physical particle production method [5]. A massive force is utilized to crush the bulk material which separates the tiny particles form it, and are called as nanoparticles. This process can also be performed by grinding, milling, chemical vapour deposition, physical vapour deposition, and lithographic cutting. Generally the top-down method is attained to produce metallic/ceramic nanoparticles, nanosheets/nanoflakes, and laminar 2-D nanostructures. During the exfoliation process, chemical reactivity and shear force are applied. Structural integrity and high crystallinity are preserved during the process [2, 5, 6, 16]. The following figure gives a brief note on the popular approaches available to process the 2-D nanomaterials by following the top-down methodology (Fig. 15.3).

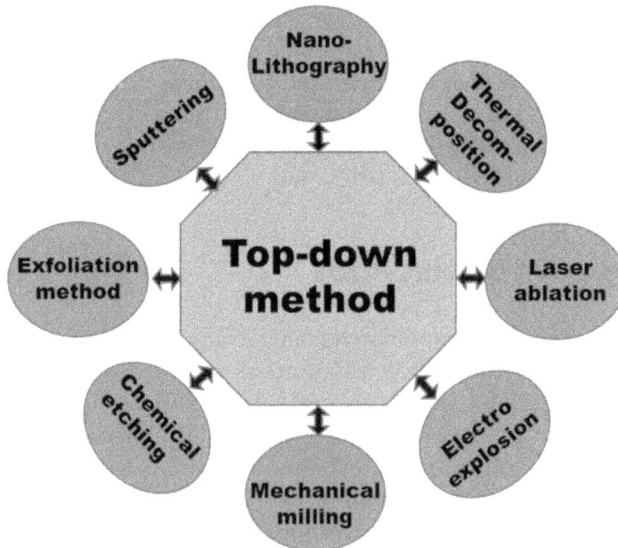

Figure 15.3 Popular top-down approaches.

15.5.2 Bottom-up Method

In case of the bottom-up method, 2-D nanomaterials are produced by permitting molecular reactions among the atoms of the nanoparticles. Atoms and molecules are the active agents for the synthesis of 2-D nanostructures throughout the process. As this process is related with the aggregation, accumulation, conglomeration, settlement, and blocking up, this method is also called a building-up method. This is based on the assembly of several small particles for the production of a complex particles, metal oxide nanoparticles, nanostructured thin films, and metal nanospheres. The popular techniques used for synthesizing the various nanomaterials in the bottom-up approach is represented in Fig 15.4.

Double-layered hydroxides, coprecipitation, ion-exchange, and reconstruction methods are normally used to produce complex 2-D nanostructures. Almost all bottom-up approaches are eco-friendly and inexpensive methods. The 2-D nanostructures obtained from bottom-up techniques, are of least defects, more homogeneous, better morphology, and proper size distribution [2, 5, 6, 16, 17].

Savants are focusing on the concept of molecular manufacturing or mechanosynthetic chemistry, which allows the atoms to self-assembled into a nanostructure. Hence, promoting and broadening up a contemporary technique by controlling the physicochemical properties has become a fresh domain of interest.

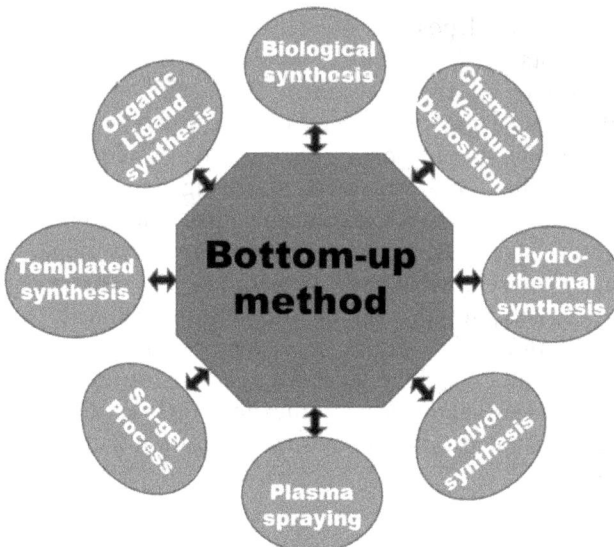

Figure 15.4 Popular bottom-up approaches.

15.6 FABRICATION TECHNIQUES

15.6.1 Thin Film Fabrication

The thin film fabrication processes are divided into five basic categories: physical vapour deposition, chemical vapour deposition, ion-assisted beam deposition, molecular beam, epitaxy, pulsed laser deposition, and chemical bath deposition [27]. Fig. 15.5 represents the various techniques and procedures involved in fabricating thin film nanostructures.

15.6.1.1 Physical Vapour Deposition (PVD)

Thermal evaporation and cathodic sputtering are the two mechanisms engaged in physical vapour deposition method of nanostructure formation. The physical vapour deposition arrangements have been presented schematically in Fig. 15.6. The rate of evaporation is significantly dependent on the factors like wettability of elements, intense heating, and heating slots. This technique comprises deposition of film on a substrate by transporting the substance from the source to the triggered position. The entire formation of nanostructure is purely because of physical reaction like, generating vapours from substances, transferring the vapours from substances to the trigged portion and condensing to a solid form, without any involvement of chemical reactions [27–29].

15.6.1.2 Chemical Vapour Deposition (CVD)

The chemical vapour deposition (CVD) method typically depends on the type of precursors used, deposition conditions, and the form of energy imposed to initiate the required chemical reaction. The schematic view of chemical vapour deposition (CVD) method is represented in Fig. 15.7. Chemical reactions and responses are the main governing mechanisms in the CVD method and various reactive agents and precursors are available to process the chemical reactions [30]. The growth of thin films of substances, compounds, alloying elements, and crystalline materials of various kinds is more flexible in CVD methods. The surface texture, size, and shape of grain and epitaxial of the deposited film is responsive to the supersaturation of vapours and temperature of the substrate. Formation of single crystalline layer film is possible, at higher substrate temperature and less vapour supersaturation, whereas multi-layered film deposition is possible at rich vapour supersaturation and at low substrate temperatures [31].

15.6.1.3 Ion-Assisted Beam Deposition

An ion beam is employed to evaporate the substance and deposit a nanostructure on the substrate; hence, it is called an ion-assisted beam

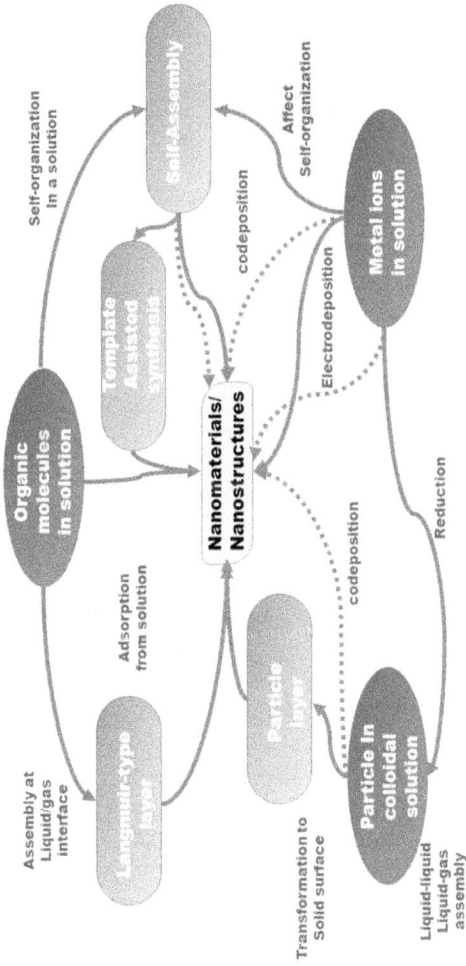

Figure 15.5 Various methods to process a thin film 2-D r.anostructure.

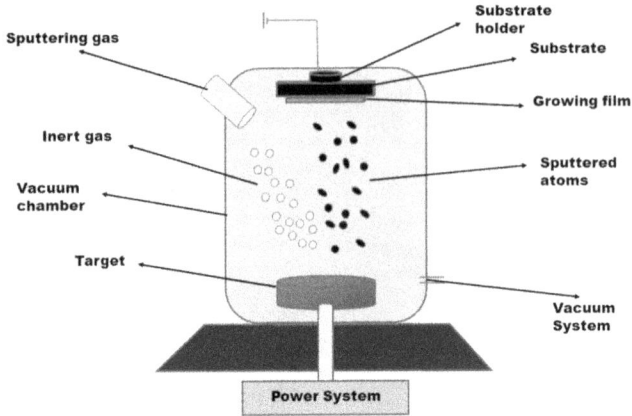

Figure 15.6 Schematic view of PVD arrangement.

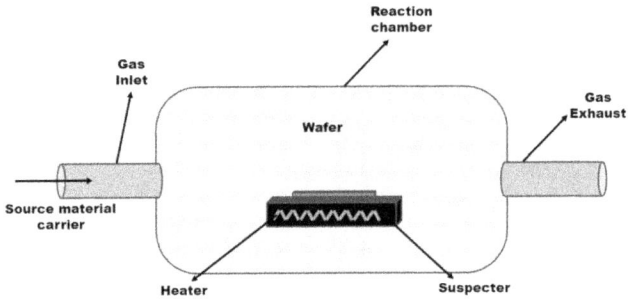

Figure 15.7 Schematic view of CVD arrangement.

deposition (IAD). Sputtering, condensation, ion implantation, thermally desorbed deposition, and crystallization come under the IAD techniques. In the IAD process, the evaporation can be done by either directing the ion beam on to the substances or by adopting sputtering system [32]. The simple IAD system mainly consists of beam source to generate and focus the ion beams on to the substances, vacuum chamber (Kauffmann type) to process the required reactions, and a substrate to deposit a solid nanostructure. By adding the adequate amount of electrons, an unbiased ion beam is generated to bombard with positively charged ions. Bombarding of positive ions helps to avoid the dielectric material charging and promotes the conduction and non-conduction recipients [32–34].

15.6.1.4 Molecular Beam Epitaxy (MBE)

The molecular beam epitaxy (MBE) is the modified version of the PVD method with enticed command on integrity of the material, establishment of interfacial, constituents of alloy, and doping concentrations. All of the above stated control is possible by the assistance of the pure environment of ultra-high vacuum within the deposition arrangements [35]. The end product of the MBE technique is more pure and highly epitaxial while comparison with other techniques. The RHEED capacity of the MBE method facilitates and accelerates reconstruction of specific surfaces with a clean substrate. The epitaxy growth in the thin film nanostructures is dependent on nature of the beam, kinetic growth along with RHEED oscillation [36]. Reproducibility, epitaxial growth, thermodynamic equilibrium, stability of metastable phases, and super-lattice are the encouraging features of the MBE technique. In recent days, MBE is also utilized to fabricate and investigate the photovoltaic substrates.

15.6.1.5 Pulsed Laser Deposition

In this technique, a laser with highly dense and confined frequency bandwidth is employed to vaporize the substances. The pulsed laser deposition (PLD) is one of the kinds of physical vapour deposition (PVD) techniques with slight modifications in vaporization arrangements [37]. When a laser beam with adequate intensity directed towards the substance material, it will elevate its temperature up to certain range. Repeating this process for a period of time gives the well-defined 2-D nanostructured films. Along with evaporating the substances, this process also facilitates the cooling of crystallites for the formation of epitaxial layered films [38]. With the increase in the temperature, rate of recrystallization improves, but at ambient temperatures takes more time than the estimated. Matrix-assisted pulsed laser deposition is the advanced technique of this kind, which is used to process the bio-material, polymers, coordinated/complex compounds, and hybrid metals.

15.6.1.6 Chemical Bath Deposition

A chemical bath deposition (CBD) is a kind of technique in which the reactive agent(s) are filled in a container and the substrate is allowed to participate the corresponding chemical reactions by dipping into it, thereby forming 2-D nanostructures. The entire process of deposition is attributed to the solubility limit of substances and the reactants. The 2-D nanostructure in the CBD method is basically precipitation product available at the end of the process. The precipitation phenomenon starts when the ionic product of the reactive agents is overreaching the solubility product, at a standard temperature [39]. Similarly, the solid state of the precipitation will dissolve into substances, when the ion product is lower than the solubility product. the precipitation and kinetic growth of the process is contingent on ion concentration, particle velocity, and nucleation of the ions.

15.6.2 Mechanical Fabrication Techniques

Multi-layer coatings are referred to as mechanical fabrication techniques of nanostructures preparative methods. These nanostructures are composed of alternatively placed metallic layers and/or various alloys. The obtained thickness in coating by placing the substance layers alternatively is termed "wavelength." In mechanical fabrication techniques only, the width will be in a nanometric scale. The nanostructures prepared under mechanical methods are more stable, and correspond to enhancement in the properties have been noted. The layers of thickness less than 1 μm can be prepared by rolling metallic foils by means of mechanical forces [40].

15.6.2.1 Multilayer Electrodeposition Using a Single Bath

In order to prepare highly purified multi-layered nanostructures, the ion concentrations must be controlled precisely when processed under single solution bath. The homogeneity of the multi-layered 2-D nanostructures depends on modulations of cathodically charged potential. During the single bath process, the electrolysis is to be prepared first, which can accelerate the deposition rate of the active-species (B), followed by the addition of required volume (approximately 1% of B) of noble-species (A). While developing the multi-layered nanostructure, with the combination of active and noble species, this forms the layers with more than 95% of purity [41,42].

15.6.2.2 Mechanical Cleavage

In this method, thin flakes can be prepared by exfoliating the layers from bulk material with the help of scotch tape. The basic intention of the exfoliation is to impose mechanical force through scotch tape to mitigate the van der Waals force among the layers of bulk material, excluding the violation of covalent bonds of every layer, so that layers can be peeled off one by one from the bulk crystal [43]. The peeling of single layer from a bulk crystal is done by attaching the adhesive side of the scotch tape and by applying shear force at the other adhesive side of the scotch tape, along with the help of ultrasonic oscillations, as shown in Fig. 15.8. Sometimes an ultrasharp diamond wedge is utilized to peel a high-ordered layered compound into an ultra-thin film. A clean surface, laterally immense size, and outstanding crystalline structure are the characteristics of the cleavage technique [44].

15.6.2.3 Other Methods

Initiating the growth of a substance inside a solution bath is an easy and cost-effective method. A solution bath is a kind of method that does not

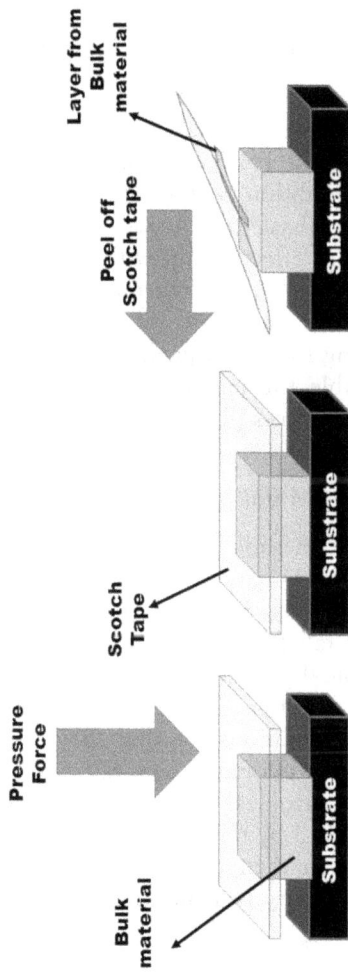

Figure 15.8 Schematic representation of scotch tape exfoliation mechanical cleavage process.

involve any complexity in both apparatus and operation. The synthesis of the solution bath technique is entirely of a chemosynthesis mechanism, which various chemical reactants are mixed together to prepare a reactive solution bath. In order to prepare a perfect reactive solution bath, the pH value of the solution needs to be controlled. By optimizing the factors like duration, temperature of the substrate, and revolution speed, a nanostructure can be obtained with outstanding quality [45].

15.6.3 Chemical/Electrochemical Fabrication Techniques

To fabricate the high-quality nanostructures, the chemical or electrochemical technique is the most suitable classical technique. The technique comprises of dissolving of the basic elements initially in a standard dissolver, followed by the addition of the deposition elements [46]. The simultaneous deposition of basic and deposition material is by the consequences of coagulational effect, outgrowth, and nucleation phenomenon. The deposition of the chemical reactants (oxidizers, reducers, and complexes) intended to form insolvable gist. The molecular reactions can be enhanced by employing the non-aqueous dissolvers. The stabilization of nanoparticles is feasible through the space method, electro-statical method, or by combined process [47].

15.6.3.1 Direct Writing of Metal Nanostructures

The colloidal lithographic method is one of the cheap and simple processes to fabricate a solid 2-D nanostructure, and comes under the plasmonic processing of materials. Fig. 15.9 represents a simple colloidal lithographic technique to fabricate a well-defined and well-organized 2-D nanostructure. The colloidal particles are closely packed together to develop the 2-D nanostructure on a substrate by the evaporation of substances under optimum conditions. During the deposition of metallic particles on to the substrate, pyramidic shapes are formed by penetration of the metallic atoms by periodical arrays. The achievability of the periodic arrays can be controlled by the diameter of the metallic atoms. The colloidal particle cover can then be removed in an easy lift-off step. Some control of the periodicity is achievable by controlling the particle diameter [48].

15.6.3.2 Phase Transformation by Electrochemical Potential

Chirality of some molecules my transform the 3-D nanostructures into 2-D nanostructures with the imposed effect of geometric characteristics. Adsorbing behaviour of the molecules is induced by reduction of symmetry superficial level and dissimilarity in interfacial charge. The obtained nanostructure by the above-stated effect is basically known as a

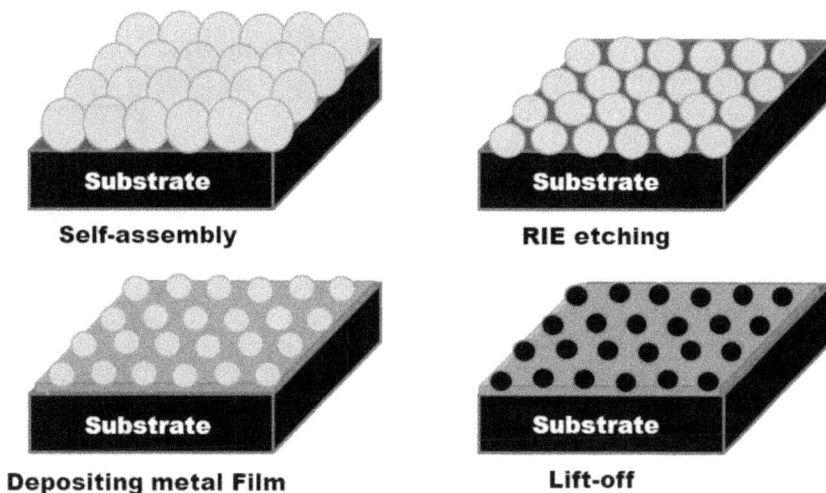

Figure 15.9 Simple colloidal lithographic method.

supramolecular nanostructure. The experimental investigations on supra-molecular nanostructures revealed that, this transformation of 3-D to 2-D nanostructures is led by the electrochemical potential [49].

15.6.3.3 Other Methods

Codeposition of the material into a fine nanostructure in aqueous suspension is attributive to the particle size, denseness of reactants, oxidation process, material addition rate, and temperature during the molecular reactions. The codeposition of the compound material is performed by reducing the chemical reactions of the cations of the corresponding material [50]. The chemical reduction of the cations is correspondence to the oxidization of the compound material. In some cases, reducing agents are used to initiate the codeposition and corresponding oxidization.

15.6.4 Physical Fabrication Techniques

Deposition of thin nanofilms (sculptured) by glancing angle is one of the popular techniques of physical fabrication methods of 2-D nanos-tructures. In the glancing angle depositing technique, the deposition is done by physical evaporation on a rotor substrate. The final end products are sculptured thin films (STFs) either in helicoidal form or in nematic form, attributed to the rate of deposition, rotational speed of substrate, and axis of rotation. Interestingly, columnar thin films (CTFs) can also be generated by simply changing the direction of rotation during the deposition [51]. These films are porous in nature and their anisotropic behaviour can be influenced by managing porousness, subsequently these

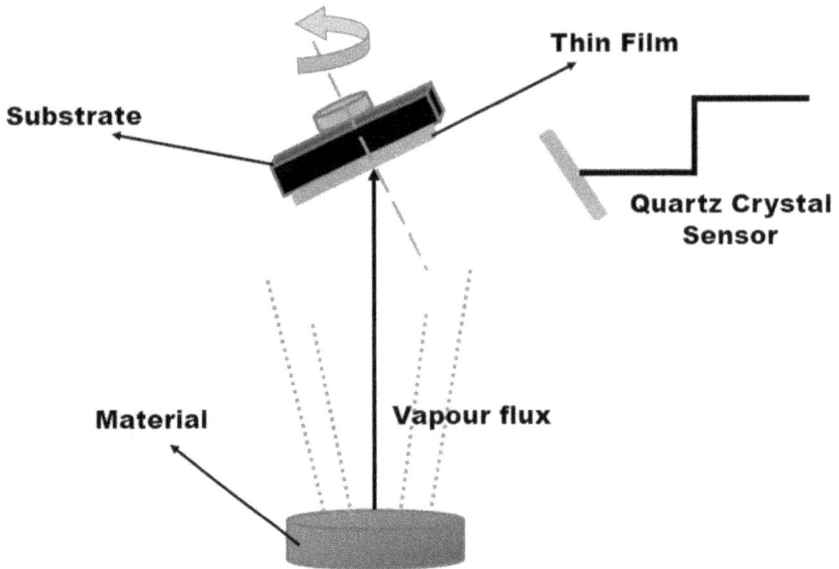

Figure 15.10 Typical glancing angle depositing technique.

films are popularly applied in the optical field. Fig. 15.10 indicates the glancing angle of a deposition technique for a sculptured thin film.

15.7 MACHINING PROCESSES

15.7.1 AFM Tip-Based Approach

The atomic force microscopy (AFM) technique consists of a typical AFM feedback arrangement, cantilever beam, AFM probe with shape AFM tip, which moves over a surface of a sample in a raster manner. Most of the AFM tips are made up of silicon or silicon nitride and incorporated at the free end of a flexible AFM cantilever beam. The linear and orthogonal movements of the AFM probe in coordinance with the surface, are under control of a piezoelectric ceramic scanner. When the AFM tip is moving over the futuristic paths, the deflection of the cantilever varies [52]. This variation in the deflection is monitored by a laser beam, focused at the back of the cantilever, which is instructed towards a position-sensing photodetector. In order to maintain an operative deflection and constant interaction force between the tip and the surface, a feedback loop is attached. By tracking the coordinates instructed by the AFM scan prob, the tip removes the material and generates a 2-D nanostructure.

The interaction between the tip and the surface texture is of basically two modes, namely static/contact mode and dynamic mode. The dynamic interaction mode is further divided into tapping/intermittent mode and non-contact mode. The basic attachments of the atomic force microscopy-assisted assembly for machining of 2-D nanostructures is illustrated in Fig. 15.11. During contact mode, the scanning probe maintains a uniform contact force between the AFM tip and the surface texture while moving over the specimen [53]. The interaction force between the AFM tip and surface texture is generally repulsive. The dominant disadvantages in contact mode process are susceptibility of AFM tips and sticking-up of eliminated portions on the top surface at ambient air. In tapping mode, the piezoelectric probe with a frequency range of 10 kHz to several kHz is employed, which oscillates when the AFM tip delicately contacts the sample at the lower end of the cantilever beam. The feedback loop maintains a constant AFM cantilever oscillation amplitude and hence a constant interaction force, whereas in non-contact mode, the cantilever beam oscillates close to or nearby resonant frequencies with a lower amplitude of 1 nm or less. The AFM tip is allowed to move along the zone of interaction of attractive forces.

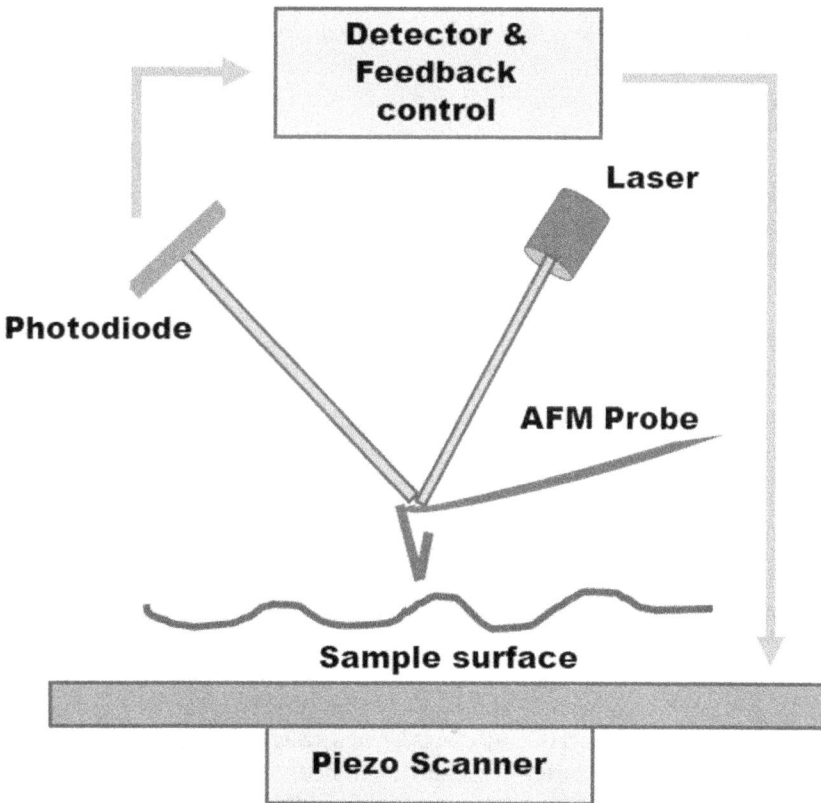

Figure 15.11 Atomic force microscopy (AFM) tip-based technique.

15.7.2 Electrodeposition and Related Techniques

The redox-active reaction helps for deposition of the 2-D nanostructures in their basic format. Electrodeposition can perform on plenty of metals via electrochemical reactions betwixt metals and precursors, suitable electrolyte and favourable conditions are the accelerating factors for the electrodeposition phenomenon. Furthermore, oblique approaches involved change the composition of the solution at the specimen through an electrochemical reaction to activate the solubility one variety in solution. Variation of pH value in aqueous solution is the noteworthy example for the above-mentioned phenomenon, like polarization to the potential at the hydrogen progression response [54].

15.7.3 Corrosion, Electroless Deposition, and Dealloying

To form a 2-D nanostructure, low noble elements are to be replaced by the high noble elements. The 2-D nanostructures are produced by depositing typical atoms utilizing the localized reduction reaction, in the case of the electroless deposition technique. Reversal of the electro and electroless processes is feasible by oxidization of elected constituents in the solid solutions and can form thin film nanostructures via dealloying [55]. Dealloying is a particular method in which enrichment of the solid phase is possible by the addition of high noble elements and dissolving the low noble elements bounded to corrosion mechanism. The corrosion phenomenon itself can process a 2-D nanostructure, with favourable conditions and top level of control. Specifically, when a thin film nanostructure is prepared by corrosion phenomenon, a metal is subjected to activate the corrosion mechanism.

15.7.4 In-situ 2-D Assembly

In self-assembly, the disorderly pre-existing nanostructures of a system turns itself into a well-ordered 2-D nanostructure with the influence of their intermolecular forces. The accomplishment of well-defined and intricately graded nanostructures with superior spatiality can be done by controlling the assemble of the 2-D substances via various efficient and effective techniques. The following are the different kinds of techniques to assemble and obtain the well-oriented nanostructures [56,57].

- **In-plane self-linking assembly:** Particular nanomaterials have self-assembling capacity in an edge-to-edge direction to form a laminar structures like sheets and plates. The self-assembling capacity of the nanomaterial is purely by the function of intermolecular reactions, which includes electrostatic force, dipole attractions, electric charges, and the environmental conditions.

- **Template-induced 2-D assembly:** Verity templates are available in hard and soft formats and are directed to build up a well-oriented 2-D nanostructures. Controlling the assemblage of various kind of templates is highly easier and amply accepted in many fields. The assembly of templates comprises the of adsorption of metallic cations and thermodynamic processing in the presence of precursor.
- **In-situ multilayer assembly:** Multi-layered assembly incorporates with stacking-up of different layers together by means of intercalated and electrostatic atomic level reactions. The anisotropic reaction in atomic level between the piled-up layers is the major obstacle for the multi-layered assembly of 2-D nanostructures.
- **Interfacial 2-D assembly:** The 2-D nanosheets can be architected by external physical forces and different kind of interfaces with superior homogeneity. Induction of interfaces with sequential order to build up a functional 2-D nanostructure even under moderate provisions.
- **Gas-liquid and liquid-liquid interface-induced 2-D assembly:** The induction of interfaces in liquid to liquid is by adsorption while interfacial energy in case of gas to liquid to build a 2-D nanostructure on a substrate. Whenever the various 2-D nanosheets are deposited in a liquid phase, then interfacial gradient occurs and epitaxially growing of 2-D nanosheets will takes place on a substrate by mismatching the lattice. The interfacial energy gradient between the two mediums (liquid to liquid or liquid to gas), trigger the dissipation instantly from high to low level of surface tension energy. This phenomenon is called the Marangoni effect, and is responsible for the formation of uniformly condensed films of 2-D nanosheets.
- **Post-synthesis multilayer assembly:** Multi-layered nanostructures can be obtained by post-synthesis assembly process in which two contradictorily charged films are compensates each other in two separate precursors when they dispersed. The alternate drenching of contradictorily charged films will disposes the 2-D nanostructure upon the substrate by the reaction of electrostatic forces. This process of deposition can also be done by means of external forces such as magnetic, electric, and shear, and is known as post-synthesis assembly driven by external forces.

15.8 CONCLUSION

The 2-D nanomaterials and 2-D nanostructures are brand new in the fields of materials and manufacturing. These structures hold unique features by the systematic changes in grains (shape, size, and concentration) and layers (agglomeration, orientation, and stacking). The molecular reactions are the major considerations in any process to build a well-defined 2-D nanostructure and their properties are related in both the atomic-level responses and precursors. A combination of a similar kind of nanosheets/nanomaterials gives some features, whereas dissimilar nanosheets/nanomaterials promote additional and enhanced properties to the final 2-D nanostructures. The fabrication techniques of 2-D nanostructures itself are considered as machining processes and they have few limitations and challenges in making the well-functioned 2-D nanostructure. The following are the challenges and constraints in machining of 2-D nanostructures:

- The grain growth and phase changes are to be in controlled manner; otherwise enlargement of the microstructure will appear and suffocates the desirable properties of the 2-D nanostructures.
- In atomic level, the grain and grain boundaries must be of the same size and phase to accomplish the well-defined 2-D nanostructure.
- While cleaving the nanostructures mechanically, crystal lattices, purity of surface, fore-and-aft of size, and orientations limit the required properties of 2-D nanostructures.
- Coexisting of flakes along the substrates are always the drawback in angulation of 2-D nanostructures.
- Single or multi-layered stacking of nanomaterials is a difficult process to control in the cases of mechanical and physical exfoliation. As a consequential result, there is an increment in magnetic properties and decrement in electrical properties (about 50%).
- In the manual stacking process, controlling the exfoliation, repeatability, and precision are the main factors to be focused on in order to fabricate a 2-D nanostructure with a balanced magnetic and electric properties.
- The establishment of nanostructures in magnetic coolers enhances their performance and output.
- In some deposition methods, epitaxial growth is possible only at low temperatures.
- By assisting the ion beam in deposition of nanostructures, only laminar growth is feasible because of rich molecular level migration and corresponding chemical reactions.
- The peeling off of single layers from a bulk crystal is feasible for the bulk materials with layered compound structure.
- The metal removal rate in AFM-tip based vibrational-assisted technique is difficult to measure, and the properties depend on various operating parameters.

REFERENCES

[1] Singh Singh, K, Kumar, S. Influence of nanomaterials on nanofluid application – A review. In AIP Conference Proceedings 2021 May 13 (Vol. 2341, No. 1, p. 040016). AIP Publishing LLC.

[2] Suryanarayana C, Koch CC. Nanostructured materials. In *Pergamon Materials Series* 1999 Jan 1 (Vol. 2, pp. 313–344). Pergamon.

[3] Cao G. *Nanostructures & Nanomaterials: Synthesis, properties & applications* 2004. Imperial College Press.

[4] Miró P, Audiffred M, Heine T. An atlas of two-dimensional materials. *Chemical Society Reviews*. 2014; 43(18): 6537–6554.

[5] Karfa P, Majhi KC, Maghuri R. Chapter 2 – Synthesis of two-dimensional nanomaterials. Khan Raju Barua Shaswat, editors. *Two-Dimensional Nanostructures for Biomedical Technology*. 2020: 35–71.

[6] Sasidharan S, Raj S, Sonawane S, Sonawane S, Pinjari D, Pandit AB, Saudagar P. Nanomaterial synthesis: chemical and biological route and applications. In *Nanomaterials synthesis*; 2019 Jan 1 (pp. 27–51). Elsevier.

[7] Yang F, Song P, Ruan M, Xu W. Recent progress in two-dimensional nanomaterials: Synthesis, engineering, and applications. *FlatChem*. 2019 Nov 1; 18: 100133.

[8] Pradeep AV, Prasad SS, Suryam LV, Kumari PP. A review on 2-D materials for bio-applications. *Materials Today: Proceedings*. 2019 Jan 1; 19: 380–383.

[9] Prasad SS, Prasad SB, Verma K, Mishra RK, Kumar V, Singh S. The role and significance of Magnesium in modern day research – A review. *Journal of Magnesium and Alloys*. 2021 Jun 24; 10(1): 1–61.

[10] Sudha PN, Sangeetha K, Vijayalakshmi K, Barhoum A. Nanomaterials history, classification, unique properties, production and market. In *Emerging applications of nanoparticles and architecture nanostructures*. 2018 Jan 1 (pp. 341–384). Elsevier.

[11] Van de Voorde M, Tulinski M, Jurczyk M. Mansfield Elisabeth, Kaiser Debra, Fujita Daisuke, Van de Voorde Marcel , editors; Engineered nanomaterials: A discussion of the major categories of nanomaterials. *Metrology and standardization of nanotechnology: Protocols and industrial innovations*. 2017 Feb 15: 49–74.

[12] Dolez PI. Nanomaterials definitions, classifications, and applications. In *Nanoengineering*; 2015 Jan 1 (pp. 3–40). Elsevier.

[13] Zhuiykov S. Nanostructured two-dimensional materials. In *Modeling, characterization, and production of nanomaterials*; 2015 Jan 1 (pp. 477–524). Woodhead Publishing.

[14] Devreese JT. Importance of nanosensors: Feynman's vision and the birth of nanotechnology. *MRS Bulletin*. 2007 Sep; 32(9): 718–725.

[15] Pokropivny VV, Skorokhod VV. Classification of nanostructures by dimensionality and concept of surface forms engineering in nanomaterial science. *Materials Science and Engineering: C*. 2007 Sep 1; 27(5–8): 990–993.

[16] Balzani V. Nanoscience and nanotechnology: The bottom-up construction of molecular devices and machines. *Pure and Applied Chemistry*. 2008 Jan 1; 80(8): 1631–1650.

[17] Teo BK, Sun XH. From top-down to bottom-up to hybrid nanotechnologies: road to nanodevices. *Journal of Cluster Science*. 2006 Dec; 17(4): 529–540.

[18] Roco MC. Reviews of national research programs in nanoparticle and nanotechnology research-Nanoparticle and nanotechnology research in the USA. *Journal of Aerosol Science*. 1998; 29(5): 749–751.

[19] Singh NB, Shukla SK. Khan Raju, Barua Shaswat , editors; Properties of two-dimensional nanomaterials. In *Two-dimensional nanostructures for biomedical technology*; 2020 Jan 1 (pp. 73–100). Elsevier.

[20] Bashir S, Liu JL, editors; Nanomaterials and their application In *Advanced nanomaterials and their applications in renewable energy*; 2015 Aug 6 (pp. 1–50). Elsevier Inc.: Amsterdam, The Netherlands.

[21] Szczech JR, Higgins JM, Jin S. Enhancement of the thermoelectric properties in nanoscale and nanostructured materials. *Journal of Materials Chemistry.* 2011; 21(12): 4037–4055.

[22] Liu JL, Bashir S. Advanced nanomaterials and their applications in renewable energy, 2015. Elsevier

[23] Ruoff RS, Lorents DC. Mechanical and thermal properties of carbon nanotubes. *Carbon.* 1995 Jan 1; 33(7): 925–930.

[24] El-Sherik AM, Erb U, Palumbo G, Aust KT. Deviations from hall-petch behaviour in as-prepared nanocrystalline nickel. *Scripta Metallurgica et Materialia.* 1992 Nov 1; 27(9): 1185–1188.

[25] Chen H, Zhou X, Ding C. Investigation of the thermomechanical properties of a plasma-sprayed nanostructured zirconia coating. *Journal of the European Ceramic Society.* 2003 Aug 1; 23(9): 1449–1455.

[26] Wang L, Li DY. Mechanical, electrochemical and tribological properties of nanocrystalline surface of brass produced by sandblasting and annealing. *Surface and Coatings Technology.* 2003 Apr 22; 167(2–3): 188–196.

[27] Helmersson U, Lattemann M, Bohlmark J, Ehiasarian AP, Gudmundsson JT. Ionized physical vapor deposition (IPVD): A review of technology and applications. *Thin Solid Films.* 2006 Aug 14; 513(1–2): 1–24.

[28] Avni R, Fried I, Raveh A, Zukerman I. Plasma surface interaction in PACVD and PVD systems during TiAlBN nanocomposite hard thin films deposition. *Thin Solid Films.* 2008 Jun 30; 516(16): 5386–5392.

[29] Knuyt G, Quaeyhaegens C, D'haen J, Stals LM. A quantitative model for the evolution from random orientation to a unique texture in PVD thin film growth. *Thin Solid Films.* 1995 Mar 15; 258(1–2): 159–169.

[30] Sherman A. Chemical vapor deposition for microelectronics: Principles, technology, and applications United States: N. p., 1987.

[31] Pierson HO. *Handbook of chemical vapor deposition: Principles, technology and applications.* William Andrew; 1999 Sep 1.

[32] Ila D, Zimmerman RL, Muntele CI, Thevenard P, Orucevic F, Santamaria CL, Guichard PS, Schiestel S, Carosella CA, Hubler GK, Poker DB. Nano-cluster engineering: A combined ion implantation/co-deposition and ionizing radiation. *Nuclear Instruments and Methods in Physics Research Section B: Beam Interactions with Materials and Atoms.* 2002 May 1; 191(1–4): 416–421.

[33] Nagase M, Takahashi H, Shirakawabe Y, Namatsu H. Nano-four-point probes on microcantilever system fabricated by focused ion beam. *Japanese Journal of Applied Physics.* 2003 Jul 1; 42(7S): 4856.

[34] Nagase T, Kubota T, Mashiko S. Fabrication of nano-gap electrodes for measuring electrical properties of organic molecules using a focused ion beam. *Thin Solid Films.* 2003 Aug 22; 438: 374–377.

[35] Liu HF, Xiang N, Chua SJ. Growth of InAs on micro-and nano-scale patterned GaAs (001) substrates by molecular beam epitaxy. *Nanotechnology.* 2006 Oct 3; 17(20): 5278.

[36] Pretorius A, Yamaguchi T, Kübel C, Kröger R, Hommel D, Rosenauer A. Structural investigation of growth and dissolution of InxGa1-xN nano-islands grown by molecular beam epitaxy. *Journal of Crystal Growth*. 2008 Feb 15; 310(4): 748–756.

[37] Wołowski J, Badziak J, Czarnecka A, Parys P, Pisarek M, Rosiński M, Turan RA, Yerci SE. Application of pulsed laser deposition and laser-induced ion implantation for formation of semiconductor nano-crystallites. *Laser and Particle Beams*. 2007 Mar; 25(1): 65–69.

[38] Youn SW, Takahashi M, Goto H, Maeda R. Microstructuring of glassy carbon mold for glass embossing – Comparison of focused ion beam, nano/ femtosecond-pulsed laser and mechanical machining. *Microelectronic Engineering*. 2006 Nov 1; 83(11–12): 2482–2492.

[39] Allouche NK, Nasr TB, Kamoun NT, Guasch C. Synthesis and properties of chemical bath deposited ZnS multilayer films. *Materials Chemistry and Physics*. 2010 Oct 1; 123(2–3): 620–624.

[40] Cho J, Joshi MS, Sun CT. Effect of inclusion size on mechanical properties of polymeric composites with micro and nano particles. *Composites Science and Technology*. 2006 Oct 1; 66(13): 1941–1952.

[41] Péter L, Kupay Z, Cziráki Á, Pádár J, Tóth J, Bakonyi I. Additive effects in multilayer electrodeposition: Properties of Co– Cu/Cu multilayers deposited with NaCl additive. *Journal of Physical Chemistry B*. 2001 Nov 8; 105(44): 10867–10873.

[42] Fei JY, Wilcox GD. Electrodeposition of zinc–nickel compositionally modulated multilayer coatings and their corrosion behaviours. *Surface and Coatings Technology*. 2006 Mar 15; 200(11): 3533–3539.

[43] Jayasena B, Subbiah S. A novel mechanical cleavage method for synthesizing few-layer graphenes. *Nanoscale Research Letters*. 2011 Dec; 6(1): 1–7.

[44] Tang DM, Kvashnin DG, Najmaei S, Bando Y, Kimoto K, Koskinen P, Ajayan PM, Yakobson BI, Sorokin PB, Lou J, Golberg D. Nanomechanical cleavage of molybdenum disulphide atomic layers. *Nature Communications*. 2014 Apr 3; 5(1): 1–8.

[45] Nahavandi M, Monirvaghefi SM. The effect of electroless bath pH on the surface properties of one-dimensional Ni–P nanomaterials. *Ceramics International*. 2020 Feb 1; 46(2): 1916–1923.

[46] Bozzini B, D'Urzo L, Mele C. Electrochemical fabrication of nano-and micrometric Cu particles: In situ investigation by electroreflectance and optical second harmonic generation. *Transactions of the IMF*. 2008 Sep 1; 86(5): 267–274.

[47] Lohmüller T, Müller U, Breisch S, Nisch W, Rudorf R, Schuhmann W, Neugebauer S, Kaczor M, Linke S, Lechner S, Spatz J. Nano-porous electrode systems by colloidal lithography for sensitive electrochemical detection: Fabrication technology and properties. *Journal of Micromechanics and Microengineering*. 2008 Sep 26; 18(11): 115011.

[48] Leggett GJ. Direct writing of metal nanostructures: Lithographic tools for nanoplasmonics research. *ACS Nano*. 2011 Mar 22; 5(3): 1575–1579.

[49] Su GJ, Li ZH, Aguilar-Sanchez R. Phase transition of two-dimensional chiral supramolecular nanostructure tuned by electrochemical potential. *Analytical Chemistry*. 2009 Nov 1; 81(21): 8741–8748.

[50] Ghodselahi T, Vesaghi MA, Shafiekhani A, Baradaran A, Karimi A, Mobini Z. Co-deposition process of RF-sputtering and RF-PECVD of copper/carbon nanocomposite films. *Surface and Coatings Technology*. 2008 Mar 15; 202(12): 2731–2736.

[51] Wang F, Lakhtakia A. Lateral shifts of optical beams on reflection by slanted chiral sculptured thin films. *Optics Communications*. 2004 May 1; 235(1–3): 107–132.

[52] Kong X, Deng J, Dong J, Cohen PH. Study of tip wear for AFM-based vibration-assisted nanomachining process. *Journal of Manufacturing Processes*. 2020 Feb 1; 50: 47–56.

[53] Yan Y, Chang S, Wang T, Geng Y. Scratch on polymer materials using AFM tip-based approach: A review. *Polymers*. 2019 Oct; 11(10): 1590.

[54] Chun S, Han KS, Shin JH, Lee H, Kim D. Fabrication and characterization of CdTe nano pattern on flexible substrates by nano imprinting and electro-deposition. *Microelectronic Engineering*. 2010 Nov 1; 87(11): 2097–2102.

[55] Sharma NC, Pandya DK, Sehgal HK, Chopra KL. Electroless deposition of epitaxial Pb1–XHgXS films. *Thin Solid Films*. 1979 May 1; 59(2): 157–164.

[56] Fang Z, Xing Q, Fernandez D, Zhang X, Yu G. A mini review on two-dimensional nanomaterial assembly. *Nano Research*. 2020 May; 13(5): 1179–1190.

[57] Xia H, Wang D. Fabrication of macroscopic freestanding films of metallic nanoparticle monolayers by interfacial self-assembly. *Advanced Materials*. 2008 Nov 18; 20(22): 4253–4256.

Index

For Product Safety Concerns and Information please contact our EU
representative GPSR@taylorandfrancis.com
Taylor & Francis Verlag GmbH, Kaufingerstraße 24, 80331 München, Germany

www.ingramcontent.com/pod-product-compliance
Lightning Source LLC
Chambersburg PA
CBHW060336220326
41598CB00023B/2726

9 781032 355481